Hydraulics

Hydraulics
Fourth Edition

Andrew L. Simon
The University of Akron

Scott F. Korom
University of North Dakota

Prentice Hall
Upper Saddle River, New Jersey   Columbus, Ohio

Library of Congress Cataloging-in-Publication Data

Simon, Andrew L.
    Hydraulics / Andrew L. Simon, Scott F. Korom. — 4th ed.
      p. cm.
    Includes bibliographical references and index.
    ISBN 0-13-213513-2
    1. Hydraulics. I. Korom, Scott F. II. Title.
  TC160.S379 1997
  627—dc20
                                       96-14293
                                           CIP

Cover photo: Flaming Gorge Dam, Utah. Last major dam built by the U.S. Bureau of Reclamation.
Editor: Ed Francis
Production Editor: Linda Hillis Bayma
Production Coordination: Gretchen K. Chenenko
Cover Designer: Russ Maselli
Cover Design Coordinator: Jill E. Bonar
Production Manager: Deidra M. Schwartz
Marketing Manager: Danny Hoyt

This book was set in Times Roman by Bi-Comp, Inc. and was printed and bound by Quebecor Printing/Book Press. The cover was printed by Phoenix Color Corp.

 © 1997, 1986, 1981, 1976 by Prentice-Hall, Inc.
Simon & Schuster/A Viacom Company
Upper Saddle River, New Jersey 07458

Printed in the United States of America

10 9 8 7 6 5 4 3 2 1

ISBN 0-13-213513-2

Prentice-Hall International (UK) Limited, *London*
Prentice-Hall of Australia Pty. Limited, *Sydney*
Prentice-Hall Canada Inc., *Toronto*
Prentice-Hall Hispanoamericana, S. A., *Mexico*
Prentice-Hall of India Private Limited, *New Delhi*
Prentice-Hall of Japan, Inc., *Tokyo*
Simon & Schuster Asia Pte. Ltd., *Singapore*
Editora Prentice-Hall do Brasil, Ltda., *Rio de Janeiro*

To My
Grandfather

Russell L. Foster
(1900–1993)

"Gramp's" formal education ended around the fifth grade when he went to work full time on the farm. He and my grandmother, E. Rozena Mathews Foster (1903), made a living in the shop behind their home sharpening shears and knives for the rubber companies in Akron, OH.

He appreciated my education; I appreciated his wisdom. He taught me that it takes more than good books to make a good engineer, it also takes experience—even if that experience comes from someone with a grade school education. Gramp's parting words were always—

"God bless ya, and keep your nose to the grindstone."
It's good advice.

Scott Foster Korom
Logging Camp Ranch, ND
March 1996

Preface

The science of hydraulics is as old as civilization itself. The control of floods, irrigation, and water supply by ancient peoples required intuitive understanding of the basic physical principles of hydraulics. Yet in spite of its age, hydraulics is still a somewhat uncertain science, full of theoretically indeterminate factors. In theoretical developments newer sciences have long eclipsed hydraulics, but its overwhelming importance in humankind's fight for survival is still unquestionable.

During the past century enormous progress has been made in the understanding of the fundamental laws of the mechanics of fluids. Powerful analytic and numerical techniques are now available for putting these fundamental principles into practice. Yet most practical hydraulics problems still defy these theoretical solutions. Practical hydraulics is perhaps as much an intuitive art as a science.

One of the reasons for the theoretical uncertainty of hydraulics is the large number of ill-defined variables that enter into even some of the simplest practical problems. The often unknown interdependence of these pertinent variables makes it impossible to develop reliable answers on the basis of the principles of fluid mechanics alone. Therefore, to consider hydraulics as simply experimental fluid mechanics is a faulty oversimplification. For a true assessment of hydraulics one may quote Professor Lofti Zadeh, father of the mathematical concept of fuzzy logic: *"As complexity rises, precise statements lose meaning and meaningful statements lose precision."* Hydraulic problems are often complex, and their solutions, to be meaningful, often lack precision.

This book provides essential information for anyone seeking practical answers for everyday hydraulic problems. No attempt is made to present the material in the framework of theoretical fluid mechanics. The topics are presented as they appear in practice and not as examples of the application of certain theoretical concepts. Derivations are included only if their understanding is necessary for the proper use of their results, and they are kept at a level that does not require calculus.

Earlier editions of this text were used in a great variety of disciplines: 4-year technology programs in agriculture, civil engineering, construction, and related fields as well as in traditional civil engineering curricula in the United States and in other English-speaking countries. The Spanish-language edition is widely used in South American schools. Because of the range of potential users, the range of topics covered is correspondingly wide. Many 4-year engineering programs have adopted this text for a course on hydraulic design. The students in those curricula had a required course in fluid mechanics, so the first two chapters were skipped. In 2-year technology programs, on the other hand, the emphasis was on the first part of the book, with later chapters selected according to the area of specialization of the program.

This new edition is the result of many fundamental changes. The whole manuscript was thoroughly rewritten. Field measurements and modeling were combined into a new third chapter titled Hydraulic Data Collection. A new chapter on coastal hydraulics was added at the end. Historical information was provided, mainly in the form of footnotes, to help the students connect this technical material to their studies of Western civilization. The new coauthor, Dr. Scott F. Korom, has reworked all example problems and generously added new ones. For increased readability, example problems are now massed at the end of each chapter.

Perhaps the most important new feature is the elimination of all computer programs from the book. While earlier editions of this book were prepared the rapidly increasing demand for "computer literacy" was the mantra of academia. By the time of this edition, general as well as specialized hydraulic software has become readily available commercially. The student is expected to be able to apply spreadsheet programs to any technical problem.

ACKNOWLEDGMENTS

We thank James Cook, West Virginia Institute of Technology; Robert Eastley, Ferris State University; and Richard Heggen, University of New Mexico, who have devoted their time and energy to review this new manuscript.

Among the many charts, tables, and formulas there are some whose creators are not known to us; hence, proper credit could not be given. We thank these anonymous engineers for their contributions to the development of hydraulics.

Mark Kennedy, University of Portland, provided insightful comments and developed new problems for the following subject areas: the physical properties of water, fluid mechanics, hydraulic data collection, and flow in pipes. We are especially indebted to Mark for his contribution to this edition.

Andrew L. Simon
Scott F. Korom

Contents

1

Physical Properties of Water

The way water behaves under various conditions encountered in practice depends primarily on its fundamental chemical and physical properties. These are actually controlled by water's molecular structure and by its internal energy. The terms and concepts introduced here will be frequently referred to in all later chapters.

1.1 INTERNAL ENERGY AND THE THREE PHASES OF WATER

Water is a chemical compound composed of oxygen and hydrogen. In each water molecule there is one oxygen atom for two hydrogen atoms. Clusters of water molecules are bonded together by electrostatic attractions between neighboring oxygen and hydrogen atoms. This type of atomic bond is called a *hydrogen bond*. The degree of hydrogen bonding—that is, the number of hydrogen bonds present—depends on the ambient temperature and pressure. Both temperature and pressure are manifestations of energy.

Energy is universally measured in joules. One *joule* (J) of energy is the potential to do work of a force of 1 *newton* (N) exerted over a distance of 1 *meter* (m). The heat required to change the temperature of a substance is one form of energy. In the conventional American system the equivalent of the joule is the *British thermal unit*, abbreviated BTU. One BTU equals 1055 joules, or approximately 1 kilojoule (kJ).

Depending on its *internal energy content* water appears in either liquid, solid, or gaseous form. Snow and ice are solid forms of water; moisture or water vapor in air is the gaseous form. The different forms of water are called its *phases*. Whether water is in its solid, liquid, or gaseous phase depends on the extent of hydrogen bonding. In solid form all hydrogen atoms are bonded; in liquid form fewer hydro-gens are bonded. There are no hydrogen bonds in the gaseous phase. Water is a stable compound; the bonds between the hydrogen and oxygen atoms do not break until the temperature reaches thousands of degrees Celsius.

The amount of energy required to raise the temperature of 1 gram (g) of a substance by 1° is a property of that substance called *specific heat.* The specific heat of ice is 1.95 joules per gram per degree Kelvin, whereas the specific heat of water is 4.19 J/g/K.[1] This means that it takes less heat to warm up ice than to warm up water; but it takes even less heat to warm up the same amount of water vapor. For water vapor the specific heat at constant pressure is 1.81; at constant temperature it is 1.35 J/g/K.

The various phases of water as a function of energy content are shown in a phase diagram in Figure 1.1*a.* The two coordinates of the phase diagram represent the two forms of energy: pressure and heat. The diagram shows three zones within which water is in solid, liquid, or gaseous phase.

For water to pass from one phase to another phase, either the pressure or the heat energy must change. Therefore, energy must be either added to or taken away from water to change its phase. The amount of energy required to change from one phase to another is called *latent heat.* Figure 1.1*b* illustrates the concepts of specific heat and latent heat as they relate to the changes in temperature of water held at constant pressure.

Melting of ice, the change from solid to liquid phase, is called *fusion.* The *latent heat of fusion* of water is 334 J/g. Freezing is the reverse process. To freeze water, 334 J of energy must be taken out of each gram of water. To change the liquid phase of water into the gaseous phase requires a *latent heat of vaporization* of 2500 J/g. The process of evaporation is complex. The heat content of a lake, the amount of heat energy entering through the inflowing waters, the velocity of wind carrying away moist air from the surface of the water, incoming solar radiation, and the temperature of the air all influence the rate of evaporation. To change water from solid to gaseous form requires 2834 J of energy per gram. This is the *latent heat of sublimation* of ice. It equals the sum of the latent heats of fusion and evaporation. In practice, we see this process when snow disappears without melting during a cold but sunny winter day. The energy input needed for this sublimation is supplied by the sun.

At sea level exposed water surfaces are under the pressure of the atmosphere. Standard atmospheric pressure is 14.7 psi at 45° latitude and 68°F, or 2116 lb/ft^2 or 33.9 ft of water, which equals 10.33 m of water. In practical computations standard atmospheric pressure is usually taken as 10 m of water, or 100 kPa. At elevations above sea level the barometric pressure may be approximated by $p = 10.33 - 0.00108H$, where H is the elevation in meters, and the pressure is expressed in meters of water.

---

[1]William Thompson Kelvin's (Ireland, 1824–1907) temperature scale is related to the Celsius scale by the expression K = °C + 273.15. Zero on the Kelvin scale refers to the theoretical lowest temperature of gases—that is, absolute zero, or −273.15°C.

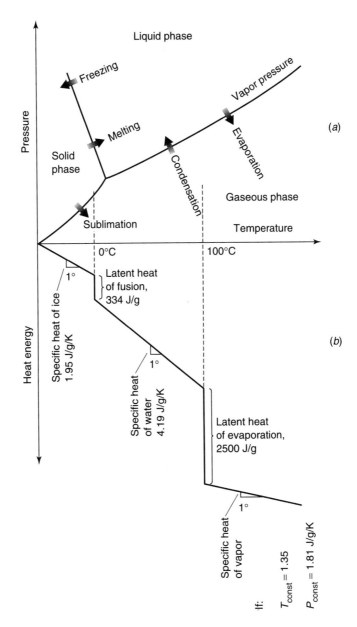

FIGURE 1.1   Phase diagram of water.

If the temperature is raised to the boiling point (100°C or 212°F), water evaporates. At higher elevations the atmospheric pressure is less; hence, water evaporates at temperatures lower than 100°C. The amount of pressure at which water changes phase from liquid to gaseous form, or begins to boil, is called *vapor pressure*. As

TABLE 1.1   Vapor Pressure
of Water

Temperature °C	Absolute Pressure[a]	
	kPa	psi
0	0.6	0.09
10	1.2	0.18
20	2.3	0.34
40	7.4	1.07
60	19.9	2.89
80	47.4	6.87
100	101.3	14.70
105	120.9	17.52

[a] Absolute pressure may be defined as the pressure on an open water surface (zero, in engineering practice) plus the pressure due to the atmosphere above it.

shown by the line representing this boundary between gaseous and liquid forms in Figure 1.1a, vapor pressure depends on the ambient pressure and temperature. Table 1.1 gives values showing this relationship.

In closed systems like pipes or pumps, water may change phase because of changes in pressure, even though the temperature remains constant. In everyday hydraulic design the vapor pressure of water is often of great importance. In the suction pipes of pumps as well as at the tip of the impellers of pumps the pressure of water often reduces to levels below vapor pressure. Water at those locations turns into vapor. As the vapor bubbles move on to points of high pressure, the bubbles collapse in a noisy and violent manner, causing considerable material damage to pump and pipe. This damaging process, called *cavitation,* can be avoided by observing correct design procedures.

In the conventional American system of weights and measures the commonly used value for vapor pressure of water is 0.34 lb/in.² at standard atmospheric pressure and normal room temperature of 68°F. In metric units it equals the pressure of a 0.24-m column of water at 20°C. This means that at sea level, if the atmospheric pressure is 10.33 m of water, 10.09 m (10.33 − 0.24) is the theoretical maximum limit on the height a pump can raise water through its suction pipe. In practice, cavitation in pumps occurs well below this theoretical maximum limit.

1.2   MASS, FORCE, AND DENSITY

The concept of *unit weight* is interpreted very differently in SI units, in the conventional American system of weights and measures, and in some regionally used metric systems. Hydraulics is a field of engineering mechanics. As such, it is based on the fundamental laws first stated by Sir Isaac Newton (England, 1642–1727). From physics we recall that Newton's second law of motion states that force equals mass times acceleration,

$$F = ma \tag{1.1}$$

The SI unit of mass is the *kilogram*. Force, then, is a secondary dimension defined by

force = kilogram × gravitational acceleration

and is expressed in *newtons* (N). Some countries with metric traditions use the term *kilogram* to express force as a primary unit. Density or mass, then, is a secondary unit given as $kg\text{-}s^2/m^4$. Although this practice is now discouraged by international standards, it is still widely used and can be found often in the technical literature.

In the conventional American system we speak of *absolute or gravitational systems*. In engineering usage the gravitational system is predominant. In the conventional American gravitational system the *pound force* is a primary unit along with *foot* for length and *second* for time. Hence, mass is a secondary unit derived from Newton's law in the following form:

pound force = slug × (1 ft/sec^2) = pound mass × gravitational acceleration

In this book the SI system will be used alongside the conventional American gravitational system. Force will be defined in newtons. One pound of force equals 4.45 N. In the SI system the *unit weight* of water is 1.0 g/cm^3 or 1000 kg/m^3. By definition, 1 liter of water is a kilogram, a quantity of mass. When we use metric terminology, weight of water is always expressed as kilogram mass.

The *gravitational acceleration* on the surface of the Earth averages between 9.78 and 9.82 m/s^2. For design the value of 9.81 m/s^2 will be used.[2] In conventional American units 32.2 ft/sec^2 will be used.

The mass of water contained in a unit volume is its *density*. Its magnitude depends on the number of water molecules that occupy the space of a unit volume. This, of course, is determined by the size of the molecules and by the degree of hydrogen bonding. The latter, as we know already, depends on the temperature and pressure. Because of the peculiar molecular structure of water and the change in its molecular structure when water takes solid form, it is one of the few substances that expands when it freezes. The expansion of freezing water when rigidly contained causes stresses in the container. These stresses are responsible for the weathering of rocks and can damage pipes or structures if their effect is not considered in the design.

Water reaches its maximum density near the freezing point, at 3.98°C. Table 1.2 gives the density of water at different temperatures. In the SI system density and unit weight mean the same. In the conventional American system density is expressed in slugs per cubic foot. The density of water in slugs per cubic foot at atmospheric pressure and at 50°F is 1.94 lb-sec^2/ft^4.

As shown in Table 1.2 the density of ice is different from that of liquid water at the same temperature. Because it is lighter then water, ice floats on the water surface.

---

[2]This value was first determined by Christiaan Huygens (Holland, 1629–1695), whose contributions to science also included the principle of centrifugal force and the law of the pendulum.

TABLE 1.2   Density of Water

Temperature °C	Density, ρ	
	kg/m³	slugs/ft³
0 ice	917	1.779
0 water	999	1.9406
3.98	1000	1.941
10	999	1.940
25	997	1.935
100	958	1.860

Because seawater contains salt, its density is greater than that of fresh water. The density of seawater is usually taken to be 1.99 slugs/ft³, about 3 percent more than fresh water. On islands or coastal areas in the groundwater zone the heavier seawater is located below a wedge or lens-shaped layer of fresh water. Heavy pumping of fresh water brings about *seawater encroachment, or intrusion*, a usually irreversible process. On the island of Malta, for example, aggressive pumping of the freshwater aquifer brought about seawater intrusion. As a result, the supply of water is increasingly dependent on desalination plants.

Density changes due to temperature influences the performance of some artesian wells discharging thermal water. If the hot water flows, its static water level may rise over the wellhead because its lower density causes the column to expand. If the well is shut off, allowing the water column to cool, the static head may subside below the wellhead. This means that pumping will be initially needed until the water column warms up again.

Density changes due to pressure are assumed to be zero for almost all hydraulic calculations. In other words, water is generally assumed to be incompressible, even though it is about 100 times more compressible than steel. However, in computations relating to shock waves in water, called *water hammer*, and in pumping groundwater from deep confined aquifers, knowledge of the elastic properties of water is essential.

The term *specific weight* is defined as a force of

$$\gamma = \rho g \tag{1.2}$$

where ρ (the Greek rho) is the density of water. In most calculations γ (the Greek gamma) is taken as 9.81 kN/m³ in SI units and 62.4 lb/ft³ in conventional American units for water at 10°C or 50°F, respectively. In the conventional American system the expression *unit weight* is also used, meaning a quantity of force.

The ratio of the change of pressure to the corresponding change of volume is called the *bulk modulus of elasticity*. In solid mechanics the modulus of elasticity (Young's modulus) is defined as the ratio of linear stresses to linear strains, as determined by tension tests. With fluids we speak of *bulk modulus* because it is determined by compression tests on volumes. An initial volume of V_0 will change by ΔV due to a change in surface pressure Δp. The formula expressing this relationship is

TABLE 1.3 Elasticity of Water

Pressure, psi	Temperature, °F				
	32	68	120	200	300
	E, Elasticity in Thousand psi				
15	292	320	332	308	—
1,500	300	330	342	319	248
4,500	317	348	362	338	271
15,000	380	410	426	405	350

$$\Delta p = -E \frac{\Delta V}{V_0} \tag{1.3}$$

As the pressure increases, the volume is reduced; hence, the negative sign in front of *E*.[3] In fluid mechanics we often refer to *modulus of compressibility K,* which is defined as the reciprocal of *E,*

$$K = \frac{1}{E} \tag{1.4}$$

Compressibility of water depends on the temperature as well as on ambient pressure. Table 1.3 gives values of *E* for water over a wide range of pressures and temperatures. Note that for all pressures water reaches its maximum elasticity (or minimum compressibility) at 120°F. The greater the pressure, the greater the elasticity. At atmospheric pressure the value of *E* for water is approximately 310,000 psi, or 2.14×10^9 Pa. Elasticities of other fluids are shown in Table 4.5.

1.3 SURFACE TENSION, ADHESION, AND CAPILLARITY

Under normal conditions molecules of water bond in three dimensions. At the surface of a container or body of water there is nothing to bond to in the upward direction, and the surface molecules have some excess bonding energy. This excess is utilized by increasing their bonds along the surface, which results in increased tension on the surface. This layer of increased molecular attraction, although only of the magnitude of one-millionth of a millimeter, has a significant influence on the physical behavior of water in a porous medium. *Surface tension* could be described as added cohesion between water molecules on the surface. Its value depends on the

---

[3]The first experimental proof of the compressibility of water was provided by philosopher Francis Bacon (England, 1561–1626), founder of the Royal Society (1660), who advocated the scientific analysis of observed facts.

TABLE 1.4   Surface Tension of Water

Temperature °C	Surface Tension, σ	
	N/m	lb/ft
0	0.075	0.00518
10	0.074	0.00508
20	0.072	0.00497
30	0.071	0.00486
40	0.069	0.00475

temperature and electrolytic content of the water. Table 1.4 shows the relationship between surface tension and temperature in water.

In the absence of other dominating forces (e.g., gravity in free fall, or in the weightlessness of space) surface tension can play an important role. For example, it causes the separation of water issued from a fire hose into water droplets. This reduces the effectiveness of the fire-stream beyond some distance from the nozzle. Firefighters use additives to reduce this effect.

Small amounts of electrolytes added to the water increase its surface tension. In groundwater, salts dissolved from adjacent soil particles tend to increase the electrolytic content and, hence, the surface tension in water. In contrast, organic substances like soaps, alcohol, and acids decrease the surface tension. The reducing effect of soap on surface tension makes it possible to stretch water film while blowing soap bubbles.

It is known that water rises in clays and fine silts to considerable heights. Clay layers become saturated with water even though the free water table exposed to the atmosphere is dozens of feet below. The cause of this *capillary rise* is complex. One of the reasons for capillary rise is the surface tension of the water. Another important contributor to the physical effect of capillary rise is the *adhesion* of water to most solid materials. Solids that have positive adhesion to water are called hydrophilic (water liking), and those that repel water are hydrophobic. The latter have negative adhesion to water.

Adhesion between fluids and solids is expressed by the *contact angle* at the edge of the contacting surfaces, as shown in Figure 1.2a. Hydrophobic materials have a contact angle with water that is larger than 90°. For example, the contact angle between water and paraffin is 107°; hence, paraffin is a good waterproofing agent. Silver, on the other hand, is neutral to pure water; its contact angle is nearly 90°. Quartz and other materials found in porous soils have a contact angle that is less than 90°; this means that they are wet well by water. The contact angle between ordinary glass and water containing impurities, for example, is about 25°. The adhesive forces between water and soil particles are so large that they can be separated only by evaporating the water. The *tensile strength* of water itself depends on the temperature and impurities present and ranges from 0.8 to 27 MPa (11,000–380,000 psi).

The capillary rise in the small pores of soils and in thin glass tubes is caused by the combined action of surface tension and adhesion. Figure 1.2b depicts the

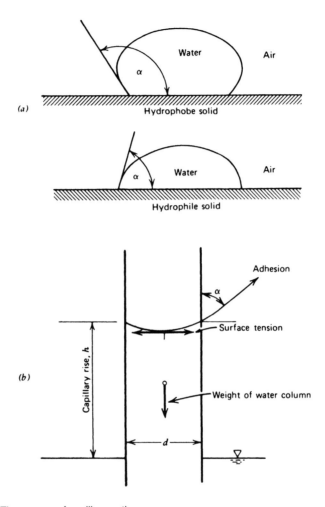

FIGURE 1.2 The causes of capillary action.

conditions present in a small-diameter glass tube in which capillary rise of water takes place. By its adhesion to the solid wall, water wants to cover as much solid surface as possible. However, by the effect of the surface tension the water molecules adhering to the solid surface are connected with a surface film in which the stresses cannot exceed the maximum possible surface tension of the water. The molecules in this surface film are joined to molecules below it by cohesive forces. As the adhesion drags the surface film upward, the film then raises a column of water filling the tube, against the force of gravity. The outcome of these factors is that the water in the small capillary tube, or in the small pores of soils, will rise upward against the force of gravity to a height at which the ultimate supporting capacity of the surface film is reached. Of course, the column of water below the surface film is under tension, which means that the water pressure in a capillary tube is below the atmospheric pressure. It is as if the water molecules hang from the surface film, held together by their molecular cohesion. The capillary rise is inversely proportional to the diameter of the tube or to the pore size in soils. Hence, the finer the soil grains, the thicker will be the capillary layer in the soil mass. Table 6.2 shows the average heights to which capillary water rises in various types of soils. In the idealized case of a small-diameter tube the height of the capillary rise, h, is

$$h = 4\sigma \frac{\cos \alpha}{d\gamma} \tag{1.5}$$

where d is the diameter of the tube, γ is the specific weight of the water, σ (the Greek sigma) is its surface tension, and α (the Greek alpha) is the contact angle characterizing the adhesion between the water and the tube. The angle α in this equation is usually assumed to be zero for water in very small, clean glass tubes; if the fluid is mercury, as in some manometers, the angle is 140°.

1.4 VISCOSITY

Perhaps the most important physical property of water is its resistance to shear or angular deformation. The measure of resistance of a fluid to such relative motion is called *viscosity*. Although viscous fluids are commonly thought of as "heavy," viscosity has nothing to do with density or unit weight. Convenient examples of fluids of different viscosity are water and honey. We define viscosity as a capacity of a fluid to convert kinetic energy, the energy of motion, into heat energy. The energy converted into heat is considered lost because it can no longer contribute to further motion. The energy either warms the fluid or is lost into the atmosphere by dissipation.

The energy required to move a certain amount of water through a pipe, open channel, or hydraulic structure is determined by the amount of *viscous shear losses* encountered along the way. Therefore, viscosity of the fluid inherently controls its movement. The effect of this viscous drag retarding the flow as it passes by piers is well demonstrated in Figure 1.3. Viscosity is due to the cohesion between fluid particles and also to the interchange of molecules between the layers of different velocities. Mathematically the relationship between viscous shear stress and viscosity is expressed by Newton's law of viscosity:

$$\tau = \mu \frac{\Delta v}{\Delta y} \tag{1.6}$$

which is merely an expression of proportionality between viscous shear resistance caused by the shear stress τ (the Greek tau) and the rate of change of velocity v in the direction perpendicular to the shear stress, as shown in Figure 1.4. The mechanism illustrated is in some ways similar to the case of a deck of cards dragged along on a table. The relative velocity between adjacent cards is Δv, the thickness of a card is Δy. The proportionality factor μ (the Greek mu) is called *absolute* (or *dynamic*) *viscosity* and has the dimension of force per area (stress) multiplied by

FIGURE 1.3 Irrigation flow control structure. Note curvature of streams as they pass the piers. (Courtesy of U.S. Bureau of Reclamation)

the time interval considered depending on the units of measurement of the velocity v. In science, viscosity is usually measured in centipoises (cp).

Conveniently, the absolute viscosity of water is 1.0 cp at 20.2°C, which is about room temperature. This fact allows us to use the absolute viscosity of water as a relative standard for viscosities of other fluids. In comparison, the absolute viscosity of air is about 0.17 cp and that of mercury about 1.7 cp. The absolute viscosity of water depends on its temperature, as shown in Table 1.5. It decreases at higher temperatures.

One hundred centipoises is a poise, which equals one gram per centimeter second, or 0.1 (N·s)/m². Accordingly,

$$1 \text{ cp} = 0.001 \frac{\text{N} \cdot \text{s}}{\text{m}^2} \tag{1.7}$$

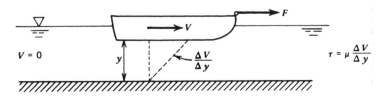

FIGURE 1.4 Interpretation of Newton's law of viscosity.

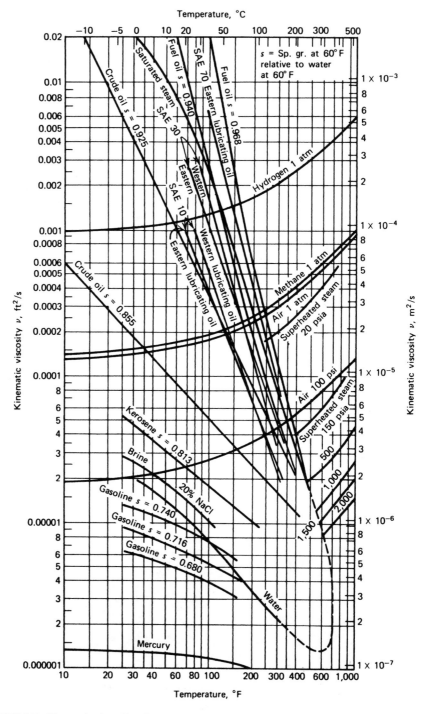

FIGURE 1.5 Kinematic viscosity of water and other fluids.

TABLE 1.5 Absolute Viscosity of Water

Temperature °C	Absolute Viscosity	
	N·s/m²	lb sec/ft²
0	1.79 × 10⁻³	0.374 × 10⁻⁴
4	1.56 × 10⁻³	0.327 × 10⁻⁴
10	1.30 × 10⁻³	0.272 × 10⁻⁴
20	1.00 × 10⁻³	0.209 × 10⁻⁴
20.2	1.00 × 10⁻³	0.208 × 10⁻⁴
30	0.80 × 10⁻³	0.167 × 10⁻⁴
50	0.549 × 10⁻³	0.114 × 10⁻⁴
70	0.40 × 10⁻³	0.084 × 10⁻⁴
100	0.28 × 10⁻³	0.059 × 10⁻⁴
150	0.18 × 10⁻³	0.038 × 10⁻⁴

In the conventional American system of units absolute viscosity is expressed in lb-sec/ft². Table 1.5 also gives absolute viscosities of water in conventional American units.

Both density and absolute viscosity are inherent in hydrodynamic equations. Thus, the absolute viscosity divided by the density of the fluid (at identical temperature) conveniently defines another property, the *kinematic viscosity*, ν, the Greek nu. The unit of kinematic viscosity in the SI system is m²/s; in the conventional American units its equivalent is ft²/sec. Figure 1.5 shows the values of kinematic viscosity for water and other common fluids and gases over a wide range of temperature. These will enable the user to apply hydraulic design formulas (e.g., flow in pipes) to fluids other than water.

Viscosity of water depends little on the pressure, but it varies considerably with the temperature. This phenomenon is often neglected, although in some cases the practical result is most noticeable. During winter the yield of wells feeding from nearby rivers may decrease by as much as 30 percent, a significant effect caused by this factor. Also, pumping warm water requires less energy than pumping cold water.

The determination of kinematic viscosity in the laboratory is a very simple procedure. Many viscosimeters have been standardized (Saybolt, Redwood, Engler, etc.), and tables are available for converting their results. One is shown in Appendix A, Table A.5. Most of these instruments operate under the principle that the amount of fluid that flows out of a hole into a cup of specified size depends on the kinematic viscosity of the fluid.

EXAMPLE PROBLEMS

Example 1.1

How much heat energy is required to melt 500 g of ice that is initially at −3°C and to raise the temperature of the water to 10°C?

Solution. The specific heat of ice is 1.95 J/g/°C. Hence, to raise the temperature to 0°

$$1.95 \times 500 \times 3 = 2925 \text{ J are required.}$$

Because the latent heat of fusion of water is 334 J/g, to melt the ice we need

$$334 \times 500 = 167{,}000 \text{ J}$$

Because the specific heat of water is 4.19 J/g/°C, to warm the water we need

$$4.19 \times 500 \times 10 = 20{,}950 \text{ J}$$

Therefore, the total heat requirement of the operation is

$$2925 + 167{,}000 + 20{,}950 = 190{,}875 \text{ J}$$

Because 1 BTU = 1055 J, this amount equals

$$181 \text{ BTUs} \qquad \qquad \square$$

Example 1.2

A pump is installed on a mountain, 3000 m high, where the atmospheric (barometric) pressure is 31 percent less than at sea level, and the ambient temperature is 10°C. What will be the theoretical maximum height of a still water column in the suction pipe of the pump?

Solution. If at sea level the standard atmospheric pressure is 10.33 m of water, the atmospheric pressure at the pump's location will be 31 percent less, that is,

$$10.33 - 0.31(10.33) = 7.13 \text{ m}$$

From Table 1.1 the vapor pressure of water is 1.2 kPa. This equals the pressure of a column of water 0.12 m high.

To find the theoretical maximum height of the still water column in the pump's suction pipe, we must reduce the ambient pressure by the vapor pressure, that is,

$$7.13 - 0.12 = 7.01 \text{ m}$$

Note that the increase in elevation decreased the atmospheric pressure. Also note in Table 1.1 that an increase in temperature increases the vapor pressure. If the pump in question were operating in hot water, the increase in vapor pressure would drastically reduce the theoretical maximum height of the water column in the suction pipe. $\qquad \square$

Example 1.3

A container weighs 3.22 lb force when empty. Filled with water at 60°F, the mass of the container and contents is 1.95 slugs. Find the weight (in pounds) and volume (in cubic feet) of the water in the container.

Solution. One lb force (lbf) is defined as the force required to give 1 lb mass (lbm) an acceleration equal to that of gravitational acceleration. By Newton's second law of motion

$$1 \text{ lbf} = \frac{1 \text{ lbm} \times 32.2 \text{ ft/sec}^2}{g_c}$$

Where g_c is a dimensional adjustment term and is given by

$$g_c = \frac{32.2 \text{ lbm-ft}}{\text{lbf-sec}^2}$$

Because a 1 lb force is also defined as the force required to give a mass of 1 slug an acceleration 1 ft/sec^2, then by definition

$$1 \text{ slug} = 32.2 \text{ lbm}$$

Therefore,

$$1.95 \text{ slugs} = 1.95 \times 32.2 = 62.79 \text{ lbm}$$

Under gravitational acceleration this equals 62.79 lb force. The weight of the container equals 3.22 lbf, so that the weight of water is $62.79 - 3.22 = 59.57$ lbf. Because the unit weight of water in water is 62.4 lbf/ft³, the volume of water in the container is

$$\frac{59.57 \text{ lbf}}{62.4 \text{ lbf/ft}^3} = 0.95 \text{ ft}^3 \qquad \square$$

**Example
1.4**

A container weighs 5 kg when empty and 87 kg when filled with water. What volume of water can it hold?

Solution. The weight of water in the container is $87 - 5 = 82$ kg. By definition 1 kg of water has a volume of 1 liter; then,

$$82 \text{ kg of water} = 82 \text{ liters}$$

Because

$$1000 \text{ liters} = 1 \text{ m}^3$$

hence,

$$82 \text{ liters} = 0.082 \text{ m}^3 \qquad \square$$

**Example
1.5**

The temperature of 0.5 m³ water is 10°C. What is the change in volume if the water is heated to 25°C?

Solution. Heating will not change the total mass; hence, we can write that

$$\text{volume} \times \text{density} = \text{mass} = \text{constant}$$

or

$$V_1 \times \rho_1 = V_2 \times \rho_2 = \text{constant}$$

Using Table 1.2 and denoting the initial state with subscript 1, we have

$$V_1 = 0.5 \text{ m}^3$$
$$\rho_1 = 999 \text{ kg/m}^3$$
$$\rho_2 = 997 \text{ kg/m}^3$$

From which

$$V_2 = V_1 \frac{\rho_1}{\rho_2} = 0.5 \frac{999}{997}$$
$$= 0.501 \text{ m}^3$$

That is, the volume will increase by 1 liter due to the increase in temperature. $\qquad \square$

**Example
1.6**

A 0.5 m³ volume of water is under 1000 Pa of pressure initially. If the pressure is increased by 9000 Pa, and the temperature remains constant, what is the reduction in volume?

Solution. Equation 1.3 is used with the following known variables:

$$\Delta p = 9000 \, \text{Pa}$$
$$V_0 = 0.5 \, \text{m}^3$$
$$E = 2.14 \times 10^9 \, \text{Pa}$$

Substituting, we have

$$9000 = -2.14 \times 10^9 \frac{\Delta V}{0.5}$$

from which

$$\Delta V = -2.1 \, \text{cm}^3$$

is the reduction in volume.   □

Example 1.7

A 100 ft³ volume of water is under 1500 psi of pressure initially, but the pressure is increased 10-fold. The temperature increases from 68 to 200°F. What is the change in volume?

Solution. Table 1.3 may be used to solve this problem. Because 1500 psi is increased to 15,000 psi, we can read off the values of the corresponding elasticities as 330,000 and 405,000 psi, respectively. Using linear interpolation, we may assume that the average value of elasticity is

$$E_{\text{average}} = \frac{405,000 + 330,000}{2} = 367,500 \, \text{psi}$$

The increment in pressure is

$$\Delta p = 15,000 - 1500 = 13,500 \, \text{psi}$$

Substituting into Equation 1.3, we get

$$13,500 = -367,500 \frac{\Delta V}{100}$$

from which

$$\Delta V = 3.67 \, \text{ft}^3$$

or, some 3.7 percent of the initial volume.   □

Example 1.8

A raft 3 × 6 m in size is dragged at a velocity of 1 m/s in a shallow channel 0.1 m deep measured between the raft and the channel bottom. Compute the necessary dragging force, assuming that Equation 1.6 is valid and that the water temperature is 20°C.

Solution. From Equation 1.6

$$\tau = \mu \frac{\Delta v}{\Delta y} = \mu \frac{1 \, \text{m/s}}{0.1 \, \text{m}}$$

Assuming that the velocity changes linearly from zero at the channel bottom to 1 m/s at the raft and obtaining from Table 1.5, μ at 20°C, we see that

$$\mu = 1.0 \times 10^{-3} \frac{N \cdot s}{m^2}$$

and

$$\tau = 1.0 \times 10^{-3} \frac{N \cdot s}{m^2} \frac{1.0\ m/s}{0.1\ m}$$

$$= 1.0 \times 10^{-2} \frac{N}{m^2}$$

The dragging force of the 3×6 m^2 raft is then

$$F = \tau A = (1.0 \times 10^{-2})18 = 0.18\ N$$

In English units 1 lb equals 4.45 N; hence, the dragging force is

$$F = 0.04\ lbf \qquad \square$$

Example 1.9

In pipe flow problems the viscosity of the water is taken into account in a dimensionless term called the Reynolds number, expressed as $\mathbf{R} = v \cdot D/v$. Compute the Reynolds number for a flow of water at 25°C in a pipe of 40 in. diameter and at a velocity of 3.3 fps.

Solution. The density of water at 25°C equals 1.935 slugs/ft^3 according to Table 1.2, and by linear approximation in Table 1.5 the viscosity is

$$\mu = 1.88 \times 10^{-5}\ lb\text{-sec/ft}^2$$

Then from

$$\mathbf{R} = \frac{vD\rho}{\mu}, \text{where } v = \frac{\mu}{\rho}$$

$$= \frac{3.3 \times 40/12 \times 1.935}{1.88 \times 10^{-5}}$$

$$= 1.1 \times 10^6 \qquad \square$$

Example 1.10

The seepage loss from a reservoir was 3 in. per month during the summer, when the average water temperature was 30°C. Assuming that the depth of water in the reservoir is constant and that the groundwater level is well below the reservoir bottom, such that it does not influence the seepage losses, estimate the seepage losses during the winter months, when the average water temperature is 4°C.

Solution. The seepage out of a reservoir depends on the coefficient of permeability of the soil on the bottom of the reservoir. Equation 6.3 shows that the permeability is proportional to the unit weight of the water and inversely proportional to the viscosity of the water. Rearranging this equation and substituting $\gamma = \rho g$ we get

$$k = \frac{(K_1 \mu_1)}{(\rho_1 g)} = \frac{(K_2 \mu_2)}{(\rho_2 g)}$$

or

$$\frac{K_2}{K_1} = \left(\frac{\rho_2}{\rho_1}\right)\left(\frac{\mu_1}{\mu_2}\right)$$

where K_1 is the summer permeability, and K_2 is the winter permeability. Substitution of the appropriate values from Tables 1.2 and 1.5 yields

$$\frac{K_2}{K_1} = \left(\frac{1000}{996}\right)\left(\frac{0.80}{1.56}\right) = 0.51$$

The winter permeability is only 51 percent of the summer permeability, so the monthly winter seepage is

$$0.51 \times 3 \text{ in.} = 1.5 \text{ in.}$$

Note that most of this change is because of a change in viscosity. Over this temperature range water density changes only (1000/996) or 0.4 percent.   □

PROBLEMS

1.1 How much heat must be removed from 10.0 m³ of water to freeze it if the initial temperature of the water is 25°C? Give the answer in joules and BTUs.

1.2 A pump is located at a site where the atmospheric pressure is only 90 percent of the standard atmospheric pressure. Determine the minimum allowable pressure at the suction side of a pump moving water at 20°C. How much does the minimum allowable pressure change if the water temperature is increased to 80°C?

1.3 What is the difference in weight of 50 gal of water at 39°F and at 77°F?

1.4 Determine the reduction of capillary rise if the contact angle representing the adhesion between water and soil material is increased from 30° to 60°.

1.5 A 1.00 m³ volume of water is heated from an initial temperature of 0°C to 100°C. Determine the change in volume.

1.6 How much pressure must be applied on water to reduce its volume by 0.1 percent (assume constant E)?

1.7 Prove that the units of absolute viscosity expressed in conventional American units (lb-sec/ft²) equal 479 poises.

1.8 Estimate the kinematic viscosity of water at 85°C, expressing it in both conventional American and metric units.

1.9 By using Equation 6.3, which gives Darcy's permeability coefficient, determine the change in permeability if the temperature of the water rises from 10°C to 85°C.

1.10 Determine the error in the discharge measured by an elbow meter that was calibrated at 25°C if the water flowing in the elbow is at 4°C (see Equation 3.13).

1.11 What is the change in Reynolds number (Equation 2.23) if the temperature of the fluid changes from 50°F to 200°F?

1.12 How many joules are required to boil away 1.0 gal of water initially at 10°C?

1.13 Refer to Example 1.8. If the same raft was dragged with the same force through water at 4°C, what would be its velocity?

1.14 A container weighs 2.06 lb. It is filled with 1.09 ft^3 of water at 25°C. What is its final weight in pounds and newtons?

1.15 Use Equation 1.5 to determine the change in h for water for a temperature increase from 4°C to 25°C (assume α and d remain constant).

MULTIPLE CHOICE QUESTIONS

1.16 The density of water depends on
   A. Temperature, pressure, and salt content.
   B. Electrolytic content, temperature, surface tension.
   C. Salt content, cohesion, and viscosity.
   D. Volume and weight.
   E. All of the above.

1.17 Modulus of compressibility of water is a function of
   A. Ambient pressure and density.
   B. Pressure change and volume.
   C. Volumetric change and ambient pressure.
   D. Volumetric change and pressure change on a certain volume.
   E. All of the above.

1.18 Elasticity of water
   A. Depends on its temperature and pressure.
   B. Is greater than that of steel.
   C. Is a minimum at 120°F.
   D. Is a negative quantity.
   E. Decreases when the pressure increases.

1.19 Surface tension in water is
   A. Larger if it contains salts.
   B. Smaller if the temperature is increased.
   C. One of the causes of capillary rise.
   D. Smaller if it is mixed with whiskey.
   E. All of the above.

1.20 Shear stresses in water depend on the
   A. Rate of change of velocity and the dynamic viscosity.
   B. Rate of change of velocity and the kinematic viscosity.
   C. Temperature and viscosity of water.
   D. Ambient pressure, temperature, and elasticity.
   E. Cohesion between water molecules.

1.21 Newton's law of viscosity states that
   A. For every action there is an equal and opposite reaction.
   B. The fluid will maintain its speed and direction unless acted upon by a force.
   C. Viscous force equals mass times acceleration.
   D. The shear stress is proportional to the rate of change of velocity.
   E. None of the above.

1.22 Kinematic viscosity and absolute viscosity are
   A. Equal but opposite in sign.
   B. Inversely proportional to each other.

    C. Directly proportional to temperature.

    D. Independent of the temperature and pressure.

    E. Related by the density of the fluid.

1.23 A phase change in water

    A. Depends on the temperature and the pressure.

    B. Requires the introduction or removal of heat energy.

    C. Requires a change in heat or pressure.

    D. Will not alter its specific heat.

    E. Is called fusion.

1.24 Sublimation is a phase change of water meaning

    A. The evaporation of ice.

    B. The opposite of fusion.

    C. A rapid loss of internal energy content.

    D. An increase in specific heat.

    E. None of these.

1.25 The vapor pressure of water is usually given in absolute pressure because

    A. The variation of height must also be considered.

    B. It equals the boiling point of water, which, in turn, depends on the atmospheric pressure.

    C. It is usually important in pumps and pipelines.

    D. Is standardized at 14.7 psi, 68°F, at 45° latitude.

    E. None of these.

1.26 Latent heat means

    A. The specific heat left in matter after freezing.

    B. The internal energy required for evaporating ice.

    C. The internal energy lost during the melting of ice.

    D. The heat of energy required to boil away water.

    E. None of these.

1.27 Water may be evaporated at constant temperature if the pressure is increased on it.

    A. True.

    B. False.

    C. It depends on the temperature.

1.28 The difference in the concept of unit weight between the conventional American system and SI is

    A. That in SI it is mass, whereas in conventional American units it is force.

    B. That in conventional American units weight is measured in slugs, whereas in SI it is measured in newtons.

    C. In magnitude only.

    D. Due to the different interpretation of pound mass and force mass.

    E. That specific weight in SI refers to cubic meters instead of cubic feet.

1.29 Unit weight divided by density gives

    A. Kinematic viscosity.

    B. Unit density.

    C. Specific weight.

    D. Gravitational acceleration.

    E. Slugs (per unit volume).

1.30 Pound force divided by pound mass equals
A. Gravitational acceleration, 32.2 ft/sec².
B. Unit acceleration.
C. One slug.
D. A dimensional adjustment constant.
E. None of these.

1.31 If the elasticity of water is 2.1 GPa and the atmospheric pressure is 101 kPa, how much will 25 m³ of water expand if the pressure is reduced by 50 percent?
A. 601 milliliters.
B. 250 liters.
C. 0.5 m³.
D. 12.5 liters.
E. Not at all.

1.32 Fifteen liters of water is frozen at −7°C. How much energy is needed to turn it into superheated steam at 110°C if the pressure is kept constant.
A. 1450 J.
B. 223 kJ.
C. 0.5 MJ.
D. 6.6 GJ.
E. 45 MJ.

1.33 The kinematic viscosity of water at room temperature is about
A. 100,000 m · m/s.
B. 9.81 m/s · s.
C. 0.000001 m · m/s.
D. 9810 Pa/s.
E. None of the above.

1.34 The absolute viscosity of water at room temperature is about
A. 0.001 Pa · s.
B. 1 Pa/s.
C. 1 poise.
D. 1.0 N · m/s.
E. 9810 kN/s.

1.35 The dimensions of the poise are
A. Force · length squared/time.
B. Mass · length/time squared.
C. Force squared · time/length squared.
D. Force · time/length squared.
E. None of the above.

2

Fluid Mechanics

The science of hydraulics is built on the foundation of engineering mechanics. Essentially, mechanics is the application of Newton's laws of motion and the laws of conservation known from physics. Knowledge of the basic physical laws controlling the motion of all fluids is essential for understanding the behavior of water. These fundamental principles are reviewed in this chapter. We will refer to them frequently in all later chapters.

2.1 HYDROSTATIC PRESSURE AND FORCE

Structures constructed for the purpose of retaining water are subject to hydrostatic forces as long as water is at rest. The hydrostatic force is the result of hydrostatic pressures acting on solid surfaces retaining the water. Hydrostatic pressure at a point is expressed by

$$p = \gamma y \qquad\qquad (2.1)$$

in which γ is the specific weight of water as shown in Equation 1.2, and y is the depth of the water over the point at which p is sought. The depth y is always measured in the direction of the gravitational acceleration.

Because water is a fluid, it cannot resist shear forces as solid materials do. For this reason the pressure p of water at a point acts in all directions. Therefore, if one wishes to represent the water pressure at one point in the form of a vector, the result is a sphere of radius p around the point. For the same reason when the pressure is evaluated along the containment boundaries of the water body, the direction of the pressure vector on the boundary must be perpendicular to the boundary surface. Also, it follows from Equation 2.1 that the total amount of water present has no bearing on the magnitude of the pressure, since it depends only on the depth vertically below the surface of the water. On the water surface the pressure is atmospheric.

If hydrostatic pressure acts on an elementary small surface area ΔA an elementary hydrostatic force $\Delta \boldsymbol{F}$ develops that may be computed by the formula[1]

$$\Delta \boldsymbol{F} = p\, \Delta A = \gamma y\, \Delta A \tag{2.2}$$

Note that in this equation both the force and the area are vector quantities; they have direction as well as magnitude. Because water cannot support shear forces without motion, the hydrostatic force must be perpendicular to the surface area on which it acts. Hence, the orientation of the area ΔA in Equation 2.2 determines the direction of the elemental hydrostatic force $\Delta \boldsymbol{F}$ that acts over it.

The total hydrostatic force $\boldsymbol{F}$ acting on a surface area $\boldsymbol{A}$ is determined by summing all elementary force components acting on all small ΔA surface components. The resultant hydrostatic force can be considered as acting at a single point of the surface. This concept is identical to the resolution of a point load from a distributed load in statics. The point on the surface is called the *pressure center*. It is placed at the center of gravity of the distributed load.

Depending on the orientation of the surface area, computation of hydrostatic forces is more or less simple. In the case of a horizontal plane, the water pressure is constant throughout. It is given by the product of the unit weight and the depth of the plane below the surface of the water. The orientation of the force depends on which side of the surface the water is located. Figure 2.1 shows two such cases. In one case the water is above the surface; hence, the force is acting downward. In the other case the water is below the surface, so the force acts upward. In both cases the pressure center is at the centroid of the area.

2.2 HYDROSTATIC FORCES ON VERTICAL AND SLANTED PLANES

If a plane surface area is not horizontal, the pressure acting on it will vary according to the variation of the depth. There are two ways to determine the resultant hydrostatic force. A pressure diagram may be constructed by computing pressures at the deepest and at the shallowest points of the area. As the pressure varies linearly with the depth, these two points may be connected by a straight line giving the pressure distribution (Figure 2.2a) along the surface. The same result is obtained if the pressure is first separated into its horizontal and vertical components, as shown in Figure 2.2b. The magnitude of the force is the sum of the horizontal and vertical components. The hydrostatic force will not act at the centroid of area $\boldsymbol{A}$ because the pressure is not distributed evenly over it; it is located at the centroid of the pressure distribution. This *pressure center* is always below the centroid of

---

[1]This principle was first recognized by Simon Stevin (Holland, 1548–1620), a mathematician who developed the triangle of forces and predicted the decimal system's use in coinage, weights, and measures.

FIGURE 2.1 Water pressure on horizontal planes.

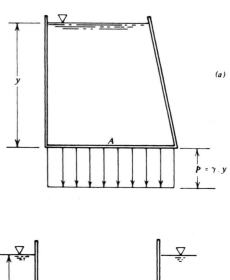

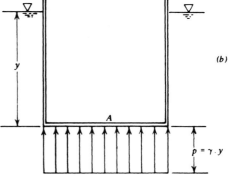

the area at a distance e, depending on the geometry of the problem. The distance e depends on the shape of the surface in question and its position with respect to the vertical coordinate.

Two important concepts of mechanics, namely, the first and second moments of areas, must be understood before the location of this pressure center can be determined. The *first moment* of a plane area, with respect to a coordinate line in the same plane, is computed by multiplying the magnitude of the area A by the perpendicular distance l between its centroid and the coordinate line, as shown in Figure 2.3. That is,

$$M = Al \qquad (2.3)$$

The *second moment* of an area, also called *moment of inertia*, is somewhat more complicated. If we are to determine I_0, the second moment of an area with respect to a coordinate line crossing its centroid, we have to sum the products of each small elemental surface component and its respective distance from the coordinate line. Generally, this can be performed by integration. However for practical work it is sufficient to refer to tables showing values of I_0 for simple areas. Table 2.1 contains

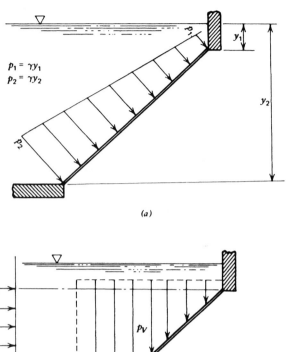

$p_1 = \gamma y_1$
$p_2 = \gamma y_2$

(a)

(b)

FIGURE 2.2 Pressure on slanted planes.

such information. For coordinate lines other than those crossing the centroid, that is, offset by a perpendicular distance c, second moments can be computed from

$$I_c = I_0 + Ac^2 \tag{2.4}$$

where I_c is the moment of inertia with respect to the coordinate line, I_0 is the moment of inertia with respect to the parallel coordinate line crossing the centroid, A is the area, and c is the offset distance between the two coordinate lines. Figure 2.4 depicts these variables.

Once the concept of first and second moments of areas is understood, we are ready to determine the location of the pressure center on the plane surface. If we refer to Figure 2.5 for our notations, the distance e between centroid and the pressure center of the hydrostatic force is given by

FIGURE 2.3 First moment of an area.

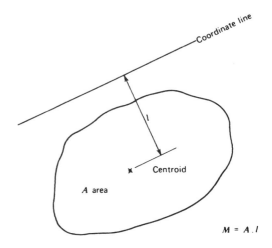

$$e = \frac{I_0}{lA} \tag{2.5}$$

where I_0 is the second moment of area A with respect to its centroid, l is the distance between the centroid and the line of intersection of the plane of A with the water level. The distance e between the centroid and the pressure center of the hydrostatic force is measured on the surface on which it acts. The magnitude of the hydrostatic force F acting on a plane surface of arbitrary orientation is the product of the hydrostatic pressure at the centroid of the plane area and the magnitude of the area, or

$$F = A\gamma y_c \tag{2.6}$$

where y_c is the depth of water above the centroid. The location where this force acts is below the centroid for all but horizontal surfaces. The reason for this is that on an inclined plane deeper locations have larger pressures.

The depth of water over the pressure center is given by

$$y_a = (l + e)\sin \alpha \tag{2.7}$$

For a vertical plane α equals 90°, in which case $(l + e)$ equals y_a. In Table 2.2 the location of the pressure center below the centroid is given for some vertical geometric arrangements often found in practice.

2.3 FORCES ON CURVED SURFACES

Hydrostatic forces acting on curved surfaces are somewhat more difficult to evaluate. In such cases a graphical approach to the computations is advantageous. Consider, for example, the segment gate shown in Figure 2.6a. Because the normal lines of

TABLE 2.1   Location of Centroid, Area, and Moment of Inertia of Common Shapes

Rectangle:		$A = b \cdot h$	$I_0 = b \cdot h^3/12$
Triangle:		$A = b \cdot h/2$	$I_0 = b \cdot h^3/36$
Circle:		$A = \pi D^2/4$	$I_0 = \pi R^4/4$
Semicircle:		$A = \pi R^2/2$	$I_0 = 0.110 R^4$
Ellipse:		$A = \pi b \cdot h/4$	$I_0 = \pi b \cdot h^3/64$
Parabolic section: $y_0 = 3h/5$ $x_0 = 3b/8$		$A = 2b \cdot h/3$	$I_0 = 8bh^3/175$

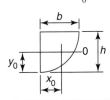

all surface elements cross at the pivot point of the gate, which is the center of the circle of which the gate is a segment, the resultant force will go through this pivot point also. Also note, however, that at each surface point the magnitude of the pressure as well as its direction varies. Thus, it is convenient to separate the horizontal and vertical components of the pressure. This results, as shown in Figure 2.6b, in two rather simple pressure diagrams: a triangular one for the horizontal pressures and a segment-shaped one for vertical pressures. For the latter, depths below the water level at various points on the gate need to be multiplied by the

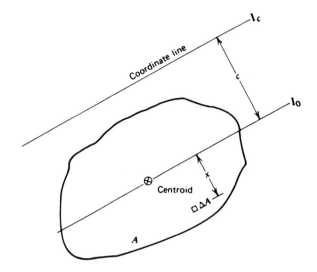

FIGURE 2.4 Interpretation of the moment of inertia. $I_0 = \sum x(\Delta A) \cdot I_c = I_0 + Ac^2$.

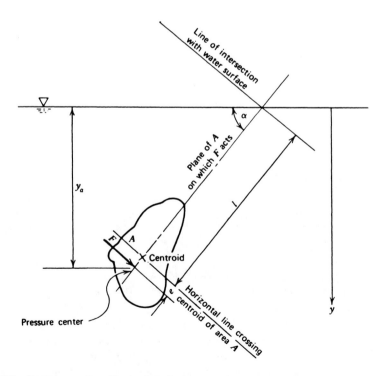

FIGURE 2.5 Pressure center of hydrostatic force.

TABLE 2.2   Location of Pressure Center of Hydrostatic Force on Vertical Gates

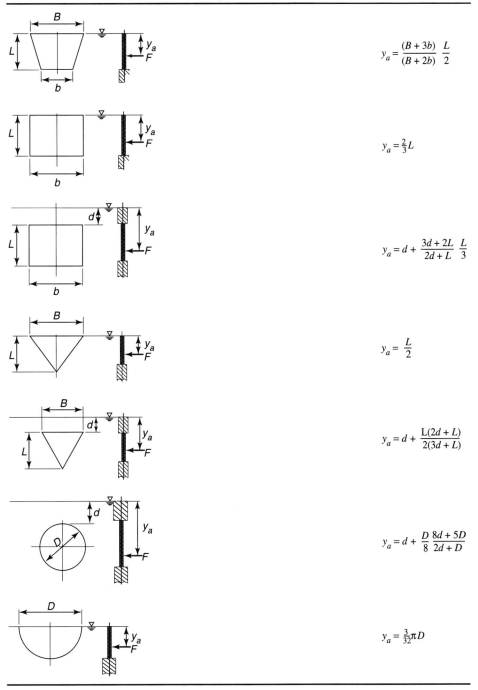

$$y_a = \frac{(B + 3b)}{(B + 2b)} \frac{L}{2}$$

$$y_a = \tfrac{2}{3} L$$

$$y_a = d + \frac{3d + 2L}{2d + L} \frac{L}{3}$$

$$y_a = \frac{L}{2}$$

$$y_a = d + \frac{L(2d + L)}{2(3d + L)}$$

$$y_a = d + \frac{D}{8} \frac{8d + 5D}{2d + D}$$

$$y_a = \tfrac{3}{32}\pi D$$

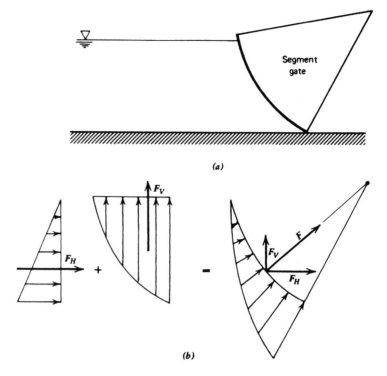

FIGURE 2.6 Hydrostatic pressure on curved surfaces.

unit weight of the water as a scaling factor in order to express vertical pressures. The magnitude of the corresponding hydrostatic forces is given by the area of the pressure diagrams. There forces act through the centroid of their respective areas. So, to find the horizontal and vertical components of the resultant hydrostatic force, we need only find the centroids of the two pressure diagrams. The sum of these two forces may be added vectorially. The resultant acts through the point of intersection of the two component forces.

2.4 BUOYANCY

The concepts developed in the previous paragraph apply also to *floating objects*. As shown in Figure 2.7, horizontal force components on the two sides of the body cancel each other. The vertical components all but cancel each other except for the volume where the floating body displaces water, resulting in a force acting upward. Accordingly, this buoyant hydrostatic force on a floating body equals the weight of the displaced water, and it acts at the center of the displaced mass. This is known as the law of Archimedes (Greece, 282–212 B.C.).[2] Of course, the floating

---

[2]Concepts outlined in his two-volume manuscript containing the analysis of hydrostatics and flotation are still valid.

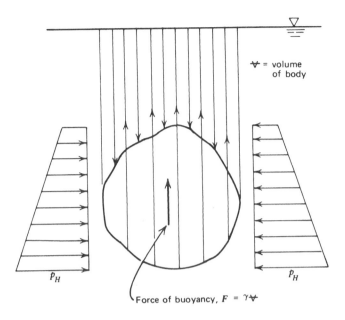

$\forall$ = volume
of body

Force of buoyancy, $F = \gamma \forall$

FIGURE 2.7 The force of buoyancy.

body itself also exerts its own weight. This force acts downward, opposing the buoyant force. Bodies floating on the water surface will submerge to a point at which the weight of the displaced water equals their own weight. Whether a body will sink to the bottom, rise to the top, or experience an apparent weightlessness in water depends on whether the hydrostatic uplift is smaller than, larger than, or equal to the body's weight. This is the principle behind the operation of submarines.

The resultant force of the hydrostatic pressure acting on the outer boundaries of a floating object, what we call *buoyant force*, acts through the centroid of the body. Depending on the weight distribution inside the floating body, its mass center may or may not coincide with the centroid. The position of the mass center relative to the centroid controls the *stability* of the floating body. For the floating body to be stable in its position, the mass center should be vertically below the centroid. Should the mass center be above or off to the side of the centroid, the body will be unstable and will turn until stability is established. Ships, one of humankind's greatest inventions, are in a special category. Even though their mass center is above the centroid of the water displaced by the hull, they will not capsize. This is because ship hulls are so designed that when tilted to some degree, their centroid shifts to the side, causing the ship to right itself.

2.5 FLOW RATE

In the science of fluid mechanics we distinguish between compressible and incompressible fluids. Laws relating to the behavior of compressible fluids like air, steam,

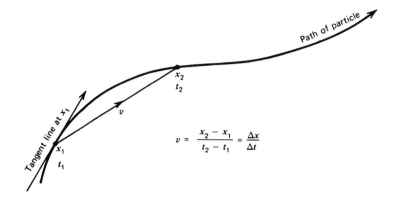

FIGURE 2.8 A velocity vector.

or gases fall within the domain of aerodynamics, or gas dynamics. Because water in its liquid form retains a constant volume for a given amount of mass for all pressures encountered in hydraulic applications, it is considered to be incompressible. The assumption of incompressibility simplifies the fundamental fluid mechanics concepts. For example, it allows us to measure amounts of water in volumetric terms instead of in mass.

Discharge, Q, meaning volume of water flowing through a certain cross-sectional area during a specified time period, is usually measured in cubic feet per second or cubic meters per second. In some specific applications other discharge units may be found. For example, discharge of pumps is quoted in gallons per minute, output of water treatment plants may be given in million gallons per day, and so forth. A table of conversion factors including the most frequently used discharge units appears in Appendix A, Table A.3.

Velocity, v, means the change of position of a water particle in the moving fluid during a certain time interval. In Figure 2.8, for example, a particle may be found at a position x_1 at time t_1. If the same particle at a later time t_2 is found at position x_2, then the velocity of the particle is defined as $(x_2 - x_1)/(t_2 - t_1)$. As we take shorter and shorter time intervals $(t_2 - t_1)$, the path of the particle between the end points x_1 and x_2 becomes less and less distinguishable from the straight line connecting the end points. For an extremely small interval (Δt) the line becomes tangent to the path at point x_1. In this case we speak of a velocity vector at that point on the path. The direction of a velocity vector shows which way the particle is moving. The magnitude of the velocity vector signifies how fast the particle is moving. Now, if we are looking at the point x_1 as a location fixed in space through which a stream of fluid particles passes, we can define another important principle: If successive water particles passing through x_1 have identical velocity vectors, we speak of *steady flow.* Conversely, if the velocity vector changes for successive fluid particles, we have *unsteady flow.* Unsteady flow is time-dependent. The assumption of steady flow simplifies the analysis of hydraulic problems.

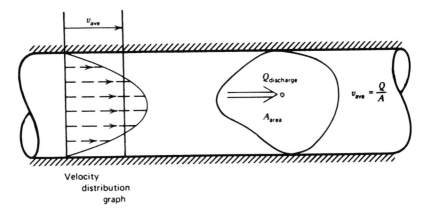

FIGURE 2.9   Average velocity.

In practice, we often find that as a discharge Q flows through a cross-sectional area A of a flow channel, the velocity of the fluid particles flowing through each point of the area is different. Usually, we find the highest velocities in the central portion of the cross section, whereas near the boundaries the velocity may be almost zero. It is convenient in such cases to define the *average velocity* as

$$v_{ave} = \frac{Q}{A} \qquad\qquad (2.8)$$

where A is considered in a direction perpendicular to the flow, as shown in Figure 2.9. The *velocity distribution graph* appears on the left side of this figure.

Acceleration of a fluid means a change in velocity. There are two kinds of accelerations: The velocity of a moving water particle can change in place as well as in time. In a pipe of enlarging diameter, as shown in Figure 2.10, the velocity

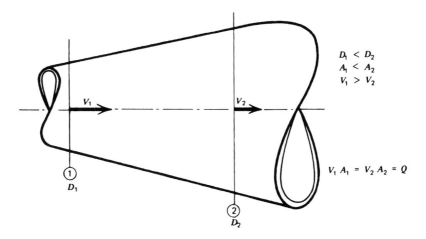

FIGURE 2.10   Continuity relationships without storage.

of the flow decreases as it passes from a section of small diameter to a section of large diameter. This form of deceleration, or negative acceleration, is called *spatial acceleration,* since the change of velocity occurs in space. If the discharge flowing through a certain cross-sectional area varies with time, we speak of *temporal acceleration.* For almost all topics considered in this book we will assume that the flow is steady, meaning that no temporal accelerations exist. Internal accelerations within the fluid result from the relative movement of the water because of external forces. By relative movement we mean the motion of water with respect to its surroundings. The external forces may be created by the gravitational acceleration, g, that acts everywhere on Earth. The driving force, in this case, can be described by Newton's law of motion. If the only acceleration acting on the fluid particles is the one due to gravity, and the water is contained, there is no cause for relative motion between the fluid particles. As long as the water is contained, it will be at rest. Forces caused by water at rest are in the domain of hydrostatics, discussed earlier. Forces caused within the moving mass of fluid, on the other hand, are in the domain of *hydrodynamics.* Hydrodynamics is the study of the behavior of water in motion. Its three basic physical laws are the law of conservation of mass, the law of energy transfer, and the law of momentum transfer.

2.6 CONSERVATION OF MASS

The law of conservation of mass is one of the three conservation laws of physics. Simply put, this law states that mass cannot be created or destroyed. This concept gives rise to the *equation of continuity:*[3] Within any hydraulic system the discharge flowing in, the stored volume within, and the discharge flowing out must be balanced. In other words, all volumetric quantities must be accounted for. Because we consider water to be incompressible, mass and volume may be used interchangeably in this equation. In mathematical form, we may write the continuity equation as

$$Q_{in} - Q_{out} = \text{change in storage} \qquad (2.9)$$

The preceding equation is often used in the analysis of reservoirs and in the routing of floods in rivers. It is important that the time scale be the same on both sides of the equation. For example, if the discharges are available in cubic feet per second (cfs), and the change in storage is desired for periods of, say, 6 hours, an appropriate conversion factor must be introduced.

In the case when no change in storage is possible, as in a pipe flowing full, the right side of Equation 2.9 reduces to zero; that is,

$$Q_{in} - Q_{out} = 0 \qquad (2.10)$$

which means that what goes in, must come out. Figure 2.11 illustrates these concepts. In certain applications, as in rivers or complex pipelines, the flow path is broken into smaller components joined together at certain points. In this case, care should

---

[3]Leonardo da Vinci (Italy, 1452–1519) is credited with the first formal statement of this law.

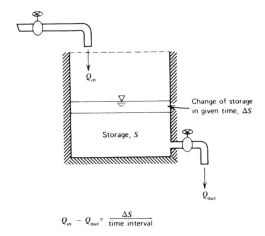

FIGURE 2.11 Continuity relationships with storage.

Q_{in}

Change of storage in given time, ΔS

Storage, S

Q_{out}

$$Q_{in} - Q_{out} = \frac{\Delta S}{\text{time interval}}$$

be exercised in selecting the proper sign for each discharge component. A common sign convention is to consider discharges flowing into the hydraulic component as positive and outflowing discharges as negative. At points where two hydraulic components are joined, the sign of the discharge changes.

2.7 FORCES AND MOMENTUM

The first step of any fluid flow analysis is to define the section of the fluid that is being considered. This portion of the space occupied by the fluid is called a *control volume*. Surfaces of a control volume are generally either perpendicular or tangential to the velocities of the flow. Tangential surfaces are either solid boundaries within which the flow takes place, such as the internal surfaces of a pipe section, or the free surface of the water flowing in a channel on which the pressure is atmospheric. The other surfaces of the control volume are normal (that is, perpendicular) to the velocities of the flow as they enter or leave the control volume. The control volume is permanently fixed in space and time. As an example, Figure 2.12a shows the control volume associated with a constricting pipe elbow. The concept is identical to the free-body diagram known in statics where components removed are replaced by equivalent forces. Once the control volume associated with a problem is well defined, the next step is to enumerate and to show all forces, pressures, and velocities that are present and acting within or on the surfaces of the control volume. These are the velocities entering and leaving the control volume, the pressures or forces acting on the surfaces normal to the flow, the shear forces and normal forces (or stresses) along the solid boundaries, and the gravitational and other body forces acting on the fluid enclosed by the control volume. Within our control volume, care should be taken to distinguish between forces created by the fluid and those acting on the fluid. The problem presented by control volumes is essentially one of statics: The control volume is fixed in space; hence, the resultant of the active forces is

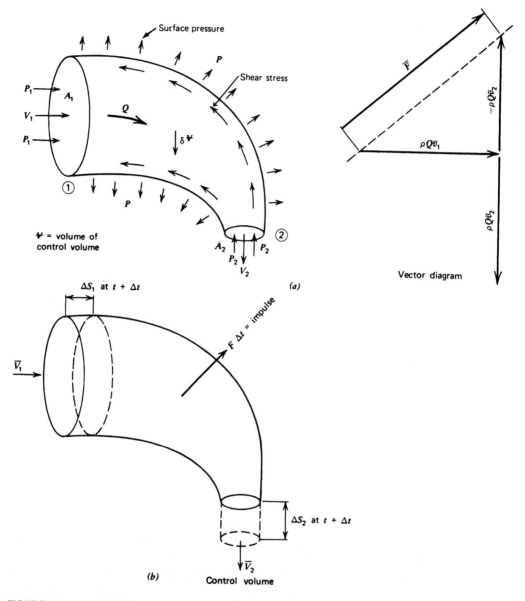

Surface pressure

Shear stress

P_1
A_1
V_1
P_1

Q

$\delta \Psi$

Ψ = volume of control volume

P

①

A_2 P_2
P_2
V_2

②

(a)

ΔS_1 at $t + \Delta t$

$\bar{V}_1$

$F \Delta t$ = impulse

ΔS_2 at $t + \Delta t$

$\bar{V}_2$

(b) Control volume

$\bar{F}$

$-\rho Q \bar{v}_2$

$\rho Q \bar{v}_1$

$\rho Q \bar{v}_2$

Vector diagram

FIGURE 2.12 Control volume.

countered by an equal but opposite reaction force. For example, in the case of a pipe elbow, the force of resistance is that force exerted by the walls of the pipe on the fluid, making it change direction.

In many applications the problem is simplified because several of the forces present may be of negligible influence. For example, in the case of the constricting pipe elbow shown in Figure 2.12a the friction forces acting between the fluid and the pipe wall are negligible. Also negligible are the weight of the fluid, the portion of the body force due to gravitational acceleration, and the resulting small difference in the normal surface pressures exerted by the fluid on the pipe walls. The sum of all external forces acting on a control volume is denoted by $\Sigma\,\boldsymbol{F}$.

Newton's second law of motion states that fluids tend to remain at rest, or continue in a state of uniform motion, unless acted on by external forces. If the velocity of a fluid changes in either magnitude or direction, or both, as it passes through a control volume, a *net* external force must be present to cause this change. In the case of steady flow there is no temporal acceleration. In this case the acceleration is the vectorial difference of the velocity flowing out of the control volume and the velocity flowing into it. For flowing fluids the mass in Equation 1.1 may be replaced by the *mass flow* rate, ρQ. Thus, Newton's second law of motion may be written as

$$\sum \boldsymbol{F} = \rho Q\,(\boldsymbol{v}_{\text{out}} - \boldsymbol{v}_{\text{in}}) \qquad (2.11)$$

The term $\rho Q\boldsymbol{v}$ can be considered a property of the flowing fluid in its own right and is called *momentum.* Because $\boldsymbol{v}$ is a vector quantity, the momentum is a vector also. The momentum of a fluid passing through a cross-sectional area on the surface of a control volume is denoted by $\boldsymbol{M}$, that is,

$$\boldsymbol{M} = \rho Q \boldsymbol{v} = \rho A \boldsymbol{v}^2 \qquad (2.12)$$

where A is the cross-sectional area and $\boldsymbol{v}$ is the average velocity perpendicular to it.

Figure 2.12b shows the position of the fluid that occupied the control volume at time t after a small time period Δt later. As we see, the surface of the fluid at the entrance has moved a Δs_1 distance away. From Section 2.5 we then can define the average velocity at the entrance as being equal to $v_1 = \Delta s_1/\Delta t$. In the case of a circular pipe this velocity is equal to

$$v_1 = \frac{\Delta s_1}{\Delta t} = \frac{4Q}{\pi D^2} \qquad (2.13)$$

and the momentum of the fluid at that surface equals

$$M_1 = 4\rho\,\frac{Q^2}{\pi D^2} \qquad (2.14)$$

One advantage of interpreting Newton's law of motion in the form of momentum changes that result in sums of acting forces is that we do not have to concern ourselves with what happens to the fluid inside the control volume. All we need to consider is the changes over the control volume's surfaces in terms of momentums and forces.

2.8 CONSERVATION OF MOMENTUM

In the previous section we learned that the *change of momentum* along the path of a flowing fluid results in force, $\boldsymbol{F}$. This force is called *impulse* force. The term

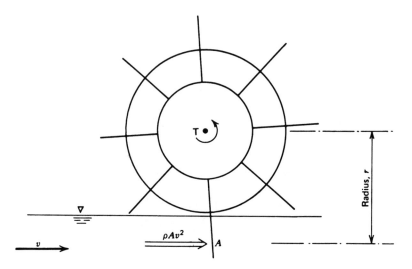

FIGURE 2.13   Torque of water on a water wheel.

impulse refers to the fact that in the development of Equation 2.11 ρQ was considered for a unit time element, Δt. Between two points (points 1 and 2 in Figure 2.12) along the path of the flow the change of momentum can be written as

$$M = M_2 - M_1$$
$$= \rho Q(v_2 - v_1) \qquad\qquad \textbf{(2.15)}$$
$$= F$$

One must keep in mind that the velocities, as well as the impulse force, are vector quantities. Often it is more convenient to solve such problems in a scalar form, separating the components of forces and momentums in the x, y, and z directions of a rectangular coordinate system.

Equation 2.15 embodies the *law of momentum conservation.* This law of nature was first recognized by René Descartes (France, 1596–1650). The law states that momentum may not be lost in a hydraulic system, although some of it may be converted into an impulse force. Hence, if the mass flow rate and the physical configuration of the flow channel causing a change in the direction of flow are known, the resulting impulse force acting on the hydraulic component or structure may be computed. The same concept may be used to measure the discharge. One example is the measuring pipe elbow in which the discharge, and consequently the velocity, is unknown and determined by measuring pressures generated by the impulse force acting on the pipe elbow. The impulse force due to momentum changes on the blades of a turbine will cause the turbine to rotate. The product of an impulse force and the radius of rotation results in the *moment of momentum* or, as it is also called, the *torque.* The concept of torque is applied in Equation 5.3 to determine the power requirements of pumps. Figure 2.13 helps explain the

FIGURE 2.14 Wave action around motor boat.

concept. Torque is the product of an impulse force causing rotation and the radius r, the moment arm.

The impulse force transferred to the water by a ship's propeller causes the ship to move. This force must be sufficient to overcome the various losses due to this relative motion. One of the losses is the momentum needed to be given to a portion of the water first to move out of the way of the ship then to move back behind it, as shown in Figure 2.14. This process causes relative movements between the water particles surrounding the ship, which in turn cause additional losses due to the viscous resistance of the water. More losses of propulsive force are indicated by surface waves created by the motion.

2.9 WORK, ENERGY, AND POWER

Force exerted over a distance represents work. For example, to push a certain amount of water through a pipeline requires that work be done. *Work*, therefore, is defined as

$$W = Fl \tag{2.16}$$

in which F is the force acting over l distance.

Power is a term closely related to work. It refers to work done per unit of time. Expressed mathematically,

$$P = \frac{W}{t} \tag{2.17}$$

where t is the time required to do the work. To perform a certain amount of work in a shorter period of time requires more power than to perform the same amount of work over a longer period of time. We use the concept of power in conjunction with pumps. Power required by electric motors to drive pumps is often expressed in *horsepower* (HP). One horsepower equals 550 foot-pounds per second in the conventional American system of measures. When dealing with electric power we use *kilowatts*. A watt is defined as 1 joule of energy per second. One kilowatt is 1.34 horsepower.

To move a discharge Q against a water pressure of (γH) the horsepower requirement is

$$HP = \frac{\gamma Q H}{550} \tag{2.18}$$

assuming that the operation is 100 percent efficient. Equation 2.18 requires that we express our variables in terms of pounds, feet, and seconds. In the metric system a horsepower is defined as 750 newton-meters per second. Hence, in SI units Equation 2.18 takes the form

$$HP_{metric} = \frac{\gamma Q H}{750} \tag{2.19}$$

in which the variables are expressed in newtons, meters, and seconds. To express the same in terms of electric power we note that

$$1\ kW = \frac{\gamma Q H}{1000} \tag{2.20}$$

in which the variables are again expressed in metric terms.

The capacity to do work is called *energy*. Energy, therefore, is simply stored work. In fluid mechanics, energy is most often expressed as energy per unit mass of fluid. There are various forms in which energy appears in nature. Put another way, work can be stored in several ways. Temperature indicates the amount of *heat energy* a substance contains. We learned in Chapter 1 that heat and pressure represent internal energy in matter. A lump of coal contains *chemical energy* that may be transformed into heat energy by the chemical reaction of burning. An object placed higher in a gravitational field has more energy by relative elevation than the same object at a lower elevation. This is called *potential energy*. A compressed spring or a stretched strip of rubber has *elastic energy*. The force maintaining their compressed (or stretched) position gives rise to *pressure energy*. Pressure energy and elastic energy may be illustrated by considering an inflated bicycle tube; the elastic rubber tube is stretched, gaining elastic energy; because the air in the tube is under pressure, it is compressed and has pressure energy with respect to the surrounding air, which itself is under pressure from the atmosphere. Moving objects contain *kinetic energy*. Their velocity may propel them to higher elevation, like water from a fire hose, in which case the kinetic energy is converted into potential energy.

As we have seen in the preceding examples, some energies can be converted freely from one form to another; others are not so reversible. Elevation or potential energy, pressure energy, and kinetic energy are reversible energies. These three are called *hydraulic energies*. In contrast, *heat energy* cannot be harnessed by hydraulic means. Once a portion of hydraulic energy is converted into heat through viscous shear action, it is lost; therefore, we refer to it as *energy loss*. Yet the heat energy converted from sunshine causes the water from the oceans to evaporate, keeping water in constant circulation through creeks and rivers.

2.10   CONSERVATION OF ENERGY

An important concept of physics is the law of energy conservation. It was first recognized by Daniel Bernoulli (Switzerland, 1700–1782), one of the preeminent founders of the science of hydrodynamics. The law states that energy cannot be lost, although it may be converted into other forms. In other words, the theorem states that in a hydraulic system the sum of all energies is a constant. Writing this total energy as E in foot-pounds of force per pounds of mass, we can sum the various forms in which hydraulic energy may appear:

$$\frac{v^2}{2g} + \frac{p}{\gamma} + y = E \qquad (2.21)$$

which is known as *Bernoulli's equation*. Each of the terms of the equation is given in "head" units, elevations in a gravitational field, so energy is expressed in feet. Indeed, water issuing upward from a garden hose will rise to a height given by the first term,[4] showing kinetic energy in terms of feet. Pressure created by a pump causes water in a pipe to rise to a height shown by the second term, defining pressure energy in terms of feet. The y term is the energy a certain mass of water may potentially have with respect to an arbitrarily selected horizontal reference level, called the *datum plane*, at an elevation $y = 0$, in feet.

Equation 2.21 does not include the heat energy term; it contains only hydraulic energies, those that are interconvertible and cannot be destroyed. Heat energy is always present, although the science of hydraulics is not concerned directly with it. As water flows from one point to another heat energy changes appear in the form of viscous shear losses, taking away some portion of the total hydraulic energy. For this reason, E is called the total available hydraulic energy at any point considered. As this total available energy decreases along the direction of the flowing

---

[4]Neglecting losses due to friction with air, etc.

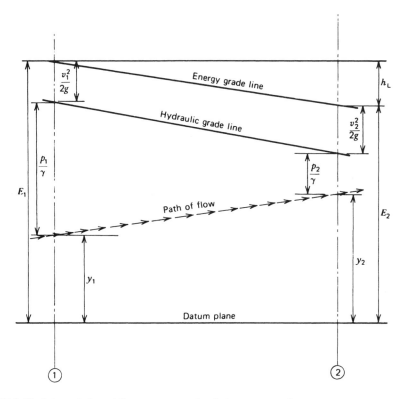

FIGURE 2.15   Interpretation of the energy equation between two points.

water between two points, as shown in Figure 2.15, the law of energy conservation has the form

$$\frac{v_1^2}{2g} + \frac{p_1}{\gamma} + y_1 = \frac{v_2^2}{2g} + \frac{p_2}{\gamma} + y_2 + h_L \tag{2.22}$$

in which subscripts 1 and 2 refer to points shown in the flow field, and the term h_L refers to the energy lost in the form of heat. This lost heat energy is actually still present in the environment in the form of an imperceptible temperature rise that is generally dissipated in the surrounding atmosphere. The h_L term is commonly called *head loss* because it is the apparent reduction of the height of E as the water moves from one point to the other.

As before, the terms in Equation 2.22 all have the dimension of elevation, measured above a datum plane.

The sum of the potential and pressure energies is sometimes called *piezometric height*. This is the height to which water would rise in a pipe with one of its ends inserted into an arbitrary point of the flow field. The line, shown in Figure 2.15, connecting several points of such piezometric measurements along the path of flow, is called the *hydraulic grade line*. It is always below the energy grade line by an amount equal to the kinetic energy head, $v^2/2g$, at the point of the piezometric measurement. In open channels, when the water surface is exposed to the atmosphere and hence is not under pressure, the hydraulic grade line coincides with the water surface.

In closed pipes flowing full the hydraulic grade line may be below the pipe, in which case the pressure there is below atmospheric pressure. The limitation on this condition is pressure that is low enough to cause cavitation. To initiate the flow in such cases requires that the energy grade line be above the pipe, but once the flow begins, it will be sustained even if the inlet and outlet energy levels are less than the height of the pipe at the middle. This case applies to pipes shaped like an inverted U and is called *siphon action.*

2.11 THE CONCEPTS OF ENERGY TRANSFER

The science of hydraulics is primarily concerned with the determination of the magnitude of the energy loss, h_L, under various circumstances. In most practical applications, as in the flow of water in pipes and in open channels, the energy loss is assumed to be proportional to the square of the average velocity of the flow, or in other words, to the kinetic energy. Seepage is one exception; here the energy loss is considered to be proportional to the velocity. This concept of proportionality gives rise to coefficients or *factors of proportionality.* Formulas for pipe flow include a coefficient *f*, which is called the friction factor; open channel flow formulas include a coefficient *n*, which is called the roughness factor; seepage has its permeability coefficient *K*; and so on. Generally, these factors are determined experimentally. Once available, these coefficients enable us to determine the rate of conversion of hydraulic energy into heat energy. This is the central question of hydraulics, but it is also the least understood one. Because of the complexity of the physical mechanisms involved, we may not necessarily assume that the influence of various parameters on the rate of conversion is linearly related to certain major variables of hydraulic systems. Indeed, because in many instances the relationships cannot be represented by a proportionality constant, these factors are themselves dependent on other variables. This gives credence to the remark made by the famous aerodynamicist Theodore von Kármán (Hungary, 1881–1963), who referred to hydraulics as "the science of variable constants." Still, because of the complexity of the physical mechanisms involved in most hydraulic applications, the influence of various parameters on the rate of conversion of hydraulic energy into heat energy is taken into account by empirical coefficients.

Many generations of hydraulicians have strived to find better means to compute the rate of this energy transfer. More mathematical and less empirical methods would give better values than the empirical coefficients, which in many instances are only slightly better than educated guesses.

During the early development of fluid mechanics theoretical findings were inconsistent with experimental results. Because of this, most mathematically derived formulas based on sound physical principles were said to apply to "ideal fluids."

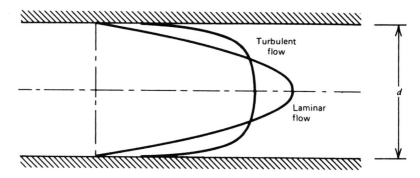

FIGURE 2.16 Comparison between the laminar and turbulent flow velocity distribution in a pipe.

These nonexistent fluids had no viscosity, and they slipped by the solid boundaries confining the flow. The principles of behavior of "real fluids" have eluded researchers for a long time. Although Louis Navier (France, 1785–1836) expressed Newton's law of motion correctly in the form of a differential equation, it was a generation later that Sir George Stokes (England, 1819–1903) included viscosity to describe shear stresses and derived the fundamental formula of hydrodynamics, called the Navier-Stokes equation.[5] One of the most important early experiments, reported by Sir Osborne Reynolds (England, 1842–1912), showed that fluid flow at low velocities is markedly different from flow at high velocities. He denoted flow conditions in a dimensionless manner by using the Reynolds number

$$\mathbf{R} = \frac{vD}{\nu} \tag{2.23}$$

where v is the average velocity, D is the pipe diameter, and ν is the kinematic viscosity. Reynolds found that up to about $\mathbf{R}$ equals 2000, dye introduced into flowing water was carried along in straight and parallel paths, but when the velocity was increased beyond this threshold level, the dye stream became mixed throughout the fluid. The former is called *laminar*, the latter *turbulent*, flow. Measuring the distribution of velocity across the pipe has shown it to be significantly different in the two conditions. Turbulent velocity distribution is more uniform across the pipe, as Figure 2.16 shows. In turbulent flow the velocity at any point varies randomly in magnitude as well as direction, thus creating a mixing effect. At the wall the velocity is always zero, indicating that real fluids stick to solid boundaries.

World War I accelerated aeronautical research. A leading scientist of Germany, Ludwig Prandtl, (1875–1953), made a major breakthrough by introducing his boundary layer theory, which explains the behavior of fluids near solid boundaries. This theory allows the determination of the velocity distribution within the layer of fluid

[5]Louis Marie Henri Navier developed the differential equation for fluid flow that included the terms accounting for viscous effects, even though he did not quite recognize their physical significance.

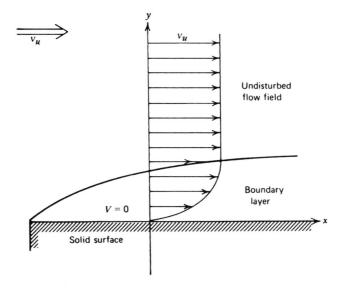

FIGURE 2.17 The development of the boundary layer on a solid surface.

influenced by the presence of a rigid boundary, as shown in Figure 2.17. The reduction of velocity within the boundary layer is directly related to a loss of momentum. Thus, aircraft designers had a way to determine the propulsive force required to maintain flight at a given velocity. Many scientists used Prandtl's theory to derive equations to express energy losses in various flow conditions other than flight. For example, Theodore von Kármán[6] applied boundary layer concepts to derive formulas for energy losses in pipes. In general, however, most practical hydraulic applications involve so many variables that theoretical methods led to unsolvable, complex mathematical problems. As a result, most theoretical formulas needed correlation and adjustment with experimental data.

Experiments have shown that the boundary layer may separate from the solid boundary, as with the stalling wing shown in Figure 2.18. Beyond the point of separation a wake may form. This is a common occurrence in hydraulics. Inside the wake a large amount of energy is dissipated by random vortices and turbulent mixing in the fluid flowing past a body. The energy lost in this way can hardly be represented in a formula in any other manner other than a proportionality constant.

Reducing the size of the wake is one way to decrease the energy loss. In everyday life we call this "aerodynamic" or "streamlined shape." In Prandtl's studies of flows past a sphere the control of the boundary layer influenced the size of the wake, hence, the total drag force. The theoretical explanation may be of significance even if it cannot be expressed mathematically. The dimpled surface of

---

[6]He was instrumental in the development of jet propulsion for the U.S. Air Force during World War II. He also introduced the "slenderness ratio" in column design.

FIGURE 2.18 Wake behind a stalled airplane wing.

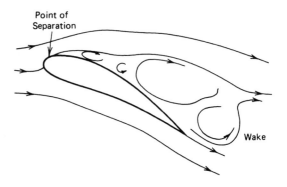

modern golf balls is a convenient illustration of the value of Prandtl's experiment shown in Figure 2.19.

The same problem was studied earlier by Stokes using traditional methods of hydraulics. Stokes considered the drag on a spherical particle, like a grain of sand falling in a jar of water. The motion is caused by the gravitational force and resisted by the force of buoyancy and the drag force. After an initial acceleration the effect of the drag will slow the particle to a constant "terminal" velocity. Stokes's equation for the drag force of a sphere is

$$F_D = (6\pi)rv\mu \tag{2.24}$$

where r is the radius of the particle, v is its velocity, and μ is the viscosity of the fluid. If the density of the particle is known, the terminal velocity of the particle may be expressed as

$$v_t = \frac{g(\rho_s - \rho_f)d^2}{18\,\mu} \tag{2.25}$$

as long as the Reynolds number is less than 1. The subscripts s and f in this equation refer to solid and fluid, respectively.

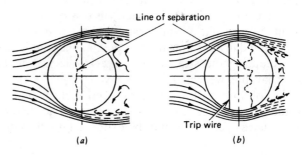

FIGURE 2.19 Flow past a sphere. (*a*) Laminar boundary layer flow separation. (*b*) Turbulent boundary layer separation point. (After L. Prandtl)

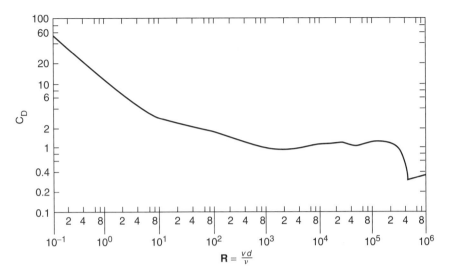

FIGURE 2.20   Drag coefficient for various Reynolds numbers of cylindrical pier.

Later studies[7] found that this formula is good only when the Reynolds number characterizing the flow is very small, that is, in the case of laminar flow. In a generalized manner the concept is expressed by the drag coefficient,

$$C_D = \frac{F_D}{\frac{1}{2}\rho v^2 A} \tag{2.26}$$

where F_D is the drag force and A is the surface area on which the drag force acts, measured perpendicular to the direction of the flow. The drag coefficient is simply the ratio of the drag force resisting the movement to the moving force caused by the kinetic energy acting on the body. Figure 2.20 shows drag coefficients for a cylindrical pier at various Reynolds numbers.

The behavior of the flow inside the wake is of particular importance in hydraulics. The turbulent eddies and secondary flow motion induced by vortices are the primary causes of erosion behind piers and other sudden changes in the flow channel. Most such information comes from studies in the field and from controlled laboratory experiments. One notable theoretical result is von Kármán's vortex trail concept, shown in Figure 2.21. Depending on the velocity of the stream, the separating boundary layer sheds vortices in an alternating sequence from either side of the solid body. The vortices travel downstream in a zigzag pattern, slowly spreading out and losing their intensity. The alternating shedding causes a vibrating transverse force on the body. On small bodies, like wires, this vibration may cause an audible

---

[7] One of these was that of Alexandre Gustave Eiffel (France, 1832–1923) builder of the Eiffel tower in Paris, who studied the drag coefficients of falling objects thrown off his tower. He also studied the drag coefficient in wind tunnels.

FIGURE 2.21   Kármán's vortex trail behind a cylindrical pier.

noise. A flag flaps in the wind for the same reason. Kármán's vortex trail–induced vibrations caused the famed collapse of the Tacoma Narrows bridge in 1940.[8]

Various formulas have been derived generally for the simplest cases of physical boundaries on the basis of the classical boundary layer theory. These give a better description of the velocity distribution curve than was available previously. Many of these equations, such as the one for pipe flow, still contain some empirical coefficients because they were adjusted to fit better the experimental results available. There are relatively few formulas that were derived for hydraulic applications on the basis of the boundary layer theory. The reason for this is the immense mathematical complexity of the theoretical formulas coupled with the physical complexity of most hydraulic problems.

One of the intrinsic laws of energy transfer is the law of *optimum energy*. In popular terms it states that "water seeks the path of least resistance." Scientifically this means that of all possible flow distributions the one and only one that may occur in nature is the one requiring the least amount of energy per mass of water. In other words, nature will deliver the most fluid for a given amount of available energy. This intrinsic law controls flows in pipe networks, open channels, and seepage, to mention only a few of the important situations. Solutions obtained by statistical methods using the so-called probabilistic approach indicate that the correct answer that satisfies this minimum energy theorem corresponds to the mean value of all possible answers. Most common methods of hydraulic analysis satisfy this basic law intrinsically. Where a minimization theorem is to be specifically stated, the computations are very complex.

In practical hydraulic design experimentally determined design coefficients—variable or otherwise—provide satisfactory solutions. These coefficients may be obtained by direct measurements in the field or by studies performed in hydraulic laboratories on models designed to represent field conditions. Concepts and techniques used in such studies are the subjects of the next chapter.

EXAMPLE PROBLEMS

Example 2.1   A 3 ft diameter manhole cover is to be placed on a section of a sewer line in which the water flows occasionally under pressure. Compute the pressure of the water at which the manhole cover will be lifted if the weight of the cover is 85 lb.

---

[8]Kármán was very apologetic about his vortex trails after that catastrophe.

Solution. The area of the manhole cover is $\pi d^2/4 = \pi(3 \times 12)^2/4 = 1018$ in.²

Therefore, the weight of cover per unit area $= 85/1018 = 0.0835$ lb/in.² This means that the manhole cover will be lifted if the water pressure exceeds 0.0835 psi or 12 lb/ft², which equals 0.19 ft of water.   □

Example 2.2

A rectangular gate 4 by 4 ft in size is set on a 45° plane with respect to the water surface; the centroid of the gate is 10 ft below the water surface (Figure E2.2). Compute the required location of the horizontal axis at which the gate is to be hinged if the hydrostatic pressure acting on the gate is to be balanced around the hinge so that there will be no moment causing rotation of the gate under the specified loading.

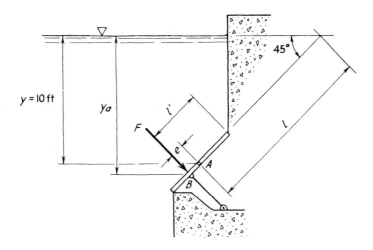

FIGURE E2.2

Solution. The centroid of the gate is 10 ft under the water surface, so that point A in the plane of the gate is at a distance l from the water surface.

$$l = \frac{10}{\sin 45°} = \frac{10}{0.707} = 14.14 \text{ ft}$$

Under static conditions the location of the hinge has to be at the level where the resultant force acts on the gate. From Equation 2.5 the distance of the pressure center from the centroid is

$$e = \frac{I_0}{lA}$$

where

$$I_0 = b \cdot \frac{h^3}{12} \quad \text{(from Table 2.1)}$$

$$= \frac{4(4)^3}{12} = 21.33 \text{ ft}^4$$

$$A = b \cdot h = 4 \times 4 = 16 \text{ ft}^2$$

Hence,

$$e = \frac{21.33}{(14.14)(16)} = 0.094 \text{ ft} = 1.1 \text{ in.}$$

This is the distance from the centroid of the gate measured in the same plane. Hence, point B is located at distance

$$l' = (2 \times 12) + 1.1 = 25.1 \text{ in.}$$

measured from the top edge of the gate. □

Example 2.3   Calculate the total hydrostatic force acting on the gate's two hinges in the previous example problem.

Solution. From Equation 2.6

$$\begin{aligned}
F &= \gamma A l \sin \alpha \\
&= 62.4(4 \times 4)(14.14)\sin 45° \\
&= 9982 \text{ lbf}
\end{aligned}$$

□

Example 2.4   Determine the magnitude and location of the hydrostatic force on the gate in Example 2.2 by constructing a pressure diagram.

Solution. A pressure diagram for the gate described in Example 2.2 is shown in Figure E2.4. The magnitude of the hydrostatic force due to the rectangular pressure prism (F_a) is

$$F_a = p_1 A = y_1 \gamma A = (10 - 2 \sin 45°)(62.4)(4 \times 4) = 8572 \text{ lb}$$

The magnitude of the hydrostatic force due to the triangular pressure prism (F_b) is

$$F_b = \frac{(p_2 - p_1)}{2} A = \frac{(y_2 - y_1)}{2} \gamma A = \frac{(4 \sin 45°)}{2}(62.4)(4 \times 4) = 1412 \text{ lb}$$

The magnitude of the resultant force F is

$$F = F_a + F_b = 8572 + 1412 = 9984 \text{ lb}$$

The location of F can be obtained by summing moments about point O

$$l' = \frac{8572(2) + 1412\,(4 \times \frac{2}{3})}{9984} = 2.094 \text{ ft} = 25.1 \text{ in.}$$

measured from the top edge of the gate. This answer is identical with that obtained in Example 2.2.

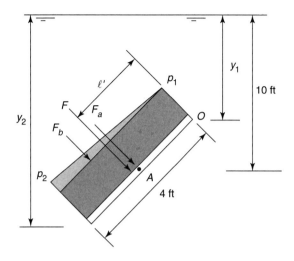

FIGURE E2.4

□

Example 2.5

A 1 m diameter flood gate placed vertically is 4 m below the water level at its highest point. The gate is hinged at the top. Compute the required horizontal force to be applied at the bottom of the gate to open it. Assume that the pressure on the other side of the gate is atmospheric pressure, and neglect the weight of the gate.

Solution. We know that

$$\gamma \text{ of water} = 9810 \text{ N/m}^3$$

$$\text{Area of the gate} = \frac{\pi}{4}(1)^2$$

$$A = 0.785 \text{ m}^2$$

The depth of the centroid of the gate is $l = 4.5$ m. Next, determine the location of F_1 acting on the gate, as shown in Figure E2.5.
From Equation 2.5

$$e = \frac{I_0}{lA}$$

From Table 2.1

$$I_0 = \frac{\pi R^4}{4}$$

$$= \frac{\pi(0.5)^4}{4} = 0.049 \text{ m}^4$$

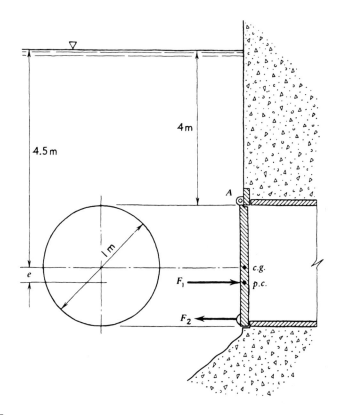

FIGURE E2.5

Therefore,

$$e = \frac{0.049}{4.5(0.785)} = 0.014 \text{ m}$$

The force F_1 acting on the gate is located at a depth $y_a = (l + e)$. This equals 4.5 m + 0.014 m, which equals

$$y_a = 4.514 \text{ m}$$

This answer could also be obtained by using the equation given in Table 2.2

$$y_a = d + \frac{D}{8}\frac{8d + 5D}{2d + D} = 4 + \frac{1}{8}\frac{(32 + 5)}{(8 + 1)} = 4.514 \text{ m}$$

The magnitude of this hydrostatic force is

$$F_1 = p_c A = 4.5 \times 9810 \times 0.785 = 34.65 \text{ kN}$$

To determine F_2 we take the moment of the two forces with respect to the hinge at A:

$$F_2 \times D = F_1 \left(\frac{D}{2} + e\right)$$

$$F_2 = \frac{34.65(0.5 + 0.014)}{1} = 17.8 \text{ kN}$$

□

Example 2.6

A cylindrical (drum) gate with a diameter of 3 m is 8 m long. The water is on one side only, reaching the top of the gate. Determine the magnitude and direction of the resultant hydrostatic force acting on the gate.

Solution. If we separate the vertical and horizontal pressure components, we obtain the pressure diagrams shown in Figure E2.6.

The horizontal force is acting at the centroid of the triangular pressure diagram, 2 m below the water surface. Its magnitude equals the area of the pressure diagram multiplied by the gate length:

$$F_H = 0.5 \times \text{gate height} \times p_1 \times \text{gate length} = 0.5(3)(3)(9810)(8) = 353 \text{ kN}$$

The vertical force is located a c distance away from the vertical centerline of the gate. The value of c may be determined from Table 2.1, in which

$$c = \frac{4R}{3\pi} = \frac{4(1.5)}{3\pi} = 0.64 \text{ m}$$

Its magnitude is the area of the semicircle multiplied by the unit weight of water and the gate length.

$$F_V = 0.5\pi R^2 \gamma \times \text{length} = 0.5\pi(1.5)^2(9810)8 = 277 \text{ kN}$$

The resultant force is

$$F = \sqrt{F_H^2 + F_V^2} = 449 \text{ kN}$$

The force acts through the intersection of its two components whose coordinates are 2 m vertical and 0.63 m horizontal with respect to the top of the water over the gate.
    The angle of the resultant force is

$$\tan \alpha = F_V/F_H = 0.78 \quad \text{and} \quad \alpha = 38.12°$$

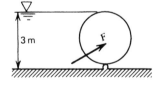 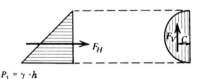

FIGURE E2.6   ☐

Example 2.7

A 100,000 deadweight ton sunken ship is to be raised from 200 ft to the water surface by pumping air into its sealed holds. How much water must be expelled from the ship to make it float?

Solution. We know that 10^5 tons equal 2×10^8 lb and that 1 ft^3 of water weighs 62.4 lb. Hence,

$$2 \times 10^8 \text{ lb of water}$$

takes up

$$3.205 \times 10^6 \text{ ft}^3$$

which is the minimum amount of air space required.

The required air pressure to expel the water from the ship's holds is thus

$$200 \text{ ft} \times 62.4 \text{ lb/ft}^3 = 12,480 \text{ lbf/ft}^2$$
$$= 86.7 \text{ psi} \qquad \square$$

Example 2.8

Compute the average velocity of water in a pipe 12 in. in diameter when the discharge of the pipe is 20 gal/min (gpm).

Solution. From Appendix A, Table A.3, 1 gpm = 2.227×10^{-3} ft³/sec (cfs); hence, 20 gpm = 0.0445 cfs. The cross-sectional area of a 12 in. or 1 ft diameter pipe is

$$A = \frac{\pi d^2}{4} = \frac{\pi}{4} 1^2 = 0.785 \text{ ft}^2$$

By Equation 2.8

$$v_{ave} = \frac{Q}{A} = \frac{0.0445}{0.785} = 0.057 \text{ fps} \qquad \square$$

Example 2.9

A 20 by 20 ft excavation is surrounded by sheet piles. In order to remove the water entering from the soil on the bottom, a pump is installed that delivers 10 gpm. The soil is loose sand with a porosity of 42 percent. Compute the average discharge velocity and the seepage velocity of the entering groundwater.

Solution. The discharge, 10 gpm, equals 0.022 cfs. The superficial area of the flow is 20 × 20, which equals 400 ft². Hence, the average discharge velocity is

$$v = \frac{Q}{A} = \frac{0.022}{400} = 5.5 \times 10^{-5} \text{ fps}$$

which, in metric units, is

$$v = 5.5 \times 10^{-5} \frac{\text{ft}}{\text{sec}} 30.48 \frac{\text{cm}}{\text{ft}}$$
$$= 1.68 \times 10^{-3} \text{ cm/s}$$

The seepage velocity, that is, the average velocity of the groundwater within the pores of the sand, is

$$v_{seepage} = \frac{v_{discharge}}{\text{porosity}}$$
$$= \frac{1.68 \times 10^{-3} \text{ cm/s}}{0.42}$$
$$= 4.0 \times 10^{-3} \text{ cm/s} \qquad \square$$

Example 2.10

The discharge of a creek was measured to be 0.5 m³/s at 9 A.M. and 0.6 m³/s at 4 P.M. on the same day. The area of the flow was 1.445 m² during the first measurement and 1.57 m² during the second measurement. Determine the average velocities of the flow and the average temporal acceleration during this time period.

Solution. At 9 A.M. the velocity was

$$v_1 = \frac{Q_1}{A_1} = \frac{0.5}{1.445} = 0.346 \text{ m/s}$$

At 4 P.M. it was

$$v_2 = \frac{Q_2}{A_2} = \frac{0.6}{1.57} = 0.382 \text{ m/s}$$

The time interval between 9 A.M. and 4 P.M. is 7 hours, or

$$T = 7 \times 3600 = 25,200 \text{ s}$$

The average temporal acceleration

$$a = \frac{v_2 - v_1}{T} = \frac{0.382 - 0.346}{25,200} = \frac{0.036}{25,200}$$
$$= 1.43 \times 10^{-6} \text{ m/s}^2 \qquad \square$$

Example 2.11

In a farm's water supply system a 10,000 gal underground storage tank is continually replenished from a well by a submersible pump delivering 2 gpm. Assuming that the reservoir is initially full, how much time will it take to empty the reservoir by using a 30 gpm pump?

Solution. By Equation 2.9

$$30 \text{ gpm} - 2 \text{ gpm} = 28 \text{ gpm} = \text{change in storage}$$

Hence,

$$\frac{10,000 \text{ gal}}{28 \text{ gal/min}} = 357 \text{ min}$$
$$= 5.95 \sim 6 \text{ hr} \qquad \square$$

Example 2.12

A 12 in. diameter pipe reduces to a diameter of 6 in., then expands to 8 in. If the average velocity in the 6 in. pipe is 16 fps, what is the discharge in the pipeline, and what are the average velocities in the other two pipes?

Solution. The cross-sectional area of the pipe is

$$A = \frac{\pi}{4} D^2$$

For a 6 in. pipe, it is

$$A_6 = \frac{\pi}{4} \left(\frac{6}{12} \right)^2 = 0.196 \text{ ft}^2$$

For an 8 in. pipe,

$$A_8 = \frac{\pi}{4} \left(\frac{8}{12} \right)^2 = 0.349 \text{ ft}^2$$

And for a 12 in. pipe,

$$A_{12} = \frac{\pi}{4}\left(\frac{12}{12}\right)^2 = 0.785 \text{ ft}^2$$

From Equation 2.8

$$Q = v \cdot A$$

Because the discharge in the pipeline is constant,

$$Q = v_6 A_6 = 16\frac{\text{ft}}{\text{sec}}(0.196 \text{ ft}^2) = 3.14 \text{ cfs}$$

The velocities in the other two pipes are then

$$v_8 = \frac{Q}{A_8} = \frac{3.14}{0.349} = 9.0 \text{ fps}$$

$$v_{12} = \frac{Q}{A_{12}} = \frac{3.14}{0.785} = 4.0 \text{ fps}$$ □

Example 2.13   Water is flowing through a fire nozzle at 60 gpm. The inlet diameter of the nozzle is 4 in.; the outlet diameter is 2 in. The nozzle is equipped with a pressure gage at the inlet. The gage registers 3.3 psi pressure. If there is no force exerted between the fire hose and the nozzle, how much force is required to hold the nozzle stationary?

Solution. Consider the nozzle to be the control volume and apply the momentum theorem:

$$\sum F = \rho Q v_{\text{out}} - \rho Q v_{\text{in}}$$
$$\rho Q = 62.4(\text{lbm/ft}^3)60/60(\text{gal/sec})\ 1/7.48(\text{ft}^3/\text{gal})$$
$$= 8.34 \text{ lbm/sec}$$

Calculating the inlet and outlet velocities, we find that

$$v_{\text{in}} = \frac{Q}{A_{\text{in}}} = \frac{1 \text{ gal/sec } (1/7.48) \text{ ft}^3/\text{gal}}{\pi/4\ (4/12)^2 \text{ ft}^2} = 1.53 \text{ fps}$$

$$v_{\text{out}} = \frac{Q}{A_{\text{out}}} = \frac{1(\text{gal/sec})\ 1/7.48\ (\text{ft}^3/\text{gal})}{\pi/4\ (2/12)^2 \text{ ft}^2} = 6.13 \text{ fps}$$

The sum of all forces acting on the control volume include the impulse force due to the momentum change between inlet and outlet and the static pressure measured by the gage times the inlet area. The pressure at the outlet is assumed to be atmospheric pressure, which is our base; hence, $p_{\text{out}} = 0$. Therefore,

$$\sum F = p_{\text{in}}A_{\text{in}} - F_{\text{impulse}} = \rho Q(v_{\text{out}} - v_{\text{in}})$$

or

$$F_{\text{impulse}} = p_{\text{in}}A_{\text{in}} - \rho Q(\boldsymbol{v}_{\text{out}} - \boldsymbol{v}_{\text{in}})$$
$$= (3.3 \text{ lbf/in.}^2)(4 \text{ in.})^2(\pi/4)$$
$$- (8.34 \text{ lbm/sec})[(\text{lbf-sec}^2)/(32.2 \text{ ft-lbm})](6.13 - 1.53) \text{ ft/sec}$$
$$- 8.34 \text{ lbm/sec} \left(\frac{\text{lbf/sec}^2}{32.2 \text{ ft-lbm}} \right) (6.13 - 1.53) \text{ fps}$$
$$= (41.5 - 1.2) \text{ lbf}$$

Therefore, the force exerted by the nozzle is

$$F = 40.3 \text{ lbf}$$

acting in the direction opposite to the flowing water.   □

Example 2.14

A 90° elbow in a 6 in. pipeline carries 3 cfs of water at room temperature (Fig. E2.14). Compute the magnitude and direction of force acting on the elbow that results from the change of momentum of the flow.

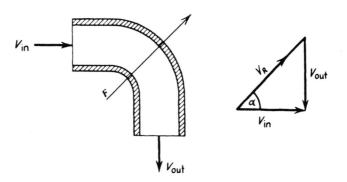

FIGURE E2.14

Solution. The momentum of flow entering the elbow is $\rho Q v$. From Equation 2.8

$$\boldsymbol{v} = \frac{Q}{A} = \frac{4Q}{\pi D^2} = \frac{4 \times 3}{\pi (6/12)^2} = 15.28 \text{ fps}$$

The direction of the resultant force vector is α, where

$$\alpha = \tan^{-1}\left(-\frac{\boldsymbol{v}_{\text{out}}}{\boldsymbol{v}_{\text{in}}} \right) = \tan^{-1}(-1) = 45°$$

The magnitude of this velocity change

$$\boldsymbol{v}_R = \boldsymbol{v}_{\text{in}} - \boldsymbol{v}_{\text{out}}$$

and

$$\cos \alpha = \frac{v_{in}}{v_R}$$

$$v_R = \frac{v_{in}}{\cos(45°)} = \frac{15.28}{0.707} = 21.61 \text{ fps}$$

From Equation 2.15

$$F = \rho Q(v_2 - v_1)$$
$$= \rho Q v_R$$
$$= \left(\frac{62.4}{32.2}\right)(3)(21.61) \left[\frac{\text{lbf sec}^2}{\text{ft}^4} \cdot \frac{\text{ft}^3}{\text{sec}} \cdot \frac{\text{ft}}{\text{sec}}\right]$$
$$= 125.6 \text{ lbf}$$

which is the magnitude of the force acting on the elbow due to the flow of water. □

Example 2.15

From a 12-in. diameter garden sprinkler two horizontal jets of water are issued in a tangential direction. The velocity of the jets is 30 fps, and the discharge of the sprinkler is 0.5 cfs. Compare the torque generated by the sprinkler that causes its rotation. (See Figure E2.15.)

FIGURE E2.15

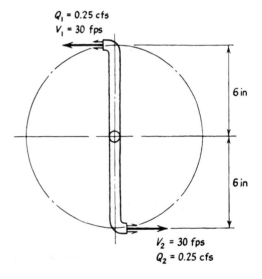

Q_1 = 0.25 cfs
V_1 = 30 fps

6 in

6 in

V_2 = 30 fps
Q_2 = 0.25 cfs

Solution. Because the total discharge = 0.5 cfs, the discharge of each jet = 0.5/2 = 0.25 cfs. Then the tangential impulse force from one arm is the product of the discharge, density, and the velocity of the jet, as given in Equation 2.11.

$$F = \rho Q v = \frac{62.4}{32.2}(0.25)(30) = 14.53 \text{ lbf}$$

This force is tangent to the radius of motion, which is 0.5 ft. There the torque from each arm is

$$14.53 \times 0.5 = 7.265 \text{ ft-lbf}$$

Hence the total torque

$$T = 7.265 \times 2 = 14.5 \text{ ft-lbf}$$ □

Example 2.16

A mountain creek discharges at 3 m³/s along a rocky bottom without apparent freezing in subzero weather. The channel elevation is reduced by a height of 80 m over a channel length of 1 km. The flow in the channel is uniform. Compute the heat generated by the moving water that prevents it from freezing.

Solution. From physics we know that power in the SI is expressed in kilowatts, where, in hydraulic terms

$$1 \text{ kW} = \frac{\gamma QH}{1000} = \frac{(\text{specific weight, N/m}^3)(\text{discharge, m}^3\text{/s})(\text{head loss, m})}{1000}$$

The specific weight of water expressed in SI units is 9810 N/m³. The discharge in the problem is 3 m³/s, and the head loss is 80 m. Hence, after substituting, we get

$$\frac{9810(3)80}{1000} = 2354 \text{ kW}$$

Distributing this total power over the channel length in a uniform manner, we have

$$\frac{2354 \text{ kW}}{1000 \text{ m}} = 2.35 \text{ kW/m}$$

which is equivalent to the power used by more than 23 light bulbs of 100 W each for each meter of channel length. This is the heat energy that keeps the water from freezing.   □

Example 2.17

The velocity in a pipe 4 in. in diameter carries 0.025 m³/s. The pipe is 12 m above the reference level. The pressure in the pipe is measured to be 7357 N/m². Calculate the total hydraulic energy in the pipe at the point in question.

Solution. The pipe 4 in. in diameter = $4 \times 2.54 = 10.16$ cm

The cross-sectional area $= \dfrac{\pi D^2}{4} = \dfrac{\pi}{4}(10.16)^2 = 81.1 \text{ cm}^2$

$$A = 81.1 \times 10^{-4} \text{ m}^2$$
$$Q = 0.025 \text{ m}^3\text{/s}$$

Then the velocity $v = \dfrac{0.025}{81.1 \times 10^{-4}} = 3.083$ m/s

The kinetic energy $= \dfrac{v^2}{2g} = \dfrac{(3.083)^2}{2(9.81)} = 0.48$ m

The pressure energy $= \dfrac{p}{\gamma} = \dfrac{7357}{9810} = 0.75$ m

From Equation 2.21 the hydraulic energy

$$E = \frac{v^2}{2g} + \frac{p}{\gamma} + y$$

$$= 0.48 + 0.75 + 12 = 13.23 \text{ m} \qquad \square$$

Example 2.18

A 1000 m long pipe is 30 m higher at the entrance point and 10 m higher at the exit point than the reference level (Fig. E2.18). The pipe diameter is constant. The velocity in the pipe is 8 m/s. The water elevation at the entrance is 12 m above the pipe. The pressure at the exit point is the atmospheric pressure. Calculate the energy loss due to the flow in the pipe.

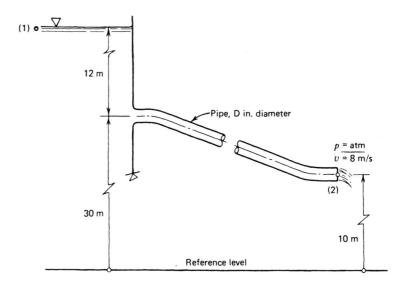

FIGURE E2.18

Solution. The pipe has a constant diameter. The kinetic energy is

$$\frac{v_1^2}{2g} = \frac{v_2^2}{2g}$$

$$= \frac{(8)^2}{2 \times 9.81} = 3.26 \text{ m}$$

From Equation 2.22, therefore,

$$\frac{v_1^2}{2g} + \frac{p_1}{\gamma} + y_1 = \frac{v_2^2}{2g} + \frac{p_2}{\gamma} + y_2 + h_L$$

$$0 + 0 + 42 = 3.26 + 0 + 10 + h_L$$

Hence, $h_L = 28.7$ m = hydraulic energy loss in the pipe due to friction and other factors.

$\square$

Example 2.19

Determine the flow and pressure at point A in the 1 ft diameter siphon shown in Figure E2.19, assuming energy losses are negligible.

Solution. Writing Equation 2.22 between the reservoir surface (1) and the siphon outlet (2) results in

$$0 + 0 + 5 \text{ ft} = \frac{v_2^2}{2g} + 0 + 0 + 0$$

Solving for velocity, we obtain

$$v_2^2 = 5(32.2)(2) = 322 \text{ ft}^2/\text{sec}^2$$
$$v_2 = \sqrt{322} = 17.9 \text{ ft/sec}$$

FIGURE E2.19

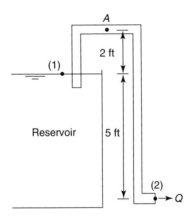

Flow can then be determined as

$$Q = v_2 A = 17.9(\pi)(0.5)^2 = 14.1 \text{ ft}^3/\text{sec}$$

Pressure at point A can be determined by writing Equation 2.22 between the reservoir surface and point A:

$$0 + 0 + 5 = \frac{(17.9)^2}{2(32.2)} + \frac{p_A}{\gamma} + 7$$

Solving for p_A gives

$$p_A = -7(62.4) = -437 \text{ lb/ft}^2 = -3.0 \text{ psi} \qquad \square$$

Example 2.20

An 8 in diameter post stands in 6 ft of deep water that flows with an average velocity of 6 fps. Determine the drag force acting on the post if the water temperature is 50°F.

Solution. The Reynolds number is

$$\mathbf{R} = \frac{vD}{\nu} = \frac{6(8/12)}{0.000015} = 0.267 \times 10^6$$

where kinematic viscosity is obtained from Figure 1.5. From Figure 2.20 the drag coefficient may be taken as

$$C_D = 1.0$$

With the density from Table 1.2 taken to be 1.94 slugs/ft³ (or 1.94 lb sec²/ft), Equation 2.26 gives

$$F_D = C_D(\tfrac{1}{2}\rho v^2 A) = 1.0\,(\tfrac{1}{2})(1.94)(6^2)(6)\left(\frac{8}{12}\right) = 140\,\text{lbf} \qquad \square$$

Example 2.21   Determine how long it will take for a spherically shaped particle of sediment to settle in 1 m of water at 20°C. The particle has a diameter of 0.1 mm and a density of 2650 kg/m³. Assume that the particle reaches terminal velocity instantaneously.

Solution. Terminal settling velocity is given by Equation 2.25,

$$v_t = \frac{g(\rho_s - \rho_f)\,d^2}{18\mu}$$

$$= \frac{9.81(2650 - 998)(10^{-4})^2}{18(10^{-3})} = 0.009\,\text{m/s}$$

Equation 2.25 is valid only for Reynolds numbers less than 1.0

$$\mathbf{R} = \frac{\rho v d}{\mu} = \frac{998(0.009)(10^{-4})}{10^{-3}} = 0.9$$

Because $\mathbf{R}$ is less than 1, the calculated value of settling velocity is correct. The time required for the particle to settle 1 m in water is

$$t = \frac{1\,\text{m}}{0.009\,\text{m/s}} = 111\,\text{s} \approx 2\,\text{min} \qquad \square$$

PROBLEMS

2.1 For a square area 4 m on a side, determine the first and second moments with respect to an axis parallel to one side and 6 m away from the centroid.

2.2 A large bridge pier 50 ft by 12 ft in cross section is built such that its base is 15 ft below water level. Determine the hydrostatic uplift force.

2.3 A 10 cm diameter plug is placed on the bottom of a tank in which the maximum water level is 2 m. What is the magnitude of the hydrostatic force for which the threads of the plug should be designed?

2.4 Water 6.0 ft deep is on the upstream side of a closed vertical sluice gate 8.0 ft wide. Determine the point of action and the magnitude and direction of the hydrostatic force on the gate.

2.5 On the gate described in Problem 2.4 determine the combined action of both upstream and downstream hydrostatic forces if the water depth at the downstream side is 5.0 ft.

2.6 A vertical circular flood gate 2.5 ft in diameter is under 11.25 ft of water with respect to its center. Determine the pressure distribution, the resultant force, and the location of the pressure center.

2.7 A semicircular channel is 2.5 m deep and is closed by a vertical gate. Determine the hydrostatic force and its point of action on the gate.

2.8 A 5 ft square gate is set on a 45° plane with respect to the water surface (see Figure E2.2); the centroid of the gate is 18 ft below the water surface. Compute the magnitude and direction of the hydrostatic force.

2.9 A 3 ft diameter pipeline will be laid underwater to a maximum depth of 10 ft above the top of the pipe. Determine the magnitude and direction of the resultant force on the pipe wall.

2.10 The floor of a watertight basement is 5 ft below the groundwater level. Assuming that the unit weight of the concrete is 162 lb/ft³, determine the required floor thickness to prevent uplift using a safety factor of 1.5.

2.11 A 12 ft² floating dock weighs 2000 lb. How many 55 gal steel drums should be placed under the platform so that they will sink to their centerlines? The weight of a steel drum is 75 lb.

2.12 In a 5 ft diameter pipe the velocity distribution is semicircular, with a maximum value of 3 ft/sec at the center. The velocity at the pipe wall is zero. Calculate the discharge and the average velocity in the pipe.

2.13 A storage reservoir is continuously filled through a pipe delivering 2.2 cfs. The reservoir is circular in shape with a diameter of 20 ft. The height of the tank is 12 ft. A pump operates for the last 5 minutes of successive 20 minute intervals, draining out 350 ft³ of water. Determine the time it takes for the tank to overflow, assuming that it is initially empty.

2.14 The discharge in a 1.0 ft diameter pipe is 2.5 cfs of water. Determine the momentum of the flow.

2.15 A 100 liter/s discharge is flowing in a 50 cm diameter, 180° pipe elbow. Determine the magnitude and direction of the momentum acting on the fixture.

2.16 A constricting 45° pipe bend is 6 in. in diameter at the inlet and 4 in. in diameter at the outlet. For a discharge of 5 liters/s determine the change of momentum through the bend.

2.17 A water wheel delivers a torque of 1450 ft-lbf at 15 rpm. The radial distance of the point of impact from the axis of the wheel is 6 ft. If the submerged area of one blade of the wheel is 15 ft², determine the average velocity imparted to the water by the blade.

2.18 A pump is delivering 60 cfs of water against a total head of 20 ft. What is the theoretical power requirement in terms of horsepower and kilowatts?

2.19 The bottom of a rectangular flume is 5 m above reference level. The depth of the water in the flume is 3 m. The average velocity is 1.7 m/s. Compute the total hydraulic energy at the surface of the water and at the bottom of the flume.

2.20 A pipeline 1200 ft long and 8 in. in diameter has at its inlet point 75 ft of available hydraulic energy with respect to the reference level. The velocity in the pipe is 5 fps; the pressure at its end is 12 psi. The elevation of the end of the pipe is 10 ft above reference level. How much energy in horsepower is lost by the water while flowing through the pipe?

2.21 A 25 cm diameter, 2 km long pipeline carries a discharge of 0.2 m³/s. The two ends of the line are at equal elevations. At the downstream end the total hydraulic energy is equal to 24 m of water above the pipe. If the highest point of the pipeline is a hill halfway along the pipeline, and if the slope of the energy grade line is 0.088 m/m, determine the maximum relative elevation of the hill before cavitation occurs and the total hydraulic energies at the inlet point needed to initiate and sustain the flow.

2.22 A horizontal pipeline is composed of a 10 cm diameter, 500 m long pipe connected to a 20 cm diameter, 700 m long pipe. The discharge in the pipeline is 0.01 m³/s. If the pipe discharges into the atmosphere, and the rates of energy loss are 0.01 m/m and 0.002 m/m, respectively, determine the required total hydraulic energy at the inlet (neglect local losses).

2.23 A pump delivers 6 cfs in an 8 in. diameter pipe over a 2000 ft length. The entrance of the pipe is 850 ft above the reference level; the exit is 1020 ft above the reference level. How much initial energy is required if the energy loss in the pipe is 120 ft, and the pressure required at the exit point equals 25 ft of water?

2.24 Is the flow in Problem 2.23 laminar or turbulent?

2.25 Compare the terminal velocities of a grain of sand with a diameter of 0.1 mm and a solid density of 2.65 g/cm³ in crude oil with a density of 0.855 g/cm³ and in water. Use a temperature of 20°C for the fluids.

2.26 A 1 m diameter bridge pier stands in 3 m of water with an average velocity of 1 m/s. What is the drag force on the pier if the water temperature is 20°C?

MULTIPLE CHOICE QUESTIONS

2.27 If all velocity vectors representing flow through an area remain the same over some time, it is called
A. Average flow.
B. Constant flow.
C. Steady flow.
D. Permanent flow.
E. None of the above.

2.28 Unsteady flow is defined in hydraulics as
A. The variation of discharge with time.
B. Flow in a channel or pipe with varying cross section.
C. Zero spatial acceleration of the fluid.
D. Spatial acceleration that is different from zero.

2.29 Spatial acceleration means
A. The change of velocity within a confined space.
B. The change of position of a moving particle over a given time period.
C. A negative acceleration, or deceleration.
D. Change of velocity between two points along a flow path.
E. The inverse of temporal acceleration.

2.30 In hydraulics the equation of continuity
A. Stems from Newton's first law of motion.
B. Is used for incompressible fluids only.
C. Means that what goes in, must come out.
D. Means that water cannot be created or destroyed.
E. None of the above.

2.31 The momentum of a flowing fluid is defined as the product of the discharge and the density.
A. True.
B. False.

2.32 The change of momentum of a flowing fluid results in
  A. An impulse.
  B. A force.
  C. A corresponding change of velocity.
  D. All of the above.
  E. None of the above.

2.33 The moment of momentum is
  A. The angular change of the impact force.
  B. Product of the rotational acceleration and the momentum.
  C. The ratio of the moment arm and the impulse force.
  D. Called the torque.
  E. All of the above.

2.34 An impulse force changes with a temperature change in a hydraulic system.
  A. True.
  B. False.

2.35 The capacity to do work is called
  A. Energy.
  B. Power.
  C. Hydraulic energy.
  D. Potential energy.
  E. Conserved energy.

2.36 Hydraulic energies are
  A. Potential and elevation energies.
  B. Potential, kinetic, and elastic energies.
  C. Those in excess of heat energy.
  D. Kinetic, potential, and pressure energies.
  E. Chemical energy and heat energy.

2.37 Hydraulic energies can be converted to internal energy.
  A. True.
  B. False.

2.38 The difference between hydraulic and energy grade lines is
  A. Constant.
  B. Dependent on the discharge and the pressure.
  C. The kinetic energy.
  D. The head loss.
  E. Zero in a gravity-free environment.

2.39 The statement that "hydraulics is the science of variable constants" was made by
  A. Bernoulli.
  B. von Kármán.
  C. Newton.
  D. Euler.
  E. Prandtl.

2.40 The law of optimum energy
  A. States that water will use most of the available energy.
  B. Means that hydraulic energies are convertible.
  C. Is valid only for laminar flows.
  D. Is a fundamental law of nature.
  E. All of the above.

2.41 Hydraulic factors of proportionality were
A. Introduced by Ludwig Prandtl.
B. Necessary to account for seepage factors.
C. Introduced to define frictional losses.
D. First used by Newton.
E. von Kármán's invention.

2.42 Bernoulli's equation is derived from
A. Hydraulic energies.
B. Newton's second law of motion.
C. Balancing energies between two points.
D. All of the above.
E. None of the above.

2.43 Hydrostatic pressure
A. Acts perpendicular to a solid surface.
B. Is proportional to the depth of the water.
C. Depends on the density of the fluid.
D. Acts equally in every direction.
E. All of the above.

2.44 On sloping surface areas the pressure center is
A. At the centroid.
B. Dependent on the angle of slope.
C. Above the centroid.
D. Defined by the formula $= \mathbf{M}/l\mathbf{A}$.
E. Independent of the slope angle.

2.45 In a 25 cm diameter, 90° pipe elbow water stands at a pressure of 11 kPa. The hydrostatic force acting on the elbow is
A. 637 N.
B. 540 kN.
C. 65.4 N.
D. 764 N.
E. 63.9 kN.

2.46 The law of Archimedes states that
A. Forces acting on the sides of a floating body cancel out.
B. The weight of a submerged body depends on the depth of the water above it.
C. The force of buoyancy equals that of the displaced water.
D. The stability of a floating body depends on the location of its centroid with respect to the mass center.
E. None of the above.

2.47 The hydrostatic force acting on a curved surface area
A. Depends on the radius of curvature.
B. Must act perpendicularly to the surface.
C. Is independent of the fluid density.
D. Is the sum of the vertical and horizontal pressures.
E. None of the above.

2.48 The momentum formula originates from Newton's second law of motion.
A. True.
B. False.

2.49 A jet of water discharged at $Q = 0.3$ cfs and 35 fps velocity hits a board perpendicularly. What is the impulse force?
  A. 13.4 lb.
  B. 544 lb.
  C. 18.5 lb.
  D. 6.87 lb.
  E. 20.4 lb.

2.50 A 0.1 m diameter pipe carries 50 liters/s of water at a pressure of 60 kPa. How much hydraulic energy is present with respect to a datum plane 15 m below the pipe?
  A. 45.4 m.
  B. 123.5 m.
  C. 16.2 m.
  D. 23.2 m.
  E. None of the above.

2.51 When 50 m³/s of water falls over a waterfall that is 42 m high, how much hydraulic energy could be generated assuming that the efficiency of the turbine is 30 percent?
  A. 6.18 MW.
  B. 54 kW.
  C. 40 MW.
  D. 9.5 kW.
  E. None of the above.

3

Hydraulic Data Collection

Whether in the field or in the laboratory, measurements of discharge and other hydraulic parameters provide the fundamental data on which hydraulic analysis and design is based. This chapter surveys the measurement techniques and formulas used in hydraulics. Reduction of such data and the formulation of dimensionless variables are introduced. The design of hydraulic laboratory models is explained.

3.1 INTRODUCTION

Measuring discharges and other flow parameters is an essential part of the analysis and operational control of all hydraulic systems. The rate of use of water by municipal customers, industrial processes, and agricultural users is usually metered. The capacity of watercourses and hydraulic structures must be determined by direct measurements. Hydrologic study of creeks and rivers is based on the statistical analysis of a long sequence of data obtained by periodically repeated or continuous measurements. For hydraulic design, formulas derived from fluid mechanics concepts need to be adjusted to practical reality by the development of hydraulic coefficients based on field measurements. Some of these data may be obtained by measurements in the field; others require the design and construction of elaborate laboratory models. For these various applications a multitude of methods and devices have been developed over the years. All are based on the fundamental physical laws of fluid mechanics.

3.2   CLASSIFICATION OF FLOW MEASUREMENTS

Flow determinations may be made either directly or indirectly. *Direct* determination of the discharge can be made by volumetric or weight measurements. The simplest way of measuring a discharge is to allow it to flow into a container of known volume and measure the time required to fill it. In laboratories, containers placed on scales allow the weighing of the water entering the container in a measured time interval. The measured weight divided by the unit weight of water and by the measured time interval gives the flow rate in volume per unit time. Municipal water meters measure the volume of water used over a period by direct measurements. They are called *positive displacement meters* because they measure the volume of water that repeatedly fills a given container. Usually, the number of fillings is counted by a register or counter, and the total quantity is obtained as the product of the volume of a chamber and the number of fillings. The registers are geared such that the counters show the total volume in gallons, cubic feet, or liters. Hence, the displacement meters may be looked on as fluid motors operating at a high volumetric efficiency under a very light load. Types of water meters include the oscillating or rotary piston meters, sliding or rotating vane meters, nutating disc meters, gear or lobed impeller meters, and other devices named according to their mechanical design. In addition to determining customers' water use, meters that register positive displacement are now also used for a broad range of hydraulic discharge measurements. These types of meters serve as *proportional secondary circuit meters* when placed in parallel with orifice or flume-type hydraulic measuring structures.

Indirect determination of the flow involves defining or establishing known flow conditions and measuring one or more parameters, such as pressure or its variation, kinetic energy, or water surface elevations. The measured parameters are used with the applicable hydraulic formulas to define the flow rate. According to the appropriate laws of fluid mechanics, the various techniques and devices may be classified by their practical applications. Thus, one may group the flow-measuring methods as they apply to

> pipe networks
> rivers
> creeks and other small watercourses
> hydraulic structures
> hydraulic laboratories

The various flow-measuring techniques will be discussed in the sequence listed.

3.3   FLOW MEASUREMENTS IN PIPES

In practice, indirect flow measurements in pipelines are made, by a great variety of commercially available measuring devices. All of them use the principle of continuity and one other equation that defines the motion. The latter most commonly used is the energy equation. Orifices, measuring nozzles and venturi tubes,

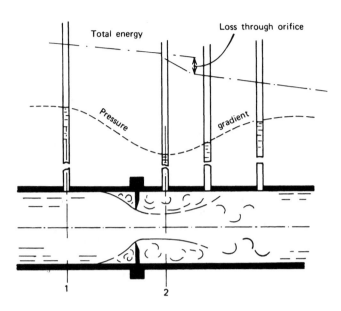

FIGURE 3.1   Flow through a circular orifice.

Pitot tubes, Prandtl tubes, and Annubars® operate on this principle. Other devices use the momentum equation. These include propeller meters and elbow meters. Other physical principles used in flow measurement include Faraday's law of electromagnetic induction, used in magnetic flow meters, and Ohm's law, utilized in hot wire anemometers measuring the rate of cooling by the flow fluid of an electrically heated resistance wire.

The simplest method of measuring discharge in a pipe is offered by a *circular orifice*. Its arrangement is shown in Figure 3.1. By the equation of continuity we may observe that

$$(VA)_{\text{pipe}} = (va)_{\text{orifice}} = Q_{\text{pipe}} = Q_{\text{orifice}} = Q \tag{3.1}$$

If we introduce the energy equation with the assumption that the pipe is horizontal, so that y_1 and y_2 cancel each other, Bernoulli's equation (Equation 2.22) takes the form

$$\frac{V_p^2}{2g} + \frac{p_p}{\gamma} = \frac{v_o^2}{2g} + \frac{p_o}{\gamma} + h_L \tag{3.2}$$

where h_L is an energy loss occurring at the orifice edges and in the turbulent exit zone behind the orifice plate. For orifices and similar flow meters the energy loss is expressed as a coefficient c, which is experimentally recorded at the time of calibration. Combining Equations 3.1 and 3.2 and rearranging the terms, we may write the ideal average velocity through an orifice as

$$v = \sqrt{(2g/\gamma)\frac{p_\mathrm{p} - p_\mathrm{o}}{1 - \left[c\left(\dfrac{a}{A}\right)\right]^2}} \tag{3.3}$$

and the discharge Q as

$$Q = \frac{ca}{\sqrt{1 - c^2(d/D)^4}}\sqrt{\frac{2g}{\gamma}\Delta p} \tag{3.4}$$

where d and D are the diameters of the orifice and the pipe, respectively, a is the orifice area, and Δp is the pressure differential.

Experimental studies indicated that the actual value of the c coefficient depends on the location where p_o is measured. In addition, it also depends on the configuration of the edge of the orifice, on the ratio d/D, and on the Reynolds number in the pipe, expressed by Equation 2.23. For Reynolds numbers above 25,000 the value of the c coefficient becomes a constant for all diameter ratios below 0.5. Because these are representative of practical cases, a good approximate formula for orifices in practical applications is the Torricelli equation,[1]

$$Q = cav = ca\sqrt{2g\frac{\Delta p}{\gamma}} \tag{3.5}$$

where v is the average velocity issued by the orifice. The value of c is to be determined by calibration. Commercially available orifice meters are supplied with a calibration chart. The discharge coefficient is usually within the range of 0.6 to 0.7. For water at constant temperature and in fully turbulent flow, the discharge equation of a given orifice can be represented by the simple formula

$$Q = K\,\Delta p^{0.5} \tag{3.6}$$

where K is a parameter lumping all variables that are fixed for the orifice in question.

To reduce the losses occurring in the constricted flow and to extend the range of the constant values of the coefficient, orifices are sometimes replaced by *flow nozzles* or *venturi tubes*, which are shown in Figure 3.2.

The derivation of the discharge formula for flow nozzles or venturi tubes is identical with that for the orifice meter. Discharge coefficients of these devices range from 0.94 to 0.98.

In metering any fluid in a pipeline, it is most important that the flow approaches the orifice, flow nozzle, venturi, or any other metering element in a normally turbulent state. The flow must not be influenced by swirls, crosscurrent, eddies, or other disturbances that create helical paths of flow. Disturbances after the flow-measuring element may introduce errors into the static pressure measurements at

---

[1]After Evangelista Torricelli (Italy, 1608–1647) who developed the concept, even though the gravity term, $2g$, was introduced more than a century later.

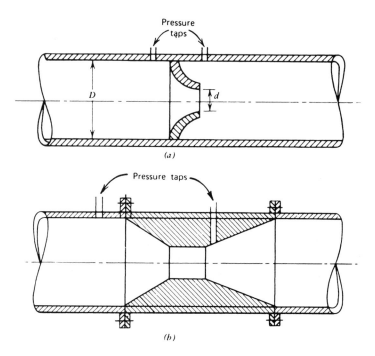

FIGURE 3.2 (a) Flow nozzle and (b) venturi tube.

that point.[2] To produce uniform conditions it is recommended that at least a length of 6 to 10 times the pipe diameter be of straight pipe in front of measuring devices. If this is not possible in some installations, then guide vanes are to be installed in the pipe to straighten the flow. After the metering device a straight length of 3 to 5 times the pipe diameter is desirable.

The *measurement of pressures* and pressure differences is performed by manometers, differential manometers, or Bourdon gages. The simplest way to measure pressure in a closed pipe is to use a transparent vertical or slanted standpipe, called a *piezometer tube*, connected to a tap on the pipe in which the pressure is to be measured. By Equation 2.1 the pressure energy in the water will force the water level up to a height at which the static pressure in the piezometric tube equals that in the pipe itself, as shown in Figure 3.3. Piezometric tubes are used for some laboratory work, but they are not feasible for practical applications. Because the height required is proportional to the unit weight of the fluid in the piezometer, large pressures can be measured more conveniently by using mercury, which is 13.5 times heavier than water. To prevent the mercury from flowing out into the pipe, the piezometer is bent into a U-shaped traplike loop. The resulting device is called a *manometer*.

---

[2]The fact that replacing an orifice plate with two cone-shaped pipes reduces or even eliminates headloss-creating turbulent eddies was first recognized by Giovanni Battista Venturi (Italy, 1746–1822).

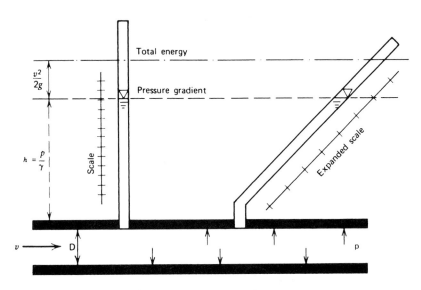

FIGURE 3.3   Piezometric tubes for pressure measurements.

The simple mercury manometer shown in Figure 3.4a allows the determination of the pressure in the pipe at point m from the relationship

$$p_m = \gamma_{\text{Hg}} z - \gamma_{\text{water}} h \tag{3.7}$$

where γ_{Hg} is the unit weight of mercury, which is 13.5 times that of water. Thus, Equation 3.7 takes the form

$$p_m = \gamma(13.5z - h) \tag{3.8}$$

In this case we assumed that the pressure at point b is atmospheric; hence, p_m is the pressure above atmospheric pressure.

If the other side of the manometer is connected to another pressure tap, as shown in Figure 3.4b, the instrument is a *differential manometer*. It is used to measure differences in pressure. Because of differences in pressure at points m and n, there exists a difference in level, z, between the two mercury columns in the U tube. By starting at m, where the pressure is p_m, and noting the changes in pressure as we proceed through the pipe to point n, we obtain the following expression for p_n:

$$p_n = p_m - \gamma x - \gamma_{\text{Hg}} z + \gamma y \tag{3.9}$$

Referring to Figure 3.4b, we see that $(x + z) = (y + h)$, and by substitution and rearranging we obtain

$$\Delta p = \gamma(12.5z + h) \tag{3.10}$$

In case the two pressure taps are at the same level, then $h = 0$ and

$$\Delta p = 12.5\gamma z \tag{3.11}$$

when mercury is used as manometer fluid. For other manometer indicating fluids, Equation 3.10 must be rewritten accordingly.

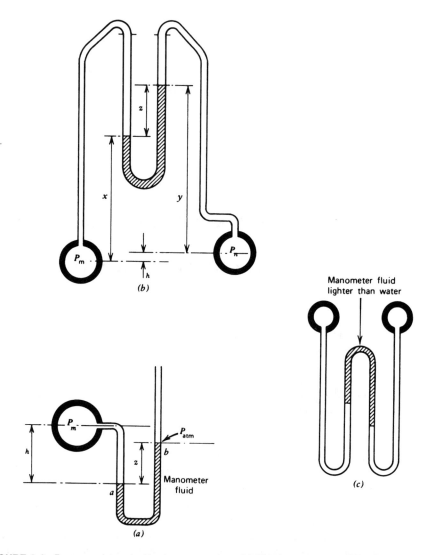

FIGURE 3.4 Pressure determination by manometers. (*a*) Simple manometer, (*b*) differential manometer, and (*c*) inverted manometer.

When the pressure difference between two measured points is very small, the elevation difference in adjacent piezometer pipes is difficult to determine. It is convenient in such cases to magnify the difference by using a manometer fluid that is lighter than water. Various manometer oils are available commercially having unit weights about 80 percent that of water. Usually these are color coded such that their specific gravity may be ascertained beyond doubt. One manufacturer's manometer fluids are listed in Table 3.1. Differential manometers designed to work with these lighter-than-water manometer fluids are of the *inverted* kind, as shown

TABLE 3.1   Manometer Fluid Properties

Trade Name	Color	Density g/cm³	Description
Meriam Red Oil	Red	0.827 at 60°F	A highly refined mineral seal oil. Noncorrosive. For use where a light fluid is required and extremes are not encountered. Useful temperature range is 40 to 120°F.
Meriam Indicating Fluid Concentrate	Green	1.000	Specially prepared concentrate for high-precision work. Noncorrosive to brass or stainless steel. Low surface tension. Hysteresis practically zero. Keeps indicating tube clean. Not for use under freezing conditions.
Meriam Unity Oil	Red	1.00 at 73°F	A mineral seal oil mixture prepared for general use from 30 to 100°F, where a fluid near the specific gravity of water is desired and water is unsuitable.
Meriam Indicating Fluid	Red	1.04 at 83°F	Specially suited to high-vacuum applications. Teflon gasketing recommended. Noncorrosive. Insoluble in water. Useful temperature range is 20 to 150°F.
Meriam Indicating Fluid	Light straw	1.20 at 77°F	A specially prepared oil particularly suited for general outdoor work. At elevated temperatures, corrosive to brass and aluminum. Useful range is 10 to 150°F.
Meriam Indicating Fluid	Blue	1.75 at 55°F	Extremely stable. For use in manometers and flowmeters. Low viscosity, nonfreezing, nonflash. Insoluble and clear interface in water. Useful range is 70 to 150°F.
Meriam No. 3 Fluid	Red	2.95 plus or minus 0.002 at 78°F	A heavy bromide. Nonflashing. All parts in contact with this fluid must be brass, glass, or stainless steel. Corrosive to steel. Useful temperature range is 40 to 100°F.
Meriam Hi-Purity Mercury	Silver	13.54 at 73°F	Specially treated for highest purity. Used in all applications where greatest density is required for maximum range and where chemical reaction is not encountered. Cannot be used in aluminum or brass instruments.

in Figure 3.4c. In using manometers care must be taken to expel all air bubbles from all connecting tubes; otherwise, the readings obtained will have little value. For correct measurements the pressure taps in the pipes must be free of burrs, and the entrance holes should be on a surface parallel to the flow in the pipe.

High pressure may also be measured with ordinary steam gages invented by Bourdon. A *Bourdon gage* consists of a coiled tube having its inner end closed and connected by a simple rack and pinion to a hand that is free to rotate over a graduated dial. The pressure in the coiled tube tends to uncoil the tube, which results in a movement of the hand. The dial is calibrated by applying a known pressure to the tube and marking the position of the hand on the dial, as shown in Figure 3.5.

A variety of modern pressure-sensing devices have been developed for use in fluids. The most common types measure the deflection of a circular elastic membrane, the diaphragm, which is under pressure from one side. The deflection is transmitted to a gage or meter either mechanically, magnetically, or electrically.

Magnetic linkage between diaphragm and dial pointer in *Dwyer Magnehelic®* gages is provided by a cantilever leaf spring that is located behind the diaphragm

FIGURE 3.5 Bourdon gage.

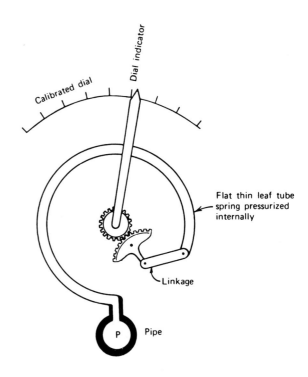

and mechanically connected to it. Deflection of the diaphragm by the fluid pressure behind it bends the leaf spring. At the end of the leaf spring a horseshoe magnet is affixed. Between the ends of this magnet there is a circular, staircase-shaped helix made of a metal of high magnetic susceptibility. The helix is pivoted at its end, so it rotates freely, moving a pointer at one of its ends. As the horseshoe magnet moves the helix tends to maintain a minimum gap between the ends of the magnet. Hence, the change of pressure on the diaphragm causes the pointer to move, indicating the pressure, as Figure 3.6 shows.

Electrical linkage between the deflecting elastic membrane and a calibrated indicating meter or a recorder is provided by *pressure transducers*. Pressure transducers are usually built with a variable resistance electric strain gage bonded to the membrane. The strain measured by the strain gage results from differential fluid pressure on the membrane. Hence, the change in measured voltage flowing through the gage is proportional to the pressure of the fluid on the diaphragm. The pressure range to be measured can be changed by simply changing the diaphragm. Another type of electric pressure transducer uses the principle of capacitance. A thin, stretched stainless steel diaphragm welded to its case forms a variable capacitor with an insulated electrode located very close to the diaphragm. The pressure change–induced variation in capacitance is translated by electronic means into a high-level, linear, DC signal. Schematic drawings of these devices are shown in

FIGURE 3.6 Dwyer Magnehelic® gage.

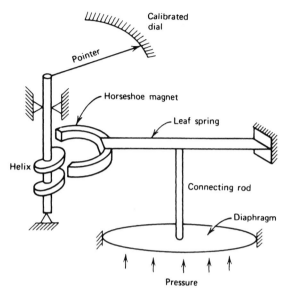

Figure 3.7. In practice, electric pressure transducers are advantageous because such signals can be amplified to attain greater precision, because they can be continuously recorded by magnetic or strip chart recorders, and because electric signals can be transmitted for remote control applications.

Velocity measurements in pipes as well as in open channels may be made by measuring the pressure corresponding to the kinetic energy of the flow.

The simplest method is to measure the pressure in an open-ended tube bent in such a way that the end is aligned opposite the velocity vector measured. Such tubes are called *Pitot tubes*.[3] The kinetic energy at the center of a Pitot tube converts into pressure energy as the flow of the fluid is halted at that point, as shown in Figure 3.8a. In pipes the pressure differential h between a Pitot tube and a nearby static tube provides the means to compute the velocity at the point measured by the Torricelli formula, where $\Delta p / \gamma$ is replaced by h. Hence,

$$v = \sqrt{2gh} \tag{3.12}$$

Moving the Pitot tube across the diameter of the pipe gives the velocity distribution, from which the discharge can be calculated. Fixed Pitot tubes may be calibrated for a range of discharges, so the discharge determination may be made directly by multiplying the measured pressure differential by a constant.

The combination of static and Pitot tubes resulted in the *Prandtl tube* shown in Figure 3.8b. Prandtl tubes are available commercially in various sizes, some as small as $\frac{1}{16}$ in. in diameter.

---

[3]Named after their inventor, hydraulic engineer Henri de Pitot (France, 1695–1771).

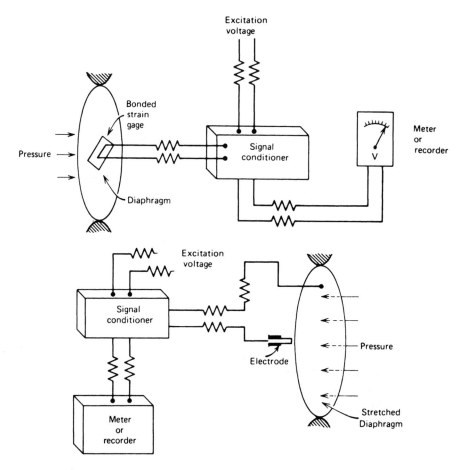

FIGURE 3.7   Pressure transducers.

Further improvement of this type of measurement is represented by the *Annubar*® shown in Figure 3.8*c*. Kinetic energy measured in the annubar is a composite of the velocities in several annuli of the pipe cross section. These measuring devices are calibrated in the factory, and they give the discharge of the flow directly from a calibration curve.

The impulse-momentum principle is the basis of velocity measurements in pipes made by *propeller meters*, although they are more commonly used in rivers and open channels. They are used for velocity measurements in hydraulic power stations, where several propeller meters are attached to a crossbar located in the conduit leading to the turbine. For very small discharges in pipes (0.15 gph to 6 gpm) there exists a flow meter in which a small rotor spins freely without any contacts with the meter housing. This bearingless flow meter, shown in Figure 3.9, is equipped with an optical readout. The rotor's light reflective marks are sensed by a photodetector

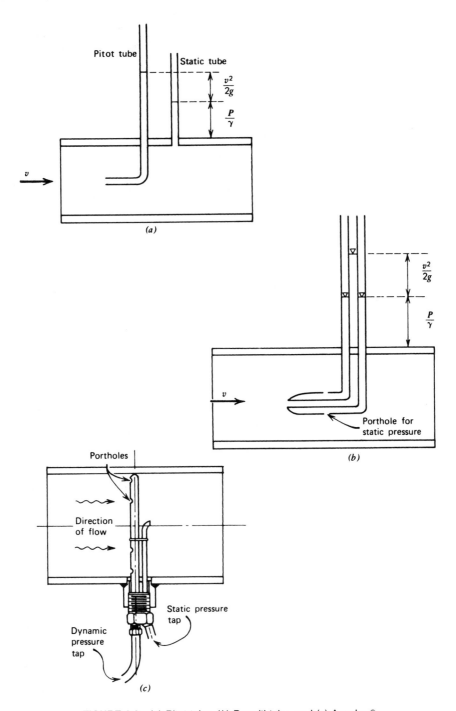

FIGURE 3.8 (a) Pitot tube, (b) Prandtl tube, and (c) Annubar®.

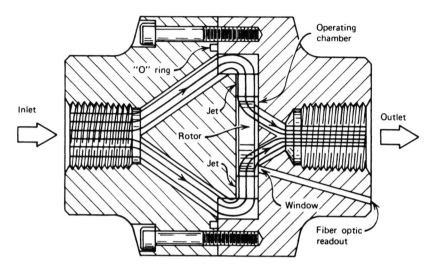

FIGURE 3.9 Bearingless flow meter. (Reproduced with permission from Bearingless Flow Meter Company, Boston, MA)

through fiber optics. The photodetector produces an electric impulse for each rotor mark passing a window and registers it by a counter.

Another flow-measuring device utilizing the momentum-impulse principle is the *elbow meter*. In a pipe elbow the direction of the velocity changes and, by Equation 2.15, an impulse force is generated acting along the outer curvature of the bend. Measuring the difference in pressure between the inner and outer curvature of the pipe bend, as shown in Figure 3.10, registers the effect of the impulse force. By proper calibration the discharge in a pipe elbow of a given diameter may be determined. An approximate discharge formula for 90° pipe elbows for which the radius of curvature of the pipe R is larger than twice the pipe diameter may be expressed as

$$Q = 0.785 \frac{D^{1.5}R^{0.5}}{\rho^{0.5}} \sqrt{\Delta p} \tag{3.13}$$

where Δp represents the pressure difference between the inside and outside pressure taps located along the 45° diagonal of the elbow. For continuous discharge measurements pipe elbows may be equipped with a standard positive displacement water meter to serve as a proportional secondary circuit meter. In such arrangements the difference in pressure between the taps will force some water to flow across the meter. This discharge, registered by the meter, will be proportional to the pressure difference, which, in turn, varies directly with the discharge in the pipe. The system, once calibrated, is a useful method for determining the output of pumping installations.

Another class of instruments for measuring discharge in piping systems are the *variable area meters*. These include the tapered tube and float meter, the multiple-holed cylinder and piston, and the slotted cylinder and piston types.

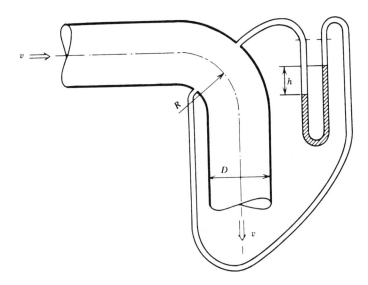

FIGURE 3.10 Pressure distribution in pipe elbows.

In all variable area meters the energy causing the flow through the meter is substantially constant for all discharges, and the discharge is directly proportional to the metering area. In the *tapered tube and float meter*, shown in Figure 3.11, the upward-flowing fluid lifts the float in the tapered tube. The float is designed in such manner that the impulse force and viscous drag force raising it, less the force of buoyancy, are constant for all positions of the float in the tapered tube. Such instruments are referred to as *rotameters*. In these the tapered tube is made of glass to make the position of the float visible. The discharge versus float elevation is determined by calibration. Because the viscous drag force lifting the float is dependent on the viscosity of the fluid, which in turn is dependent on the temperature, for different fluids or calibration fluids at different temperatures the actual readings of the meter must be reduced to calibration data. The actual discharge q_a of a fluid with a unit weight of γ_a corresponds to calibration discharge q_c as

$$q_a = q_c \sqrt{\frac{\gamma_c}{\gamma_a}} \tag{3.14}$$

where γ_c is the unit weight of the fluid for which the meter was calibrated. The *slotted cylinder* type of variable area meter utilizes a cylinder with a slot on its side through which the flow exits the meter. The varying rates of flow cause a piston in the cylinder to rise or fall. The position of the piston is seen through a sight tube in front of a calibrated scale. An example of the slotted cylinder type meter is shown in Figure 3.12.

With direct visual display the variable area flow meters give an instantaneous discharge value. One such device overcomes this shortcoming by transforming the float position in its tapered tube into an electric signal. This is accomplished by a

FIGURE 3.11 Rotameter.

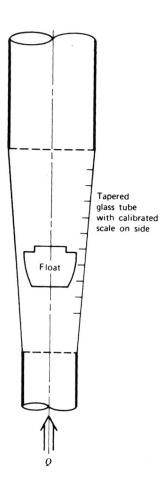

Tapered
glass tube
with calibrated
scale on side

Float

Q

linear variable differential transformer (LVDT) connected to the float by a vertical stem. The principle of an LVDT is shown in Figure 3.13. It consists of an electromagnet excited by a primary ac current. The position of the core element is dependent on the mechanical input—in the case of the flow meter it is connected to the float. The relative location of the core is sensed by two secondary coils by the induced currents generated in them according to Faraday's principle of electromagnetic induction. When the core is centrally located, the electromagnetically induced currents in the two secondary coils are equal. If the core is offset, one secondary coil will have more induced current than the other one. By electric instrumentation the position of the core, hence, the connected measuring float, can be determined and recorded. The output voltage is calibrated to the discharge of the flow meter. Faraday's principle is used directly for discharge measurement in *magnetic flow meters* (Figure 3.14). The principle of operation involves a magnetic field generated

FIGURE 3.12 Slotted cylinder meter.

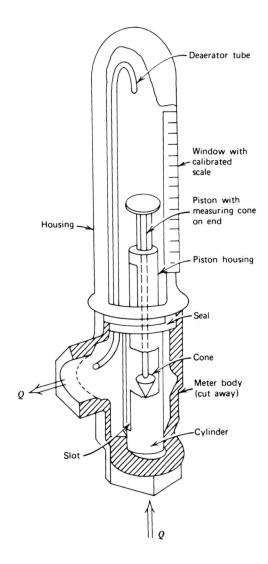

by an electromagnet placed perpendicular to the pipe in which the flow takes place. As the fluid flows through the magnetic field, an electromotive force is induced in it. The strength of this electromotive force is proportional to the velocity of the fluid and may be measured by a galvanometer connected to two electrodes attached to the pipe.

Other new pipe flow-measuring devices include the *electrostatic flow meter*, based on two dissimilar metal pipe sections connected by a nonconducting pipe section. The voltage change between the two metals relates to the flow rate of the fluid. Also under development is the *thermal wave flow meter*, which correlates the

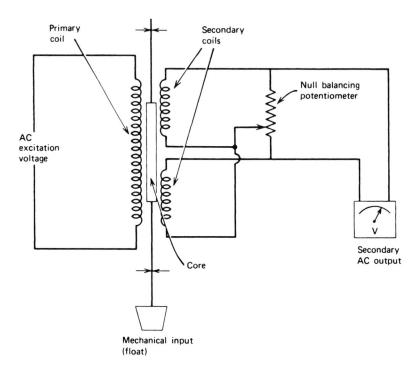

FIGURE 3.13 The principle of operation of an LVDT.

time of flow of a periodic thermal wave between an electric heater element and a sensor.

3.4 FLOW MEASUREMENTS IN RIVERS

Direct determination of river discharge is usually done by the concurrent measurements of depth and velocity distributions at a suitable cross section of the stream. The best location to perform such measurements along a stream is where the direction of the velocity is substantially perpendicular to the section, cross currents are minimal, and the cross section is reasonably uniform and free from vegetable growth. Bridges are ideal locations for such measurements; otherwise, guy cables and boats are required. Figure 3.15 shows the usual arrangement of such measurements.

In shallow rivers where the depth does not exceed 3 to 4 m and the velocity is less than 0.5 m/s, graduated *sounding or wading rods* are used for measuring the depth. Deeper channels are measured by *cables with weights* at their end. The size of these weights depends on the maximum velocity expected in the stream. For velocities ranging up to 0.5 m/s a 5 kg weight is sufficient; for 2 m/s a 50 kg weight is necessary. For high velocities up to 4 m/s, 100 kg weights are recommended.

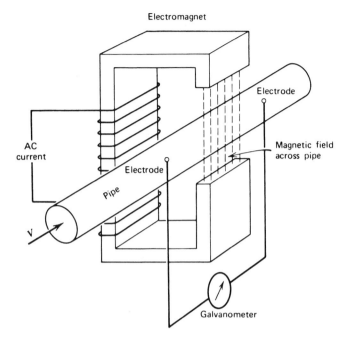

FIGURE 3.14   Magnetic flow meter.

Motorized winches are used to handle these large weights. To account for the deflection of the cable due to the velocity of the water, the approximate formula

$$y = l \cos \alpha \qquad (3.15)$$

may be used, in which the true depth is y, the length of the submerged cable is l, and α is the initial angle of the cable with the vertical. This equation is reasonably valid as long as the initial angle is less than 30°.

More precise measurements of depth may be performed by *ultrasonic devices*. These measure the time required for a sound pulse emitted by the instrument to bounce back from the bottom to a receiver in the manner in which sonar operates. *Echographs* and *fathometers* plot the results of continuous measurements across the stream.

Lead lines and rods, tag lines stretched across the channel, positioning the measuring boat by using two theodolites based on the shore, and other traditional hydrographic tools have been replaced by modern electronic systems. Depths of water in streams, lakes, estuaries, and coastal areas are measured by single- or multiple-transducer acoustic measuring devices. Laser techniques were developed to measure greater depths. Microwave systems, more like radar, give shorter and more differentiated pulses that enable hydrographers to measure the thickness of the sediment on the bottom and give indications of the type of the rock below. Side-scanning sonar technology in the hands of an experienced analyst provides

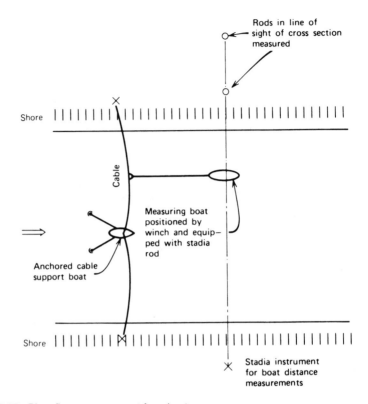

FIGURE 3.15 River flow measurement from boats.

information on the ripples and dunes as well as rocks on the bottom that may be as small as 10 cm in diameter. Hydrographic survey boats are equipped with on-board electronic positioning devices, which are coupled with the depth measuring equipment via digital computers that develop and plot all necessary survey data instantaneously. The results are increased precision and considerably greater efficiency than those of the traditional methods used in hydrographic surveying. Yet the traditional techniques are still very much in use, as they are more practical in the case of smaller projects.

To determine the stream discharge in a known channel cross section the *flow velocity* is measured along a number of verticals. For medium-size rivers at least 10 to 15 verticals are measured, with the measuring positions distributed evenly across the channel. At each vertical, velocities are measured at several depths. For shallow depths one point measurement at $0.6y$ is sufficient. For depths up to 0.6 m the average of two points measured at $0.2y$ and $0.8y$ is generally sufficient. In three-point measurements, at depths of $0.15y$, $0.5y$, and $0.85y$, the average velocity is still the arithmetic mean of the three measured values. For uneven velocity distribution along a deep vertical the measurements are taken at five points. In addition to the measured velocity values near the surface, v_s, and near the bottom, v_b, measurements

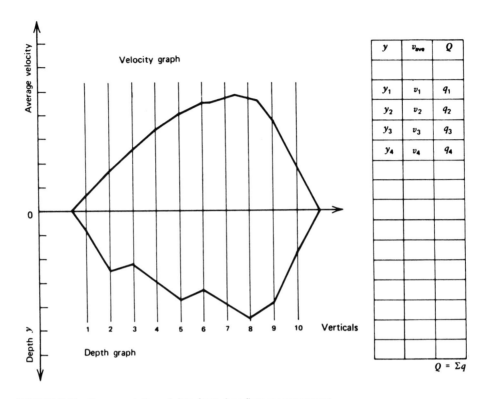

The table in the figure:

y	v_{ave}	Q
y_1	v_1	q_1
y_2	v_2	q_2
y_3	v_3	q_3
y_4	v_4	q_4
		$Q = \Sigma q$

FIGURE 3.16   Representation of data from river flow measurement.

are taken at $0.2y$, $0.6y$, and $0.8y$. The average velocity in this case is computed by

$$v_{ave} = \frac{v_s + 3v_{0.2} + 2v_{0.6} + 3v_{0.8} + v_b}{10} \qquad \textbf{(3.16)}$$

Results of discharge measurements in a river can be graphically represented in a plot similar to the one in Figure 3.16. The discharge Q may be determined from the data by computing the discharge q in each strip between two adjacent verticals. In this case the discharge of a strip between two verticals is

$$q = \left(\frac{v_1 + v_2}{2}\right)\left(\frac{y_1 + y_2}{2}\right) \cdot b \qquad \textbf{(3.17)}$$

where b is the width between the adjacent verticals, and subscripts 1 and 2 refer to the measurements taken in the verticals. The total discharge is then the sum of q values determined.

A more precise result may be obtained by graphical solution. In this method, the velocities are plotted on the cross section where they were measured and where points of equal velocity are connected by contour lines. The areas of known velocities

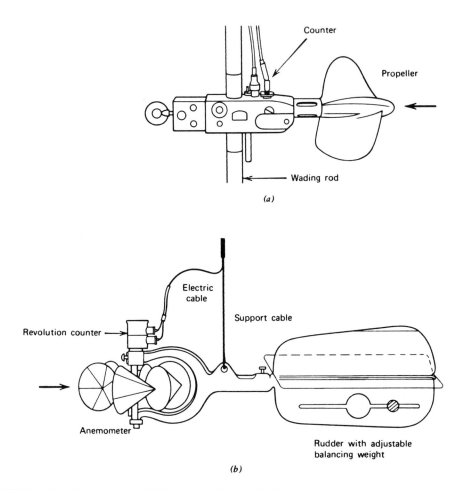

FIGURE 3.17   Current meters. (*a*) Propeller meter and (*b*) Price current meter.

may be measured by a planimeter. The sum of all areas multiplied by their respective velocities gives the total discharge of the stream.

Performing the discharge measurement of a river by the direct method just described is a time-consuming process. Because the flow in a river is rarely constant but either rises or falls, the measured velocities must be adjusted to an average value, valid for the duration of the measurement. This adjustment must be carried out if the average depth of the stream changes by more than 1 percent during the measuring time. To establish the change of stage it is recommended that the water level be marked by a stake at the shore before and after the measurement and that the change of average water level during this time be recorded.

The commonly used velocity-measuring device in rivers is the *propeller meter*, shown in Figure 3.17*a*. There are various propeller meters available commercially. A typical propeller meter consists of a propeller runner, a revolution counter, a shaft, and a tailpiece. The propeller is usually 7 to 12 cm in diameter. The pitch of the propeller is dependent on the average velocity for which the meter is designed. Normal propellers are designed to operate within the velocity range of 0.03 to 10 m/s. Care is taken to develop propellers such that they react to velocity components in their axial direction only. This enables the operator to measure the true velocity passing through the river cross section perpendicularly, holding the meter by a rigid rod. The counter gives an electric signal audible through an earphone after each revolution or after every 5, 10, or 20 revolutions. The tailpiece acts as a rudder, holding the meter in the direction of the flow, and is used only if the current is known to be perpendicular to the cross section measured. Experience indicates that the directional error with some self-aligning velocity meters can be as much as 20 percent.

The main difference among the various propeller meters is in their propeller design. A notable development in this field is the *Ott Kempten* propeller meter made in Germany. In this meter the propeller turns a small electric generator, and the velocity measured is proportional to its electric output.

The *Price current meter,* widely used in American practice, uses a cup-type anemometer rotating around a vertical shaft, as shown in Figure 3.17*b*. It is manufactured in various sizes for different applications and may be used either on a wading rod or on a cable with a weight attached. The direct-reading current meter includes a six-pole permanent magnet and a reed switch that sends electric signals to a direct readout unit on the surface, which shows the velocity of the water on a dial indicator.

All current meters, whether of propeller or anemometer configuration, operate on the momentum-impulse theory. To use them one must have a calibration table. Calibration of these instruments is done in towing tanks. The instruments are attached to a carriage that is pulled over the towing tank at various velocities during which the rates of revolution of the meters are recorded. Under careful maintenance and proper operation the attainable precision with propeller-type current meters ranges between 5 and 10 percent, depending on the rate of pulsation in the current. The larger the pulsation in the stream, the greater the possible error.

Magnetic flow meters, mentioned earlier, were also developed for measuring velocities in rivers, lakes, and estuaries, as well as in sewers and flumes. The electromagnetic current meters usually measure two perpendicular velocity components up to 10 ft/s.

3.5 FLOW MEASUREMENTS IN SMALL OPEN CHANNELS

Shallow rocky rivers, creeks, or other small open channels cannot be measured in the same manner as rivers because of their small size. In canals like those supplying water for irrigation permanent flow-measuring weirs are built. One such weir is shown in Figure 3.18. In the absence of hydraulic structures that may be utilized as points of measurement, either *portable weirs or flumes* are utilized, or they may be constructed at the site, or the discharge is determined by chemical means.

A V-notch–type portable weir, made of plywood, used in the field to measure seepage discharge is shown in Figure 3.19. With proper calibration such structures

FIGURE 3.18 Contracted weir controlling flow in irrigation ditch. (Courtesy of U.S. Bureau of Reclamation)

FIGURE 3.19 Portable V-notch weir used to measure seepage discharge behind leaky earthen dam. (Courtesy of Ohio Department of Water Resources)

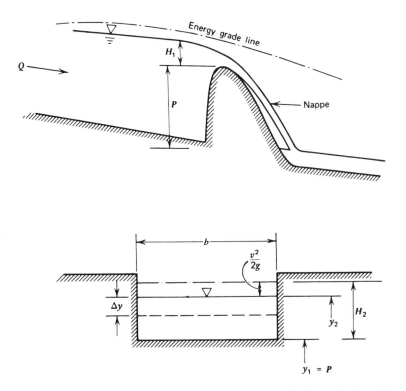

FIGURE 3.20 Notations for the weir formula. *P* represents crest height, and *b* is weir length.

may give reliable information on the flow rate, but their shortcomings are significant. If the soil in which they are placed in the channel is pervious, the prevention of seepage around them is almost impossible. The requirement of free overfall makes it necessary to dam up the upstream water, resulting in considerable pressure on the upstream side of the structure, which requires significant structural supports.

The discharge formula for water flowing over weirs is derived from the Bernoulli equation. For a rectangular weir with freely falling water the velocity at any point over the crest can be determined by the energy equation. The variables of the problem are shown in Figure 3.20. The total available energy on the upstream side is

$$E_{\text{upstream}} = H_1 + \frac{v_1^2}{2g} \tag{3.18}$$

where H_1 is the depth of water on the upstream side over the crest, back at a point where the water level is unaffected by the surface curvature, and v_1 is the approach velocity in the channel. When the P height of the crest is significant, as in most hydraulic structures, the approach velocity is relatively small; hence, the kinetic energy on the upstream side may be neglected. It was recognized by early eighteenth-century hydraulicians[4] that the depth of the water over the weir is two-thirds of the upstream depth, that is

$$y_2 - y_1 = \tfrac{2}{3} E_{\text{upstream}} \qquad (3.19)$$

or, neglecting the approach velocity and considering P equals y_1 as the datum plane,

$$y_2 - y_1 = \tfrac{2}{3} H_1 \qquad (3.20)$$

In modern hydraulics we know that this depth is the so-called critical depth, to be introduced later in connection with flow in open channels. However, this phenomenon is one example of how the minimum energy theorem inherently controls nature. The energy available for the flow will deliver the maximum discharge over the weir, at optimum velocity.

The velocity at any point between y_2 and y_1 will be determined by the energy available at that point. For example, in a thin horizontal strip just below y_2 the velocity equals

$$v = \sqrt{2g(H_1 - y_2)} \qquad (3.21)$$

according to the Torricelli equation, Equation 3.5. For the area of the thin strip of width Δy, shown in Figure 3.20, the elementary discharge is

$$\Delta q = b(\Delta y)v \qquad (3.22)$$

Integrating over the whole area of the nappe between y_1 and y_2 results in the common form of the weir formula,

$$Q = c\, b \sqrt{2g}\, H_1^{3/2} \qquad (3.23)$$

in which Q is the total discharge over the rectangular weir of b width under a head of H_1. The parameter c in this formula is the *discharge coefficient* of the weir. The discharge coefficient is generally determined by calibration.

For small discharges, 90° or 60° V-notch weirs, the so-called *Thomson weirs*, are often used. The discharge formulas for these weirs in conventional American units are

$$Q_{90°} = 2.28\, H_1^{5/2}$$
$$Q_{60°} = 1.31\, H_1^{5/2} \qquad (3.24)$$

where Q is in gallons per minute and H_1 is in inches. The nomograph shown in Figure 3.21 is useful in solving Equations 3.24. These equations are valid for smooth, knife-edge weir plates. Experience shows that the V-notch weirs are very sensitive to any change in the roughness of the weir plate; hence, the equations can be assumed to give only approximate values under practical conditions.

---

[4]This formula was first derived by Giovanni Poleni (Italy, 1683–1761).

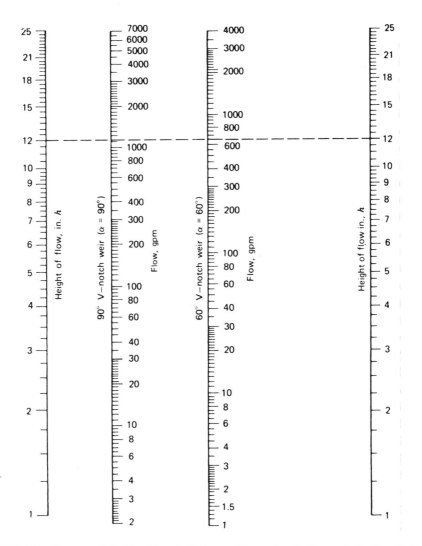

FIGURE 3.21 Nomograph for V-notch weir discharge. (Reproduced with permission from F. E. McJunkin and P. A. Vesilind: Practical Hydraulics for Public Works, *Public Works Magazine*, Vol. 99, No. 9, 10, and 11, 1968.)

For larger discharges the trapezoidal *Cipoletti weir* is used with a base width of b and steep side slopes with 4 to 1 rise. Cipoletti's discharge formula in conventional American units (cfs and ft) is

$$Q = 3.37 \, b \, H_1^{3/2} \tag{3.25}$$

Figure 3.22 provides a graphical solution of the discharge formula for Cipoletti weirs in SI units.

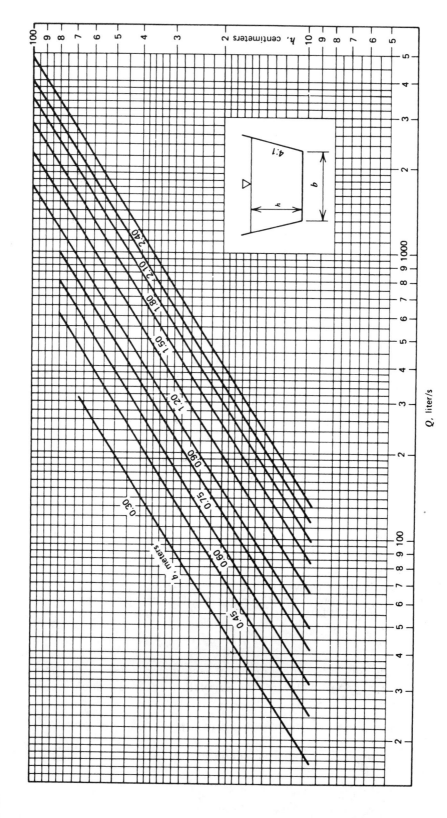

FIGURE 3.22 Discharge nomograph for Cipoletti weirs. (Reproduced with permission from F. E. McJunkin and P. A. Vesilind: *Practical Hydraulics for Public Works, Public Works Magazine*, Vol. 99, No. 9, 10, and 11, 1968.)

94

Chemical discharge measurements are most valuable for small streams where a large part of the flow passes through the rocks and gravel of the stream bottom. The common principle of the chemical discharge measuring techniques is the continuity equation. The methods involve the introduction of a known amount of tracer chemical into the water, which, after traveling in the stream over a certain length, will completely diffuse in the water. The concentration of the tracer chemical in the water after complete mixing is proportional to the discharge of the stream. Hence, by determining the concentration of tracer in the water, the discharge may be computed.

The water is sampled, and the concentration of the chemical is determined by various methods. Depending on the chemical used, these procedures may involve chemical, colorimetric, flame photometric, radioactive, or electric conductivity measurements. Because of the relative simplicity of electric conductivity measurement in the field, only this method will be discussed here.

There are two ways to introduce tracer chemicals into the stream. One is the continuous introduction of a constant tracer discharge Q_t of a known concentration C_t (in milligrams of tracer material per liter of water). The discharge of the stream Q then is determined by measuring the concentration C of the tracer in the stream and using the equation of proportionality

$$Q = \frac{C_t}{C} Q_t \tag{3.26}$$

Another way of measuring the discharge is to use a tracer wave. In this method a given amount of tracer material is mixed in a container of water, which then is dumped into the creek. At the point of sampling the concentration is measured continuously or at frequent intervals (15 or 30 s) until the concentration of the passing wave of tracer material becomes insignificant. The wave of the passing tracer at the sampling point may be plotted as shown in Figure 3.23, and the area under the tracer concentration versus time curve is determined. The area A under the curve (in seconds · milligrams/liter) defines the discharge Q (liter/second) of the creek as

$$Q = \frac{E}{A} = \left(\frac{\text{mg}}{\text{s} \cdot \text{mg/liter}} \right) = \frac{\text{liters}}{\text{s}} \tag{3.27}$$

where E is the tracer input in milligrams. Of all materials used as tracers, by far the most convenient for practical use is sodium chloride, ordinary table salt.[5] This allows the use of commercially available portable *electric conductivity meters.*

The maximum concentration of salt in water is 360g/liter; hence, a 5 gal bucket of water will easily dissolve a 10 lb bag of salt. The electric conductivity of salt solutions has been determined experimentally and is expressed by the formula

$$\text{conductivity in } \frac{1}{\text{ohm}} = 2.14 \left(\text{concentration in } \frac{\text{mg}}{\text{liter}} \right)^{0.95} \tag{3.28}$$

---

[5]The salt-velocity method was developed by Charles Metcalf Allen (United States, 1871–1950) at the Worcester Polytechnic Institute.

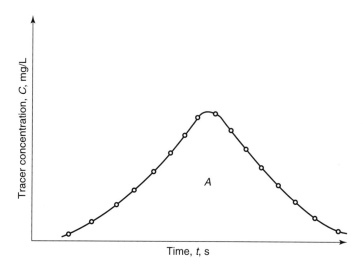

FIGURE 3.23 Discharge determination by tracer wave.

This allows the conversion of the readings on the electric conductivity meter into concentrations in milligrams per liter. The *1/ohm* term in this equation is referred to as *mho*. Although salt is often found in nature, this base amount can be eliminated from the procedure by starting the recording before the tracer is dumped and ending the recording when the readings return to near the original base conductivity. This way the base readings can be excluded from the area under the tracer curve.

The environmental effects of tracer materials should be carefully considered. In small creeks a measurable tracer wave may require less salt than is ordinarily encountered during winter road salt runoff. Tracer use in larger creeks could raise more concern.

Simple but approximate information on the velocity in an irrigation or drainage canal may be obtained by *surface floats*. For streams of medium size sometimes 15 to 25 floats are used simultaneously. These are equally distributed along the starter channel cross section. The times of arrival of the floats are recorded at a location a distance away from the starter section. The distance between the two sections should provide a travel time of several minutes, unless the velocity of the stream is rapid and the channel is meandering. Because the float velocity is representative of the surface velocity of the stream, the measured velocities should be reduced by a factor ranging from 0.75 to 0.9, with the larger value being applied to deeper and faster (more than 2.0 m/s) streams. Somewhat more precise results may be obtained with *adjustable submerging* floats—thin aluminum tubes closed at both ends and filled with enough weight at one end so that they float. Float lengths are selected so that the floats will submerge to a depth of about 10 to 15 in. above the bottom and stick out over the surface by about 2 to 4 in. The observed velocity of a submerging float, v_f, allows the determination of the average stream velocity, v_{ave}, along its path from the experimental formula

$$v_{ave} = v_f \left(0.9 - 0.116 \sqrt{1 - \frac{h_f}{y}} \right) \qquad (3.29)$$

where h_f is the length of the float and y is the depth of the stream. In small channels repeated use of the float used for fishing may give acceptable results.

3.6   DISCHARGE MEASUREMENTS IN HYDRAULIC STRUCTURES

Hydraulic structures are ideal locations for discharge measurements, particularly if they are of simple hydraulic design and if the water passes through them in a known condition. At *control structures*—weirs, gates, manholes, and the like—the discharge rating curve can be developed by a few independent discharge measurements in the stream, using the methods described in the previous sections.

In installations where the amount of passing water must be known, as in irrigation, drainage, or sewage networks, permanent measuring structures are often built. One of the most successful structures developed for the measurement of small open channel flows is the *Parshall flume,*[6] shown in Figure 3.24. Although more expensive than weirs, the Parshall flume has found considerable use in irrigation, water supply, and waste water treatment plants because of its inherent low permanent head loss and its self-cleaning capacity. Its main feature is its rapidly converging and dropping throat section in which critical flow conditions develop unless a hydraulic jump[7] forms downstream and backs the water level up into the throat section. For free flow in the throat the discharge in the flume can be computed by measuring the upstream water level in the stilling basin located on the outside of the converging side wall. The approximate formula for the discharge in this case is

$$Q_0 = 4\,w\,H_a^{1.522w^{0.026}} \qquad (3.30)$$

where w is the throat width and H_a is the upstream water level. The nomograph shown in Figure 3.25 facilitates the solution of Equation 3.30. For larger discharges, when the hydraulic jump moves upstream and completely submerges the throat section, the critical flow condition is eliminated. In this case, expressing the submergence by the ratio H_b/H_a, in which H_b is measured at the throat section, we may adjust the discharge by a correction factor given by the graph shown in Figure 3.26.

Many other *venturi-type flumes* were developed for various applications, each claiming certain advantages over the Parshall flume. For example, the *Palmer-Bowlus* flumes, widely used in California and elsewhere for measuring sewage flows through manholes, have a flat bottom; hence, they are easier to place in existing installations. In some irrigation systems venturi flumes, different from the Parshall

---

[6]Named after Ralph Leroy Parshall (United States, 1881–1960).
[7]The concept of hydraulic jump will be explained in detail in Chapter 8.

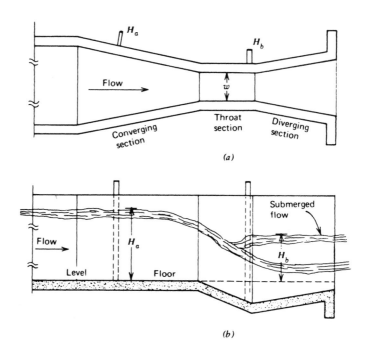

FIGURE 3.24 Parshall flume configuration. (*a*) Plan and (*b*) section. (Reproduced with permission from F. E. McJunkin and P. A. Vesilind: Practical Hydraulics for Public Works, *Public Works Magazine*, Vol. 99, No. 9, 10, and 11, 1968.)

flume, are used with configurations that are cheaper to construct. Trapezoidal-shaped venturi flumes of various standard designs are used in Europe and elsewhere.

Regardless of design, the general requirements of well-designed discharge measuring structures are that their discharge rating curve be independent of the channel slope and of downstream changes, such that their discharge can be determined by a single depth measurement; that the changes in upstream and downstream channel configurations do not influence their discharge characteristics; that their size and configuration allow the designer to use established theoretical principles for preliminary design; that their discharge be calibrated by only a few measurements in the field; and that their structure remain unchanged for a long period of time, hence, make their rating curve reasonably permanent.

Measuring structures are generally installed on small streams. Because changes in flow in small streams are usually much more rapid than in large rivers, the occasional (daily or twice daily) reading of a graduated stage may introduce large errors in the continuous discharge data. This consideration suggests the necessity of installing *continuous recording devices* to measure the stage. Float-type continuous recorders are available commercially. These recorders are either of the mechanical, clock motor driver type, like the *Stevens Total Flow Meter,* or are electric strip chart recorders. The popular Stevens flow meter is equipped with interchangeable

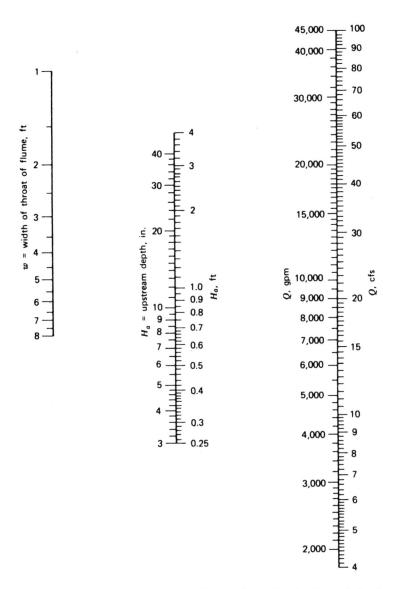

FIGURE 3.25 Discharge nomograph for Parshall flumes. (Reproduced with permission from F. E. McJunkin and P. A. Vesilind: Practical Hydraulics for Public Works, *Public Works Magazine*, Vol. 99, No. 9, 10, and 11, 1968.)

gears and cams to convert them to different flow ranges and includes a paper-carrying roller on which a pen marks the instantaneous position of the float. The roller is connected to a counter that integrates the "run" of the roller and, hence, provides a reading that relates directly to the total volume of the flow. Newer

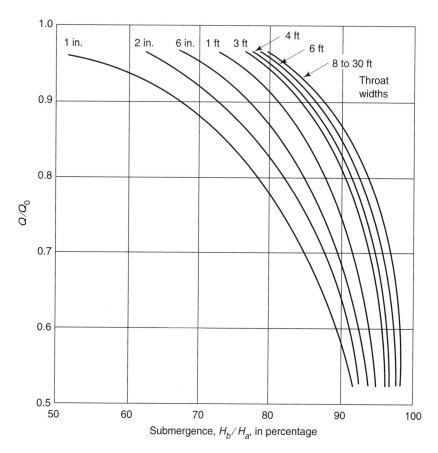

FIGURE 3.26 Correction factor for submerged Parshall flume's discharge determination. (Reproduced with permission from F. E. McJunkin and P. A. Vesilind: Practical Hydraulics for Public Works, *Public Works Magazine*, Vol. 99, No. 9, 10, and 11, 1968.)

devices on the market incorporate a magnetic tape casette recorder to collect data.

3.7 REDUCTION OF EXPERIMENTAL DATA

Hydraulics is an experimental science. Its development over the past two centuries has been marked by the assemblage of an immense pile of experimental data from both field and laboratory measurements. Hydraulics is also a complex subject. Most phenomena involving the movement of water are known to be dependent on many variables—geometric characteristics, fluid properties, and flow characteristics. Even some of the most commonplace hydraulics problems depend on perhaps a dozen variables. Early books on hydraulics were replete with endless pages listing "raw" experimental data. Design consisted of selecting parameters from tables showing data of previous projects "that worked." Much of these raw experimental data were incomplete, ambiguous, and incorrectly taken. Attempts to represent previous experiments by plotting them on charts and expressing the plots by empirical formulas increased the chances for errors by not stating the limits within which data were available, leading the users to incorrect extrapolations. To overcome manipulative difficulties inherent in handling large sets of experimental data, techniques of dimensional analysis and the formation of dimensionless groups of variables were introduced at the end of the nineteenth century. Dimensional analysis helps the experimenter in the systematic study of his or her experimental data. Dimensionless grouping reduces the number of variables that have to be processed.

In physical systems, including hydraulic ones, certain quantities such as mass or force, length, time, and temperature are considered five *fundamental quantities* that cannot be expressed in simpler terms. These fundamental quantities are said to have dimensions. We measure these dimensions in various units; for instance, length may be measured in meters, light-years, feet, or fathoms. All other physical quantities are expressed in terms of the fundamental quantities: for example, velocity is expressed as length per time. These are the *derived quantities.*

Because force and mass are related to each other by Newton's second law of motion, only four of the five fundamental quantities are independent of each other. In hydraulics, temperature is rarely used; hence, we speak of three fundamental quantities only. To describe motion we use either the force, length, and time (FLT) system or the mass, length, and time (MLT) system. In either case, the mass or force becomes a derived quantity. Either set may be used for dimensional analysis of physical systems, but each must be used exclusively. Table 3.2 lists the derived dimensions of the most common hydraulic parameters for both sets of fundamental quantities.

In order to organize best a set of raw experimental data or to plan an experimental study, the first step is to list all physical variables that pertain to the problem. This requires that the analyst consider all the fundamental theoretical concepts that play a part in the physical process studied. For example, hydraulic energy must be looked on as foot-pounds of energy per pound of fluid instead of its common representation as "feet of head." Terms not obviously playing a part in the problem must be listed if they inherently influence the physical processes considered. For example, the gravitational acceleration is essential in the analysis of weir flow, even though its value remains a constant; as such the careless experimenter may tend to neglect it.

Glancing over the formulas and charts in the previous pages of this book, we notice the advantage of dimensionless representation of data. Dimensionless groups of variables are independent of the units of measurement. Dimensional homogeneity is a fundamental requirement in all computations, since all variables have to be expressed in a consistent system of measures.

A dimensionless number may be formed by taking the ratio of two corresponding terms of identical dimensions. For instance, the height of the average roughness element on the inside wall of a pipe divided by the pipe diameter furnishes a

TABLE 3.2   Dimensions of Hydraulic Quantities

	FLT	MLT
Geometric characteristics		
Length	L	L
Area	L^2	L^2
Volume	L^3	L^3
Fluid properties[a]		
Mass	FT^2/L	M
Density (ρ)	FT^2/L^4	M/L^3
Specific weight (γ)	F/L^3	M/L^2T^2
Kinematic viscosity (ν)	L^2/T	L^2/T
Absolute viscosity (μ)	FT/L^2	M/LT
Elastic modulus (E)	F/L^2	M/LT^2
Surface tension (σ)	F/L	M/T^2
Flow characteristics		
Velocity (v)	L/T	L/T
Angular velocity (ω)	$1/T$	$1/T$
Acceleration (a)	L/T^2	L/T^2
Pressure (Δp)	F/L^2	M/LT^2
Force (drag, lift, shear)	F	ML/T^2
Shear stress (τ)	F/L^2	M/LT^2
Pressure gradient ($\Delta p/L$)	F/L^3	M/L^2T^2
Discharge	L^3/T	L^3/T
Mass flow rate	FT/L	M/T
Torque and moment	FL	ML^2/T^2
Work or energy	FL	ML^2/T^2
Work or energy per unit weight	L	L
Work or energy per unit mass	L^2/T^2	L^2/T^2

[a] Density, viscosity, elastic modulus, and surface tension depend on temperature; therefore, temperature is not considered a property in the sense used here.

dimensionless geometric parameter describing one geometric aspect of the flow in a pipe.

Newton's second law of motion states that any change from static condition or from linear steady motion comes about by external effects. The force resisting change in motion is the force of inertia,

$$F = ma$$

The active forces resisting the maintenance of this inertia force are

gravity force $= mg$
viscous force $= \tau A = (\mu \Delta v / \Delta L)A$
pressure force $= \Delta p A$
elastic force $= (-E \Delta V / V_0)A$
surface tension force $= \sigma L$

A in these equations represents the surface area on which the stresses act. L is length, and V is volume. A number of dimensionless ratios may be derived by dividing these terms into the inertial force. For example,

$$\frac{\text{inertia force}}{\text{gravity force}} = \frac{a}{g}$$

From Table 3.2 the dimension of a is L/T^2. Substituting this value into the preceding relationship and multiplying by L/L (or 1) yields $(L/T^2)\,L/(gL)$. But from Table 3.2 L/T is the dimension of velocity v. Thus,

$$\frac{\text{inertia force}}{\text{gravity force}} = \frac{v^2}{gL} = \mathbf{F}^2 \quad \text{or} \quad \mathbf{F} = \frac{v}{\sqrt{gL}} \qquad (3.31)$$

$\mathbf{F}$ is an important characteristic of free surface flow. It is called the Froude number after William Froude (England, 1810–1879), who built the first towing tank for the study of ship models. Other common dimensionless numbers follow:

$$\frac{\text{inertia force}}{\text{viscous force}} = \frac{\rho L v}{\mu} = \mathbf{R} \qquad (3.32)$$

$\mathbf{R}$ is the dimensionless number for flow in pipes called the Reynolds number. It was defined earlier as Equation 2.23.

$$\frac{\text{inertia force}}{\text{pressure force}} = \frac{\rho v^2}{\Delta p} = \mathbf{E} \qquad (3.33)$$

is a ratio used quite commonly, although implicitly, to express resistance coefficients such as friction factor in pipes, and drag coefficients of bodies. $\mathbf{E}$ is called the Euler number after Leonhard Euler (Switzerland, 1707–1783), one of the greatest mathematicians of his time.[8]

$$\frac{\text{inertia force}}{\text{elastic force}} = \frac{v}{\sqrt{K/\rho}} = \mathbf{M} \qquad (3.34)$$

in which K is the modulus of compressibility and the denominator is equivalent to the speed of sound. $\mathbf{M}$ is called the Mach number after Ernst Mach (Austria, 1838–1916), who studied ballistics.[9]

$$\frac{\text{inertia force}}{\text{surface tension force}} = \frac{v}{\sqrt{\sigma/\rho L}} = \mathbf{W} \qquad (3.35)$$

where $\mathbf{W}$ is called the Weber number in honor of Moritz Weber (Germany, 1871–1951), who developed it under the name "capillary parameter."

Other ratios may be formed to describe relationships important in the description of certain motions. For example, in seepage the ratio of viscous shear forces

---

[8]In his eulogy it took 50 pages to list the titles of his scientific papers.
[9]Mach was a philosopher who made major contributions in the development of "logical positivism," striving to rid science of metaphysical assumptions that were prevalent in his days.

to the gravity forces is the most important parameter. This is equivalent to the ratio of the Froude number and the Reynolds number.

In the following chapters most hydraulic parameters are expressed by dimensionless parameters. Most numerical flow coefficients, or proportionality factors, are dimensionless combinations of variables that characterize the flow. When there are large numbers of variables, often there is no explicit mathematical solution method available. In such cases a formal, albeit less intuitive, method of composing dimensionless groups of variables is used. It was introduced in the United States by Edgar Buckingham as the Π (Greek capital pi) theorem.[10] It can best be learned by working out an example.

As an example of dimensional analysis, let us consider the case of flow in pipes. The pressure drop in a unit length of pipe, $\Delta p/L$, is dependent on the pipe diameter D, the pipe roughness e, the average flow velocity v, the density of the fluid ρ, and its kinematic viscosity μ. Hence, we may write

$$\frac{\Delta p}{L} = f(D, e, v, \rho, \mu) \qquad (3.36)$$

If we replace these variables with their dimensions in the force, length, and time system, using Table 3.2, we get

$$\frac{F}{L^3} = f\left[L, L, \left(\frac{L}{T}\right), \left(\frac{FT^2}{L^4}\right), \left(\frac{FT}{L^2}\right) \right] \qquad (3.37)$$

Observing that the formula involves all three fundamental quantities, force, length and time, we write $m = 3$. Because there are six derived quantities in our problem, we write $n = 6$. The theorem of dimensional analysis states that *the minimum number of dimensionless units that can be composed from these variables equals* $(n - m)$, which in our example is three.

The procedure of forming meaningful dimensionless variables consists of the following steps:

First, a set of m variables are selected from the ones in Equation 3.37 under the following three rules:

1. Among the three variables all the fundamental quantities are involved.
2. One of the three variables is a geometric characteristic, another is a fluid property, and the third is a flow characteristic.
3. The three variables do not include the dependent variable (pressure loss per unit length in this case).

Next, the variables to be included in the $(n - m)$ dimensionless terms are collected so that each includes all previously picked variables plus one of the remaining

---

[10]Actually, it was first published in 1911 by Dimitri Pavlovich Riabouchinsky (Russia, 1882–1962) in one of his 200 or so scientific articles. He is also known as the original inventor of the bazooka. Buckingham failed to give credit to the original authors of some of his papers.

ones. In our example we may select D, ρ, and v as *repeating variables*; hence, the dimensionless terms will be composed of the following:

Π_1 will include $\Delta p/L$ and the three repeating variables.
Π_2 will include e and the repeating variables.
Π_3 will include μ and the repeating variables.

The next step is to substitute the dimensions for the symbols from Equation 3.37 and to write exponential formulas for each Π term:

$$\Pi_1 = D^x \rho^y v^z \left(\frac{\Delta p}{L}\right)^a = (L)^x \left(\frac{FT^2}{L^4}\right)^y \left(\frac{L}{T}\right)^z \left(\frac{F}{L^3}\right)^a \tag{3.38}$$

and so on for the other two dimensionless terms. To make Π_1 dimensionless we select the exponents x, y, and z so that Equation 3.38 equals 1. Because the zeroth power of any number is 1, Equation 3.38 may be written as

$$\Pi_1 = F^0 L^0 T^0 = 1 \tag{3.39}$$

Because F, L and T are fundamental quantities, Equations 3.38 and 3.39 may be split up in the form of

$$F^y F^a = F^0$$
$$L^x L^{-4y} L^z L^{-3a} = F^0 \tag{3.40}$$
$$T^{2y} T^{-z} = T^0$$

In the next step we recognize that there are three equations and three unknowns here:

$$y \qquad + a = 0$$
$$x - 4y + z - 3a = 0 \tag{3.41}$$
$$2y - z \qquad = 0$$

Solving Equation 3.41 we find that $y = -a$, $z = -2a$, and $x = a$. Introducing these results into Equation 3.38, and taking a as unity, we obtain

$$\Pi_1 = \frac{D \, \Delta p}{L \, \rho v^2} \tag{3.42}$$

which is the product of the dimensionless geometric ratio D/L and the reciprocal of the Euler number, Equation 3.33. In the next chapter we will find that the Euler number corresponds to the friction factor f in pipe flow.

By a similar derivation one can determine the other two dimensionless terms:

$$\Pi_2 = \frac{e}{D} \tag{3.43}$$

which is called ε, the relative roughness of a pipe, and

$$\Pi_3 = \frac{\mu}{v D \rho} \tag{3.44}$$

the reciprocal of the Reynolds number. Because the reciprocal, square, square root, or any other function of a dimensionless number is also a dimensionless number, we can conclude that Equation 3.36 expressed by dimensionless variables is

$$\frac{\Delta p}{L} = f(\varepsilon, \mathbf{R}, f) \tag{3.45}$$

or in other words, the pressure loss in a unit length of a pipe can be formulated in terms of the three dimensionless numbers: the relative roughness, the Reynolds number, and the friction factor. The actual relationship is embodied in the results of experimental studies on pipe flow, shown in Figure 4.2.

In practice, it is often unnecessary to carry out a detailed dimensional analysis to present experimental results in a proper manner. Let us take an example from open channel flow. One of the earliest open channel flow formulas relates the velocity v to the depth y and the slope of the surface S as

$$v = C\sqrt{yS} \tag{3.46}$$

in which C is a coefficient. To determine the proper dimensions of C we may substitute dimensions for the variables, which gives

$$\frac{L}{T} = C\sqrt{L(L/L)} \tag{3.47}$$

Rearranging to solve for C we obtain

$$C = \sqrt{\frac{L}{T^2}} = \sqrt{\text{acceleration}} \tag{3.48}$$

This result suggests that by including the square root of the gravitational acceleration in C we may represent our experimental results in regard to Equation 3.46 by a new dimensionless coefficient k, where

$$k = \frac{C}{\sqrt{g}} \tag{3.49}$$

By dimensional analysis we reduce the number of variables that need to be handled, but we do not find a *solution* for the problem. That can be found only by controlled experimental or field measurements of hydraulic data. However, after dimensionless grouping of our variables, we eliminate our dependence on the units of measurement.

3.8 HYDRAULIC MODELS

Hydraulic experiments may be made using prototypes or full-scale or reduced-scale models.[11] For example, in the development of a new type of fire-fighting nozzle,

---

[11]The first hydraulics laboratory in the United States was founded in 1887 at Lehigh University by Mansfield Merriman (United States, 1848–1925).

FIGURE 3.27   Hydraulic model of New York Harbor. (Courtesy of U.S. Army Corps of Engineers)

experiments would be made at full scale. But in finding the most effective shape for a ship's hull or the best location of breakwaters to protect a harbor from excessive wave action, full-scale experimentation would be economically prohibitive. The primary requirement in the design of a reduced-scale model is that it be similar to the prototype. Perfect similitude requires not only that all corresponding linear dimensions be reduced by the same scaling factor but that all other parameters of the prototype, such as velocities, forces, and material properties, be reduced by known scaling factors as well. As we shall see in the next example, trying to achieve perfect similitude is an impossible task.

Hydraulic models of a full-scale prototype (that is, the real thing) are designed such that the motion and forces in them are similar to those in the prototype. To attain similarity the designer should be familiar with the concept and laws of geometric, kinematic, and dynamic similarity.

Geometric similarity means that all corresponding lengths in the model and prototype are of the same ratio. The geometric scale of a model is selected so that it can be built and operated conveniently in a hydraulic laboratory and still be big enough to measure velocities in it with sufficient accuracy. Model scales of spillways, conduits, and similar structures range between 1:15 and 1:50. Models of rivers and harbors range from 1:100 to as small as 1:2000. Figures 3.27, 3.28, and 3.29 show typical hydraulic models. The first is a model of New York City's harbor, for the study of currents, the second is a river sedimentation experiment for the most efficient placement of spur dikes, the last one is a model of the spillway of Crystal

FIGURE 3.28 Model study of river training. Proper location of spur dikes allows sediment deposition at desired locations and rectifies stream alignment. (Courtesy of U.S. Army Corps of Engineers, Waterways Experiment Station)

Dam, studied at the U.S. Bureau of Reclamation's hydraulic laboratory in Denver, Colorado.

With such small scales as mentioned, water depths in the model would be so small that effects of surface tension and roughness could he overbearing. In such cases the vertical scale may differ from the horizontal scale, and we speak of *distorted models*. These will be discussed in the next section.

To satisfy the fundamental requirement that the model is to operate in a manner similar to the prototype, one has to recognize that in each the flow follows Newton's

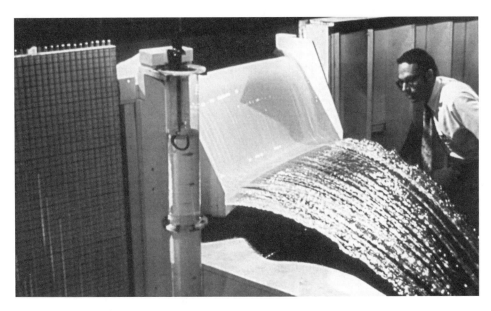

FIGURE 3.29 Model study of the spillway for Crystal Dam. (Courtesy of U.S. Bureau of Reclamation, Hydraulic Laboratory)

second law of motion. For the two systems

$$F_p = m_p \cdot a_p \tag{3.50}$$
$$F_m = m_m \cdot a_m$$

where subscripts p and m represent prototype and model. The acceleration is the time rate of velocity change

$$a = \frac{\Delta v}{\Delta T} \tag{3.51}$$

and the mass is

$$m = \rho L^3 \tag{3.52}$$

where L is the unit length, corresponding to either the prototype, L_p, or the model, L_m. To have geometric similarity between the two systems, the ratio of lengths must be constant everywhere; hence,

$$L_R = \frac{L_m}{L_p} \tag{3.53}$$

This is called the model's *scaling ratio*.

To correspond to their prototypes, models must also operate in *kinematic similarity.* This means that the ratio of corresponding velocities in prototype and model must be constant throughout, that is,

$$v_R = \frac{v_m}{v_p} = \frac{L_m/T_m}{L_p/T_p} = \frac{L_R}{T_R} \tag{3.54}$$

Where T_R is the time ratio between times in the prototype T_p and times in the model, T_m. Equation 3.54 suggests that to have kinematic similarity the

$$T_R = \frac{T_m}{T_p} \tag{3.55}$$

ratio must be maintained.

Returning our attention to Equations 3.50 and considering the formulas following it, we may conclude that to have *dynamic similarity* between model and prototype, the ratio between corresponding inertia forces should be

$$F_R = \frac{F_m}{F_p} \tag{3.56}$$

Substituting Equations 3.51 and 3.52 we get

$$F_R = \frac{\rho_m}{\rho_p} \left(\frac{L_m}{L_p}\right)^3 \Delta\left(\frac{L_R}{T_R}\right)/\Delta T_R \tag{3.57}$$

where L_R/T_R is the velocity ratio, Equation 3.54. Thus,

$$F_R = \rho_R L_R^3 \frac{L_R}{T_R^2} = \rho_R A_R v_R^2 \tag{3.58}$$

in which ρ_R is the ratio of fluid densities in the prototype and model, and A_R is the ratio of the corresponding areas.

Equation 3.58 is the fundamental equation to assure dynamic similarity between *inertia forces* of prototype and model.

In practical hydraulic structure design problems we are aware that the flow is influenced by a variety of forces; inertia, gravity, viscous shear, capillarity, and elasticity all play a part. Usually the force of gravity or viscous shear forces are the most dominant. Because of the combined effect of the various forces acting in the prototype, it is practically impossible to design a hydraulic model that is a true prototype for all forces present. It is therefore necessary judiciously to neglect forces of lesser importance and concentrate one's efforts on satisfying modeling requirements for the most important forces. For only one dominant force, neglecting all other forces, one should be able to build a model at a selected scale and operate it in such a manner that the effect of nondominant forces is negligible. In this way model studies can give reasonably accurate information on the expected hydraulic behavior of the prototype. If there are only two types of forces considered, Equation 3.58 may be satisfied by selecting a model fluid different from water, the prototype fluid.

Most hydraulic models utilize water as a model fluid. Hence, the density ratio in Equation 3.58 is unity. In some cases other fluids may be used to obtain better dynamic similarity. Properties of common fluids that may be used in laboratory

TABLE 3.3   Properties of Various Liquids Used in Hydraulic Modeling

	Temperature (°F)	Specific Weight, γ		Density, ρ		Viscosity, μ	
		lb/ft³	kN/m³	slugs/ft³	kg/m³	lb-sec/ft²	N·s/m²
Ethyl alcohol	68	49.3	7.74	1.530	789	0.0000249	0.119
Benzene	68	54.9	8.62	1.705	879	0.0000136	0.065
Carbon disulfide	68	78.7	12.37	2.446	1,261	0.0000077	0.037
Carbon tetrachloride	68	99.5	15.64	3.095	1,595	0.0000200	0.096
Glycerin	68	78.7	12.36	2.445	1,260	0.0313	149.9
Mercury	68	845.3	132.80	26.282	13,546	0.0000328	0.157
Oil, castor	68	60.0	9.41	1.862	960	0.0206	98.6
Oil, linseed	60	58.3	9.16	1.812	934	0.000292	3.31
Oil, olive	59	57.3	9.00	1.781	918	0.0023	11.0
Oil, sperm	60	54.9	8.63	1.707	880	0.000877	4.2
Turpentine	61	54.5	8.56	1.694	873	0.0000334	0.16
Water, fresh	60	62.4	9.80	1.938	999	0.0000236	0.113
Water, sea	60	64.0	10.05	1.989	1,025	—	—

work are listed in Table 3.3. Once the density ratio is fixed, Equation 3.58 will be a function of two independent variables only. These are L_R and T_R. One provides for geometric similarity and the other for kinematic similarity.

Most hydraulic models represent either open channel flow, spillways, and weirs, for example, or pipe and other closed conduit flows. The former are characterized by the work of gravity forces; the latter, by viscous forces.

To assure that the Froude numbers in model and prototype are the same, one may first write the ratio of *gravity forces* as

$$\text{gravity force} = \gamma_R L_R^3 \qquad (3.59)$$

Because the Froude number is the ratio of gravity forces to inertia forces, Equation 3.59 should be equated with Equation 3.58, because their ratio must be unity. This results in

$$\gamma_R L_R^3 = \frac{\gamma_R}{g_R} L_R^3 \frac{L_R}{T_R^2} \qquad (3.60)$$

Simplifying and rearranging this equation to express the *time ratio* that defines how time measured in the model corresponds to prototype (real) time, results in

$$T_R = \sqrt{\frac{L_R}{g_R}} \qquad (3.61)$$

This ratio must be maintained if the Froude numbers in model and prototype are to be similar. Because the model is under the same gravitational acceleration as the prototype, g_R is always unity. Hence, for models of open channels and free surface flow structures the time ratio will take the form

$$T_R = \sqrt{L_R} \tag{3.62}$$

and the velocity ratio, by Equation 3.54, will be

$$v_R = \sqrt{L_R} \tag{3.63}$$

Equations 3.62 and 3.63 represent the *Froude model law for time scale.*

When the flows are dominated by viscous shear, the Reynolds numbers in the model should correspond to those in the prototype. Writing the ratio of *viscous forces* in model and prototype as

$$\text{viscous force} = \mu_R \frac{L_R^2}{T_R} \tag{3.64}$$

and equating it with Equation 3.58, since their ratio must be unity, we obtain

$$\rho_R \frac{L_R^4}{T_R^2} = \mu_R \frac{L_R^2}{T_R} \tag{3.65}$$

Simplifying this formula, we obtain

$$T_R = L_R^2 \frac{\rho_R}{\mu_R} = \frac{L_R^2}{v_R} \tag{3.66}$$

which is the *time scale for the Reynolds model law.*

If we use the same fluid in both the model and the prototype, this formula means that in a model built on a $\frac{1}{10}$ length scale, under the Reynolds model law the time ratio is $\frac{1}{100}$.

Once the time scale for a model law is derived, all other hydraulic parameters of dynamically similar models can be derived. For example, let us express the ratio of corresponding accelerations in model and prototype designed to satisfy the Reynolds model law:

$$a_R = \frac{L_R}{T_R^2} \tag{3.67}$$

Substituting the time ratio using Equation 3.66, we obtain

$$a_m = \left(\frac{L_R \mu_R^2}{L_R^4 \rho_R^2}\right) a_p = \left(\frac{\mu_R^2}{L_R^3 \rho_R^2}\right) a_p \tag{3.68}$$

This formula may be referred to as the *transfer formula for acceleration* between model and prototype. For the two main model laws these transfer formulas were

TABLE 3.4   Transfer Formulas for Froude and Reynolds Modeling Laws

Quantity	Modeling Ratios[a] for:	
	Froude Law	Reynolds Law
Length	L	L
Area	L^2	L^2
Volume	L^3	L^3
Mass	$L^3\gamma g^{-1}$	$L^3\gamma g^{-1}$
Density	γg^{-1}	γg^{-1}
Time	$L^{0.5}g^{-0.5}$	$L^2\nu^{-1}$
Velocity	$L^{0.5}g^{0.5}$	$L^{-1}\nu$
Acceleration	g	$L^{-3}\nu^2$
Angular velocity	$L^{-0.5}g^{0.5}$	$L^{-2}\nu$
Angular acceleration	$L^{-1}g$	$L^{-4}\nu^2$
Force and weight	$L^3\gamma$	$\nu^2\gamma g^{-1}$
Pressure	$L\gamma$	$L^{-2}\nu^2\gamma g^{-1}$
Impulse and momentum	$L^{3.5}\gamma g^{-0.5}$	$L^2\nu\gamma g^{-1}$
Discharge, volume/sec	$L^{2.5}g^{0.5}$	$L\nu$
Discharge, weight/sec	$L^{2.5}\gamma g^{0.5}$	$L^{-2}\nu^3\gamma g^{-1}$
Energy and work	$L^4\gamma$	$L\nu^2\gamma g^{-1}$
Power	$L^{3.5}\gamma g^{0.5}$	$L^{-1}\nu^3\gamma g^{-1}$
Torque	$L^4\gamma$	$L\nu^2\gamma g^{-1}$
Absolute viscosity	$L^{1.5}\gamma g^{-0.5}$	$\nu\gamma g^{-1}$
Kinematic viscosity	$L^{1.5}g^{0.5}$	ν
Surface tension	$L^2\gamma$	$L^{-1}\nu^2\gamma g^{-1}$

[a] All variables are shown without subscript R.

determined for a number of hydraulic parameters that may be measured in prototype and model. Table 3.4 lists transfer formulas for Froude and Reynolds model laws.

The roughness of a model built under either the Froude or the Reynolds model law presents another design problem. The scaling ratio of the roughness coefficient between model and prototype is determined by

$$n_R = g_R^{-1/2} L_R^{1/6} \qquad (3.69)$$

For most models in which roughness may be a controlling factor the proper model of roughness can be obtained by a trial-and-error procedure that should be carried out before the actual experimentation in the model begins. This procedure involves observations of prototype behavior and adjustment of model roughness until the time ratio of the model and prototype corresponds to its desired value. River and harbor models are the most sensitive to these effects. Hence, their geometric distortion is usually controlled by their relative roughness.

3.9 DISTORTED MODELS

For the uninitiated the true model is the one that has identical horizontal and vertical scales. Such models rarely represent the true physical conditions in which some natural occurrences must be analyzed. Quite often the model must include more than two dominant forces. Some examples are the movement of a flood wave in a river, the effect of friction in outlet structures of dams, sedimentation and erosion of a river channel, the diffusion of heated discharge or waste effluent in a lake, river, or estuary, and the erosion or shoaling of a coast line. In such situations it is necessary to determine the physical phenomena present at the prototype scale and how the proposed reduction to model scale may influence them.

River models, for example, require a large horizontal reduction because of the large distances encountered in nature. The corresponding reduction in flow depth results in velocities so small that laminar flow occurs in the model. Also, undistorted river models have slopes so flat that they cannot be measured adequately. To overcome these difficulties one may distort the model by arbitrarily increasing the slope by tilting the model, using a different scale vertically, increasing the discharge scale, or increasing the roughness of the model. Such changes are fully justifiable, since the ultimate purpose of the model study is to obtain flow phenomena comparable and transferable to the natural prototype scale.

Erosion or *sediment deposition* studies in a river require the model bed to be built using grainy material that exhibits the same behavior as that of a natural erodible channel bed. Scaling down the river, again, involves the use of the Froude model law. However, the transportation of sediment in water depends on the rate of turbulence, which is controlled by the Reynolds number and the relative settling velocity of the particles (Equation 2.25), a factor that depends on the density, the size, and the shape of the grains used. Finely ground coal, plexiglass, polyvinyl chloride (PVC), pumice, treated wood, cornmeal, and nylon are some of the materials that have been found useful as model representatives of sand, silt, and gravel. The use of such lighter materials reduces the need for extreme distortion in the model and permits the completion of experiments in less time than would be required otherwise. But often the friction in a model with correctly modeled sediment material does not represent the natural conditions well. On some occasions, when both Froude numbers and Reynolds numbers must be satisfied simultaneously, one may find the necessary condition by equating the velocity ratios for the two. Considerable experience is required to handle such modeling problems.

An essential part of modeling is the trial-and-error procedure required to adjust the performance of a model to best represent the natural conditions of the prototype. Without proper *verification* of the model studies, the experimental results may have only a faint qualitative resemblance to the physical phenomena appearing in the prototype. Correctly verified model studies, on the other hand, are of immense value in the design of an expensive hydraulic structure. The cost of hydraulic model experiments is minimal when compared with the cost of the actual project. Successive alterations of the different design features of the prototype based on results of small-scale modeling may bring about considerable savings and operational benefits in the completed structure.

EXAMPLE PROBLEMS

Example 3.1

A 12 in. diameter pipe is equipped with a 4 in. diameter orifice plate (Figure E3.1). By calibration the discharge coefficient was determined to be 0.63. What is the discharge in the pipe if the differential pressure registered during the flow is 12 psi?

FIGURE E3.1

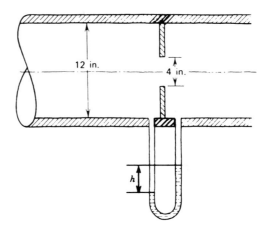

Solution. From Equation 3.5

$$Q = ca \sqrt{\frac{2g}{\gamma}(\Delta p)}$$

where

$$c = 0.63$$

$$a = \frac{\pi}{4}\left(\frac{4}{12}\right)^2 = 0.087 \text{ ft}^2$$

$$\gamma = 62.4 \frac{\text{lb}}{\text{ft}^3}$$

$$\Delta p = 12 \times 144 = 1728 \frac{\text{lb}}{\text{ft}^2}$$

Then

$$Q = 0.63 \times 0.087 \sqrt{\frac{2(32.2)(1728)}{62.4}}$$

$$= 2.3 \text{ cfs} \qquad \qquad \square$$

Example 3.2

A flow nozzle of 6 in. diameter is placed in an 18 in. diameter pipe. During calibration the pressure differential was measured to be 10 psi for a flow of 7.2 cfs, and 17 psi for a flow of 9.3 cfs. Determine the discharge coefficient of the measuring nozzle (Figure E3.2).

Solution. From Equation 3.5,

$$Q = ca \sqrt{\frac{2g}{\gamma}(\Delta p)}$$

$$a = \frac{\pi}{4}\left(\frac{1}{2}\right)^2 = \frac{\pi}{16} \text{ ft}^2$$

$$c_1 = \frac{7.2}{\frac{\pi}{16}\sqrt{\frac{2 \times 32.2}{62.4} \times 10 \times 144}}$$
$$= 0.951$$

$$c_2 = \frac{9.3}{\frac{\pi}{16}\sqrt{\frac{2 \times 32.2}{62.4} \times 17.0 \times 144}}$$
$$= 0.942$$

$$c = \tfrac{1}{2}(c_1 + c_2) = \tfrac{1}{2}(0.951 + 0.942) \simeq 0.946$$

FIGURE E3.2

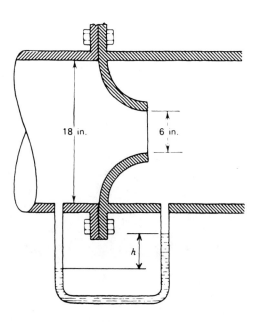

18 in.   6 in.

h

Example 3.3

The results of a calibration study for an orifice meter are summarized in the following table. Develop a discharge equation for this orifice meter using the calibration data.

Pressure Drop across Orifice Meter (psi)	Flow (cfs)
1.3	0.19
2.7	0.26
4.0	0.30
5.3	0.34
6.6	0.37
8.0	0.41
9.2	0.43
10.5	0.47
11.9	0.50
13.0	0.52
14.4	0.55
15.6	0.56
16.3	0.58

Assuming turbulent flow, the discharge through the orifice meter can be represented by Equation 3.6

$$Q = K \, \Delta p^{1/2}$$

Solution. The value of K can be determined by plotting flow versus the square root of pressure drop and determining the slope of the resulting best-fit line, as depicted in Figure E3.3.

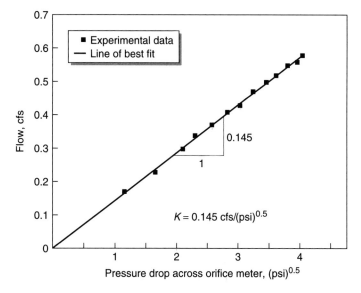

FIGURE E3.3

From Figure E3.3 the slope of the line gives a K value of 0.145 cfs/(psi)$^{0.5}$. Thus, the discharge equation for the orifice meter is

$$Q = 0.145 \, \Delta p^{1/2} \qquad \square$$

Example 3.4

An orifice meter has a 2 in. orifice plate and is placed in an 8 in. diameter pipe. The meter is to be placed into a straight pipe 8 ft in length. Where should the meter be located?

Solution. The most efficient spot to place the meter is where the flow is uniform and undisturbed. To locate it properly, a distance at least 6 to 10 times the pipe diameter from the inlet flow is recommended. For 8 ft of pipe, 10 × 8 in. = 80 in. ≈ 6 ft from the inlet point. In this case 2 ft remains between the meter and the end of the pipe, which equals 3 diameters. This distance is still sufficient to eliminate downstream influences. $\square$

Example 3.5

The maximum working length (scale length) of a manometer is 16 in. Calculate the maximum pressure differential that may be measured if the indicating fluid is Meriam Blue.

Solution. From Table 3.1 we find that the unit mass of Meriam Blue indicating fluid is 1.75 g/cm^3.

Assuming that the pressure taps are at the identical level, h in Equation 3.10 is zero. Rewriting Equation 3.10 for $\rho_{indicator} = 1.75$, we have

$$\Delta p = \gamma z (1.75 - 1) = \gamma z (0.75)$$

Observing z on Figure 3.4b, we note that

$$z_{max} = l = 16 \text{ in.}$$

Therefore, the maximum pressure differential is

$$\Delta p_{max} = 62.4 \frac{\text{lbf}}{\text{ft}^3} \left(\frac{16}{12} \text{ ft} \right) 0.75 = 62.4 \text{ lbf/ft}^2$$

If the pressure is expressed in the form of height of a water column h, then

$$h_{max} = \left(\frac{\Delta p_{max}}{\gamma} \right) = \left(\frac{62.4 \text{ lbf/ft}^2}{62.4 \text{ lbf/ft}^3} \right) = 1 \text{ ft} = 12 \text{ in.} \qquad \square$$

Example 3.6

A manometer pressure differential in two adjacent ($h = 0$) pipelines is 15 psi. Select a manometer fluid for a differential manometer such that the length of the manometer does not exceed 36 in.

Solution. From Equation 3.9

$$\Delta p = \gamma m z - \gamma z = \gamma z (m - 1)$$

$$15 \times 144 = 62.4 \frac{36}{12} (m - 1)$$

$$m - 1 = 11.5$$

$$m = 12.5$$

where m is the specific gravity of the manometer fluid; thus, the fluid to be used has to have a unit mass larger than

$$\rho = 12.5 \text{ g/cm}^3$$

suggesting that the use of mercury is required. ☐

Example 3.7 The pressure differential measured on a Prandtl tube inserted into a flowing stream was 6.5 in. of water (Figure E3.7). What is the velocity of the flow?

FIGURE E3.7

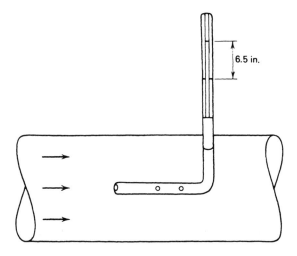

6.5 in.

Solution. By Equation 3.12

$$v = \sqrt{\frac{2}{\rho} \Delta p} = \sqrt{2gh}$$

where

$$h = 6.5 \text{ in.} = 0.54 \text{ ft}$$

Therefore,

$$v = \sqrt{32.2 \times 2 \times 0.54} = 5.9 \text{ fps} \qquad ☐$$

Example 3.8 Velocity and depth measurements for a small creek are depicted in Figure E3.8. Velocity measurements are in units of feet per second. Determine the discharge.

Solution. Because the average velocities at stations 1 and 2 are 0 and 1.0 fps, respectively, the flow in the first section of the creek is

$$q = \left(\frac{0 + 1.0}{2}\right)\left(\frac{0 + 2}{2}\right)(2) = 1.0 \text{ cfs}$$

FIGURE E3.8

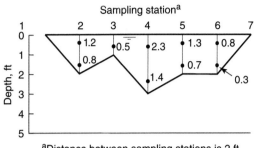

Sampling station[a]

aDistance between sampling stations is 2 ft

Calculations are summarized in the following table

Station number	y ft	v_{ave} fps	Q cfs
1	0	0	—
2	2	1.0	1.0
3	1	0.5	2.25
4	3	1.85	4.70
5	2	1.0	7.13
6	2	0.55	3.10
7	0	0	0.55
		Total flow =	18.73

Example 3.9

A Cipoletti weir of a base width of 12 in. and sides of 4 to 1 slope carries water at an overflow depth of 8 in. What is the discharge measured?

Solution. From Equation 3.25

$$Q = 3.37 b H_1^{3/2}$$

where

$$b = 1 \text{ ft}$$

$$h = \frac{8}{12} \text{ ft}$$

Then,

$$Q = 3.37 \times 1.0 \times \left(\frac{8}{12}\right)^{3/2}$$

$$= 1.8 \text{ cfs}$$

Example 3.10

A salt wave measurement in a small creek was performed by dumping brine containing 9.8 lb of salt (4.45×10^6 mg NaCl) into a stream. At 200 ft downstream from the dumping point, a portable conductivity bridge read every 30 s gave the following results:

Time(s)	$\dfrac{1}{\text{ohm}}(\times 10^2)$	Time(s)	$\dfrac{1}{\text{ohm}}(\times 10^2)$
30	5.1	450	6.0
60	5.0	480	5.8
90	5.0	510	5.5
120	5.0	540	5.6
150	5.0	570	5.4
180	7.5	600	5.4
210	10.0	630	5.4
240	8.5	660	5.4
270	7.5	690	5.3
300	7.0	720	5.2
330	6.8	750	5.1
360	6.5	780	5.0
390	6.1	810	5.2
420	6.0	840	5.1

Determine the discharge in the creek.

Solution. A plot of concentration versus time for the salt wave measurement is shown in Figure E3.10. Concentrations were calculated using Equation 3.28. The baseline salt concentration is 311 mg/liter. The area under the salt wave from a time of 180 to 840 seconds is calculated to be 52170 mg · s/liter using the method of trapezoids. From Equation 3.27 the flow is then

$$\frac{4.45 \times 10^6}{52170} = 85.3 \text{ liters/s}$$

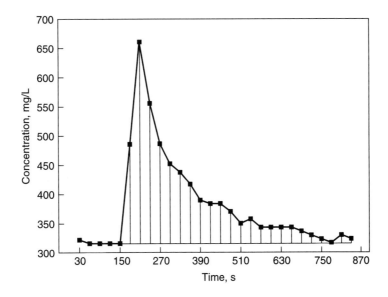

FIGURE E3.10

Example 3.11

The discharge over a rectangular weir was determined by repeated volumetric measurements to be 6.5 m³/s on one occasion and 1.5 m³/s on another. The width of the weir was 6 m. On the first occasion the depth of the water nappe over the weir was 57 cm; the second time the depth was 21.5 cm. Design a rating curve for the weir.

Solution. From Equation 3.23 and letting $M = C\sqrt{2g}$ gives

$$M_1 = \frac{Q_1}{bH_1^{3/2}} = \frac{6.5}{6(0.57)^{3/2}} = 2.52$$

$$M_2 = \frac{1.5}{6(0.215)^{3/2}} = 2.51$$

$$M_{ave} = \tfrac{1}{2}(2.52 + 2.51) = 2.515$$

Then

$$Q = (2.515 \times 6)H_1^{3/2} \qquad Q = 15.1\,H_1^{3/2}$$

Plotting Q versus H_1 will provide the rating curve desired. ☐

Example 3.12

Determine the discharge through a Parshall flume with a throat width of 4 ft if the upstream depth is 1.15 ft and the flume is flowing under 90 percent submergence.

Solution. From Equation 3.30 (or using Figure 3.25),

$$Q_0 = 4wH_a^{1.522w^{0.026}} = 4(4)\,(1.15)^{1.522(4)^{0.026}} = 19.9 \text{ cfs}$$

From Figure 3.26, at a submergence of 90 percent

$$\frac{Q}{Q_0} = 0.83$$

Thus, flow through the flume is

$$0.83(19.9) = 16.5 \text{ cfs} \qquad ☐$$

Example 3.13

Apply Buckingham's Pi theorem to determine the form of the discharge equation for V-notch weirs of angle ϕ. The discharge, Q, is a function of the upstream depth above the bottom of the notch, H_1, the approach velocity, v_1, and the acceleration due to gravity, g.

$$Q = f(\phi, H_1, v_1, g)$$

Solution. Using Table 3.2 to reduce each variable to its fundamental dimensions, we get

$$\frac{L^3}{T} = f\left[\phi, L, \left(\frac{L}{T}\right), \left(\frac{L}{T^2}\right)\right]$$

Only two of the fundamental quantities are involved, so $m = 2$. There are five derived quantities, so $n = 5$. Thus, there are three dimensionless units that can be composed from these variables. We select H_1 as a geometric characteristic and g as a flow characteristic. There is no fluid property in this problem. Because ϕ is considered dimensionless, it will be our first dimensionless term. Our list then includes:

$$\Pi_1 = \phi$$
$$\Pi_2 = f(H_1^x, g^y, Q^a) = f[(L)^x, (L/T^2)^y, (L^3/T)^a]$$
$$\Pi_3 = f(H_1^x, g^y, v_1^b) = f[(L)^x, (L/T^2)^y, (L/T)^b]$$

The equations derived from the exponents of Π_2 are

$$\text{For } L: \qquad x + y + 3a = 0$$
$$\text{For } T: \qquad -2y - a = 0$$

From these we find $y = -a/2$ and $x = -5a/2$. Choosing $a = 1$, we get

$$\Pi_2 = \frac{Q}{H_1^{5/2}\sqrt{g}}$$
$$\text{or } Q = CH_1^{5/2}$$

where C is a constant based on experimental results. Note the similarity of this result of Equation 3.24.

In like manner we find

$$\Pi_3 = \frac{v_1}{\sqrt{gL}}$$

which is the Froude number (Equation 3.31).

In this chapter it was stated that if the height of the weir is significant, then the approach velocity, v_1, is negligible. In such cases only the dimensionless terms defined by Π_1 and Π_2 need to be considered. $\square$

Example 3.14

For the structure described in Example 9.17 design a hydraulic model for which the geometric ratio is 1:10. Develop transfer formulas for discharge and roughness such that the model will truly represent the prototype in question.

Solution. With gravity as the dominant force the model must satisfy the Froude model law. From Example 9.17,

$$L_p = 150 \text{ ft}$$
$$n = 0.024$$
$$S_0 = 0.02$$
$$H_p = 6 \text{ ft}$$
$$Q_p = 112 \text{ cfs}$$

we have $L_m = L_p \cdot L_R = 150 \times 0.1 = 15$ ft (Equation 3.53) for the length of our model. The diameter of the model culvert pipe is similarly computed as

$$L_m = 4 \text{ ft } (L_R) = 4(0.1) = 0.4 = 4.8 \text{ in.}$$

To satisfy the dynamic similarity conditions, we note that since $g_R = 1$, then from Equation 3.62, $T_r = \sqrt{L_R}$. Thus, the discharge transfer formula is

$$Q_R = A_R v_R = (L_R)^2 \frac{L_R}{T_R} = \frac{L_R^3}{L_R^{1/2}} = L_R^{2.5} = 0.1^{2.5} = 0.0032$$

Since

$$Q_R = \frac{Q_m}{Q_p}$$

the model discharge will be

$$Q_m = 0.0032 Q_p$$

This means, for example, that if the optimum discharge in the prototype is $Q_p = 168.6$ cfs, in the model similar conditions will be attained at 0.540 cfs, assuming that all other modeling characteristics are satisfied. The most important of these is the proper modeling of the pipe roughness. From $n = 1.49 R^{2/3} S^{1/2}/v$, where S is dimensionless,

$$n_R = \frac{n_m}{n_p} = \frac{[R^{2/3} S^{1/2}]_m}{[R^{2/3} S^{1/2}]_p} \cdot \frac{v_p}{v_m} = \frac{L_R^{2/3}}{v_R} = \frac{L_R^{2/3}}{L_R/T_R}$$

and by introducing the time scale of the Froude law, Equation 3.61

$$n_R = \frac{\sqrt{L_R}}{\sqrt{g_R}} \frac{L_R^{2/3}}{L_R} = \frac{L_R^{1/6}}{g_R^{1/2}}$$

which is Equation 3.69. Since $g_R = 1$ and $L_R = 0.1$, then $n_R = (0.1)^{1/6} = 0.68$.

Because the roughness of the prototype is stated in Example 9.17 to be

$$n_p = 0.024$$

then the roughness in the model should be

$$n_m = 0.68(n_p) = 0.68(0.024)$$
$$= 0.016$$

Checking in Table 8.2, which shows roughness coefficients, we note that the most commonly used model pipe material, lucite, has a roughness coefficient of 0.009, which is almost half the roughness factor required for the model. Therefore, the pipe in the model should be artificially roughened to attain the required conditions. □

Example 3.15

For the broad-crested weir flow problem described in Example 9.6 design a hydraulic model of a scale $L_R = 1:5$ and write the transfer formulas for all major variables. The purpose of the model study is to determine proper discharge coefficients.

Solution. From Example 9.6

$$Q_p = 477 \text{ cfs}$$
$$b_p = 30 \text{ ft}$$
$$h_p = 2 \text{ ft}$$

Hence,

$$h_m = \frac{2}{5} = 0.4 \text{ ft} = 4.8 \text{ in.}$$

Because the flow is essentially two dimensional, the width of the spillway may be reduced at will as long as side friction in the model is negligible. Usually, the width of the available laboratory flume establishes the value of b. Assuming that the flume width available is 3 ft, the discharge is to be adjusted accordingly. It is advisable, to avoid confusion, to use the discharge per unit width, q.

$$q_p = \frac{Q}{b} = \frac{477}{30} = 15.9 \text{ cfs/ft}$$

Because the problem is one of gravity flow, the Froude model law will apply with $L_R = 0.2$. From Table 3.4 the transfer formulas are

$$\text{discharge} = L_R^{2.5} \cdot g_R^{0.5} = 0.0179$$

$$\text{energy} = L_R^4 \cdot \gamma_R = 0.0016\gamma_R$$

$$\text{absolute viscosity} = L_R^{1.5} \cdot \gamma_R \cdot g_R^{-0.5} = 0.089\gamma_R$$

Because $g_R = 1$, it will not enter the problem. Also, since both model and prototype will use the same fluid, water, γ_R will be unity.

From the transfer equations we have

$$q_m = 0.0179(q_p) = 0.0179(15.9)$$
$$= 0.285 \text{ cfs/ft}$$

or

$$Q_m = 0.285 \times 3 = 0.855 \text{ cfs}$$

for the 3 ft wide modeling flume.

We now recall that the head H_1 over the weir was introduced as a measure of the driving energy; hence, it is to be modeled as such. As long as the model fluid is water, the modeling ratio for the head is as given above for energy, that is,

$$H_m = 0.0016H_p$$
$$= 0.0032 \text{ ft}$$
$$= 0.038 \text{ in.}$$

which is an impractically small quantity, subject to influences by surface tension and other factors. Disregarding the energy consideration, and simply using the geometric scaling ratio, we would get

$$H_m = 0.2(2) = 0.4 \text{ ft} = 4.8 \text{ in.}$$

Otherwise, we may consider using another fluid, in which case since

$$\gamma_p = 62.4 \text{ lb/ft}^3$$

and for the energy transfer formula

$$E_R = \frac{H_m}{H_p} \gamma_R \text{ to equal } L_R = 0.2$$

$$= 0.0016\frac{\gamma_m}{\gamma_p} = 0.0016\frac{\gamma_m}{62.4} = 0.2$$

$$\gamma_m = \frac{0.2(62.4)}{0.0016} = 7800 \frac{\text{lb}}{\text{ft}^3}$$

an improbably heavy model fluid to find.

To complicate the problem further, we may consider the discharge coefficient to be in some way dependent on the viscosity. In that case the ratio of the viscosities of the fluids in the model and prototype is defined by

$$\mu_R = \frac{\mu_m}{\mu_p} = 0.089\gamma_R$$

And since the unit weight and viscosity of the prototype water is known, we get

$$\mu_m = 0.089\frac{\gamma_m}{62.4}(\mu_p)$$

and if μ equals 1 cp (water at 20°C), then

$$\mu_m = 0.0014\gamma_m$$

The γ_m was previously determined to be 7800; hence,

$$\mu_m = 11.125 \text{ cp}$$

which is an exceedingly high value for model fluids.

Accordingly, in order to perform this simple model experiment in an absolutely correct manner, one would have to either develop a yet unknown and improbable model fluid or move to another planet where the gravitational acceleration is markedly different. The other alternative is to relax the stringent requirements for modeling and accept a degree of uncertainty about the behavior of the model. This shortcoming may be overcome by full-scale proof testing and comparing the two results.   □

PROBLEMS

3.1 A circular orifice in an 8 in. pipe has a diameter of 3 in. Determine the velocity in the pipe if the discharge coefficient is 0.65 and the pressure differential measured is 18 psi.

3.2 Determine the K constant in Equation 3.6 for the orifice described in Problem 3.1.

3.3 A 4 ft long vertical differential mercury manometer registers $z = 15$ in. How much is the pressure difference measured in feet of water if the manometer is connected to the two pressure taps of an orifice meter?

3.4 A Pitot tube is inserted into a 6 in. diameter pipe. Assuming that at the center where the measurement is made the velocity is 25 percent larger than the average velocity, how high will the water rise in a vertical standpipe connected to the Pitot tube for a discharge of 1.5 cfs?

3.5 A 6 in. diameter, 90° elbow meter with a radius of curvature of 4 ft measures a discharge of 1.8 cfs. How much is the corresponding pressure difference between the inside and outside pressure taps?

3.6 Compare the areas of flow through 90° and 60° V-notch weirs for a given head, H_1. Does the difference between the areas explain the difference in the discharge coefficients for the weirs given by Equation 3.24?

3.7 A float submerged to a depth of 20 cm moves in a small river with a velocity of 75 cm/s. The depth of the river is 1.3 m. Estimate the average velocity of the flow.

3.8 The measured water depth in a Parshall flume is 2 ft. The throat width of the flume is 15 in. How much is the discharge?

3.9 If the Parshall flume in Problem 3.8 was flowing under 80 percent submergence, what would be the discharge?

3.10 Develop a dimensionless ratio for gravity and viscous forces.

3.11 Develop a dimensionless ratio for viscous and pressure forces.

3.12 Determine the Reynolds number for the flow through the pipe in Example 2.17.

3.13 A rectangular channel 5 ft wide is delivering a discharge of 106 cfs. If the water depth in the channel is 4.2 ft, what is the Froude number of the flow?

3.14 Perform a dimensional analysis for the drag force on the raft described in Example 1.8 using Buckingham's theorem. Choose the density of the fluid, the raft velocity, and the fluid depth as the m variables.

3.15 A hydraulic model designed under Froude's law has a $\frac{1}{100}$ geometric scale. The model fluid is olive oil. Develop the pressure ratio and the momentum ratio between the prototype and the model.

3.16 Derive an equation for the time scale if a model is designed to maintain the ratio of inertia to elastic forces.

3.17 Derive an equation for the length scale if a model is designed to maintain the ratio of inertia to surface tension forces.

3.18 Rework Example 3.14 using a geometric scale of $\frac{1}{12}$. Which scale leads to a model that is more practical to implement, $\frac{1}{10}$ or $\frac{1}{12}$?

3.19 Rework Example 3.15 using glycerin as a model fluid. Is glycerin a satisfactory model fluid for this problem?

MULTIPLE CHOICE QUESTIONS

3.20 A typical direct measurement of discharge is the use of
A. Sharp-crested weirs.
B. Venturi meters.
C. Municipal water meters.
D. Bourdon gages.
E. Electric conductivity.

3.21 A rotameter measures water by
A. A tapered tube and float.
B. The rotational speed of a propeller.
C. Impulse force in an elbow.
D. A bearingless rotor.
E. Flow past a Prandtl tube.

3.22 An Annubar measures the
A. Hydrostatic pressure in a pipe.
B. Voltage drop in a transducer.
C. Stretch of a membrane in a pressure meter.
D. Average kinetic energy of the flow.
E. Deformation of a pressurized leaf spring.

3.23 A Price current meter is used in determining
   A. Electric current in a stream.
   B. Conductivity in creek measurements.
   C. Flow velocity in rivers.
   D. Discharge in flumes.
   E. Sewage flows.

3.24 The advantage of a Parshall flume is
   A. Small loss of head.
   B. Universality.
   C. Portability.
   D. Precision.
   E. Simplicity of use.

3.25 A slotted cylinder meter is a type of
   A. Magnetic flow meter.
   B. Thermal wave flow meter.
   C. LVDT.
   D. Rotameter.
   E. Velocity meter.

3.26 Concentration of a tracer chemical is used to measure
   A. Flow in pipes.
   B. Seepage velocity.
   C. Discharge in creeks.
   D. Pollution.
   E. Magnetic flows.

3.27 In a mass-length-time system the dimensions of absolute viscosity are
   A. M/LT.
   B. ML/T.
   C. T/M.
   D. MT/L.
   E. None of the above.

3.28 In the force-length-time system the dimensions of mass flow rate are
   A. FL.
   B. T/FL.
   C. FT/L.
   D. L/T.
   E. None of the above.

3.29 How many independent dimensionless terms may be needed to represent a hydraulic problem described by 15 variables that include all three fundamental dimensions among them?
   A. 5.
   B. 15.
   C. 10.
   D. 12.
   E. Less than 5.

3.30 What is the advantage of using Buckingham's theorem in the solution of hydraulic problems?
   A. It provides manageable functional relationships between variables.
   B. It decreases the number of variables to be dealt with.

   C. It eliminates the dimensions of all variables.

   D. It facilitates SI to English conversions.

   E. None of the above.

3.31 Distorted models are used when

   A. A true model does not fit the space available.

   B. A normal reduction has unwanted force effects.

   C. One particular effect should be emphasized.

   D. The prototype is expected to change later.

   E. The laboratory pump discharge is limited.

3.32 What is the reason that reduced-scale hydraulic models cannot fully represent their prototypes?

   A. Gravity cannot be reduced for the model on Earth.

   B. No ideal model fluid exists.

   C. Viscosity and density cannot be conveniently reduced.

   D. All of the above.

   E. None of the above.

3.33 A hydraulic model is to be designed to study the flow between piers of a bridge. Which modeling law should be used?

   A. Buckingham's theorem.

   B. Bernoulli's law.

   C. Reynolds' law.

   D. Froude's law.

   E. None of the above.

3.34 Why should Froude's law not be used in the design of a model to represent a piping system?

   A. Because viscous forces dominate.

   B. Because the flow is driven by gravity.

   C. Because it neglects surface tension.

   D. It invalidates Buckingham's theorem.

   E. None of the above.

3.35 What is the basic variable represented by transfer formulas in hydraulic modeling?

   A. Velocity.

   B. Length.

   C. Density.

   D. Force.

   E. Acceleration.

4

Flow in Pipes

The great majority of industrial and municipal applications of hydraulics involve flow in pipes. After an introduction that puts the development of pipe flow hydraulics into historical perspective, this chapter reviews the various methods used in the computation of energy losses in pipes. Following this the design of pipe networks is explained. The fundamental concepts of water hammer are introduced. Elemental considerations relating the moving of slurries through pipe lines are explained.

4.1 INTRODUCTION

During the Roman civilization water was led from springs in the mountains to nearby forts and palaces by gravity flow in aqueducts. But lead and clay pipes were also used by the Romans. In these, water often moved under pressure. To supply water to palaces, pipelines were built sporadically from the Middle Ages on. With the development of iron and steel manufacturing and after the introduction of efficient pumps more than 150 years ago, the building of pipelines became common practice. The amount of energy required to push the desired amount of discharge through a pipeline became a paramount question for designers. Because the discharge and the pressure lost through a pipe were easily measurable quantities, a great deal of experimental and field data was collected on pipe flow. A natural scientific development to put these data into the form of design formulas followed quite early. Most of the fundamental concepts related to flow of water in pipes were developed in the early part of the nineteenth century. Researchers in major countries developed different formulas for use in their respective regions. Even today in Western Europe engineers use the Kutter-Ganguillet equation; in Russia, the Pavlovsky formula. The American equivalent of these is the Hazen-Williams formula. Many of these concepts were first developed for other scientific questions and were then adopted by hydraulic problems. For example, Jean Louis Poiseuille (France, 1799–1869) was a physician who derived a formula for the flow of blood through veins and capillary blood vessels, which gave rise to the analysis of laminar flow in pipes. Much of modern-day pipe hydraulics stems from aeronautical developments driven by military research.

4.2 THE EMPIRICAL METHOD

At first, all information used in the design of pipelines was developed by collecting data on similar installations. Plotting such data on graphs and developing empirical equations was one of the ways to simplify design. The validity of most of these early design formulas is questionable because of the dubious origin of their underlying experimental data. However, at least one of these early equations is still widely used in American engineering practice. This is the so-called Hazen-Williams equation. In the conventional American system Equation 4.1 is written as

$$Q = 0.285 \, C \, D^{2.63} \, S^{0.54} \qquad \textbf{(4.1)}$$

in which Q is the discharge in gallons per minute and D is the pipe diameter in inches. S is the slope of the energy grade line, that is, the loss of energy per unit length of pipe. Figure 4.1 is a graphical representation of Equation 4.1 to facilitate computations.

In the SI system the Hazen-Williams formula is as

$$v = 0.355 \, C \, D^{0.63} \, S^{0.54} \qquad \textbf{(4.2)}$$

where v is the average velocity in the pipe in meters per second, and D is in meters.

The coefficient C in the Hazen-Williams formula is a *roughness coefficient*, a constant to represent the smoothness of the pipe walls. It is not a dimensionless constant; its actual dimension is length to the -0.13th power. Table 4.1 provides a list of C coefficients for various types of pipe walls. Generally, the magnitude of the coefficient is taken as 100 for average conditions. Its value can be as low as 50 for badly corroded pipes and as high as 150 for very smooth plastic or glass pipes.

The Hazen-Williams formula does not contain any terms related to the physical properties of the fluid because it is understood by its users that it applies to water only. Although the formula is incorrect from several theoretical points of view, it still seems to give acceptable results in practice. The reason for this is the built-in uncertainty in the determination of the C coefficient, which is really nothing more than a fudge factor. For these reasons the Hazen-Williams formula should be used only for rough preliminary estimates.

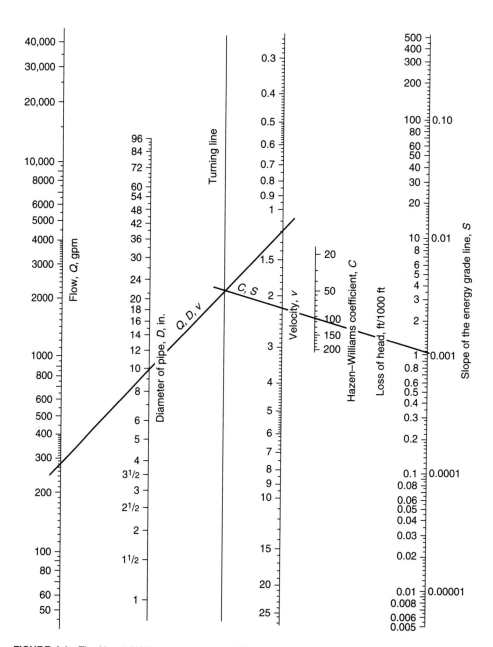

FIGURE 4.1 The Hazen-Williams nomograph. (Courtesy of Public Works Magazine)

TABLE 4.1   Values of C for the Hazen-Williams Formula

Type of Pipe	C
Asbestos cement	140
Brass	130–140
Brick sewer	100
Cast iron	
New, unlined	130
Old, unlined	40–120
Cement lined	130–150
Bitumastic enamel lined	140–150
Tar-coated	115–135
Concrete or concrete lined	
Steel forms	140
Wooden forms	120
Centrifugally spun	135
Copper	130–140
Fire hose (rubber lined)	135
Galvanized iron	120
Glass	140
Lead	130–140
Plastic	140–150
Steel	
Coal-tar enamel lined	145–150
New unlined	140–150
Tin	130
Vitrified clay	100–140

There are many other empirical formulas for pipe flow similar in form to the Hazen-Williams formula. Most were originally developed for open channel flow, and they all share the shortcomings of the Hazen-Williams formula.

4.3 THE SCIENTIFIC METHOD

Scientific speculations concerning the physical relationships controlling pipe flow date to the middle of the nineteenth century. One of the most satisfactory from a conceptual standpoint was the approach named after Henri Darcy (France, 1803–1858) and Julius Weisbach (Germany, 1806–1871). Darcy was the first to recognize the importance of pipe roughness on the flow in pipe, whereas Weisbach introduced engineering mechanics concepts into the analysis of pipe flow, particularly the use of Bernoulli's equation. The formula named after the two,

$$h_{\mathrm{L}} = f\left(\frac{L}{D}\right)\frac{v^2}{2g} \qquad (4.3)$$

states that the head loss is proportional to the kinetic energy $v^2/2g$ of the flow and to the length of the pipe and is inversely proportional to the diameter. The other influencing parameters related to fluid properties and the physical influence of the condition of the pipe wall were lumped into variable f, which was named the *friction factor*. This factor was subjected to much theoretical and experimental research. It was found that generally the value of f is dependent on the Reynolds number of the flow and the relative roughness of the pipe. The relative roughness, ε, is the magnitude of the average roughness elements of the pipe wall divided by the pipe diameter,

$$\varepsilon = \frac{e}{D} \qquad (4.4)$$

In practice there is no way to define the average roughness size, so this term is, again, an indefinite theoretical abstraction. Because the actual measurement of e is obviously not possible, its value is computed backward from experimental measurements in which all other variables are controlled. We shall return to this matter soon. The Reynolds number, defined in the previous chapter, is essentially the energy causing the motion divided by the energy resisting it, a dimensionless term. As Reynolds found in his experiments, for flows of very small velocity, or for fluids of high viscosity, or for very small pipe diameters, the Reynolds number may be less than 2300. In this case the flow is smooth and ordered, the path lines in the fluid are parallel, and dyes injected into the fluid travel along their respective flow lines without mixing into the whole fluid. The friction factor in laminar flow is computed from

$$f = \frac{64}{\mathbf{R}} \qquad (4.5)$$

which shows an inversely linear relationship between the Reynolds number and the friction factor. In practice one rarely designs for slowly moving fluid.

In flows at higher Reynolds numbers turbulence sets in. When the flow is turbulent, the fluid particles move about in the pipe in a random manner, completely mixing with the dye injected into the flow. The higher the Reynolds number, the higher the rate of turbulence. This random mixing of fluid particles consumes a considerable amount of energy in the flow. Hence, proportionally more energy is needed to move a turbulent flow than a laminar one. In turbulent flow the influence of the roughness of the pipe wall becomes dominant.

A great step forward in the determination of the effect of pipe roughness on the flow was taken by Johann Nikuradse (Russia, b. 1894), who after the Russian revolution in 1917 worked with Ludwig Prandtl in his research laboratory in Göttingen, Germany. Nikuradse introduced the concept of *equivalent sand roughness*. For his experiments he coated the inside of pipes of different diameters with sands of various grain sizes and defined the relative roughness of the pipe as the ratio of the sand grain diameter to the pipe diameter. Table 4.2 lists e, the equivalent sand roughness value for various commercially available pipe materials. Results of Nikuradse's experiments indicated that when the roughness of the pipe fully influences the flow at high Reynolds numbers, the friction factor is independent of the Reynolds number. A representative graph of Nikuradse's experimental results is shown in Figure 4.2. It is interesting to observe in this graph that the friction factor

TABLE 4.2   Equivalent Sand Roughness for Pipe Materials

Commercial Pipe Surface, New	Equivalent Sand Grain Roughness, e	
	ft	m
Glass, drawn brass, copper, lead	Smooth	
Wrought iron, steel	1.5×10^{-4}	0.46×10^{-4}
Asphalted cast iron	4.0×10^{-4}	1.22×10^{-4}
Galvanized iron	5.0×10^{-4}	1.52×10^{-4}
Cast iron	8.5×10^{-4}	2.6×10^{-4}
Concrete	$10\text{--}100 \times 10^{-4}$	$3\text{--}30 \times 10^{-4}$

rises in the transitional range between the fully turbulent flows and the laminar flows.

To express Nikuradse's experimental results in a mathematical form, Theodore von Kármán[1] used Prandtl's boundary layer formulas—originally developed for aeronautical design—to derive equations for turbulent flow within rough and smooth pipe boundaries. The concept of "rough boundary" is a theoretical abstraction rooted in boundary layer theory; it has nothing to do with the pipes being

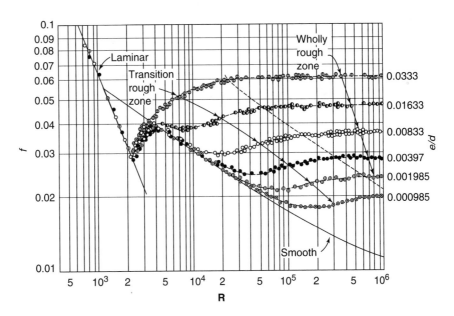

FIGURE 4.2   Nikuradse's experimental results.

---

[1]The correct pronunciation of "von" is like *phone*, a German reference to a Hungarian title of nobility.

rough or smooth to the touch. The resulting equation for the value of the friction factor within rough boundaries is

$$\frac{1}{\sqrt{f}} = 1.14 + 2\log\frac{D}{e} \qquad (4.6)$$

The constant in the equation, 1.14, was adjusted from its theoretical value to correspond to the experimental data available. Kármán's other formula, relating to flow within "smooth" pipe walls, or slower but still turbulent flows, considers the influence of the Reynolds number only; it does not contain the relative roughness term.

Subsequently, C. F. Colebrook and other investigators mathematically combined Kármán's two equations, into a complex, implicit mathematical formula, ostensibly expressing Nikuradse's experimental results in mathematical form. Lewis F. Moody (United States, 1880–1953) plotted these equations, along with Equation 4.5 of Reynolds on a logarithmic graph similar to the one reproduced in Figure 4.3. After its publication in 1944, accurate sizing of pipes for different fluids on a scientific basis became possible. Comparison of Figures 4.2 and 4.3 reveals that the

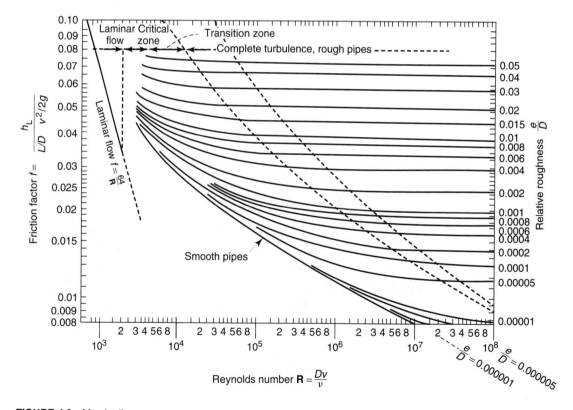

FIGURE 4.3 Moody diagram.

uncertain experimental results in the transitional range are artificially smoothed by the mathematical formulas. From a practical point of view this is rather immaterial because almost all designs concern flow well in the turbulent range.

With the Moody diagram available, frictional losses in pipes can be computed with relative ease and reasonable accuracy as long as the head loss is sought. For a given discharge and a known pipe (i.e., Q, D, e, and L are known), the Reynolds number and the relative roughness can be readily determined. These allow the value of the friction factor f to be read directly from the Moody diagram.

If the discharge Q is the unknown, or when the pipe diameter is to be determined for a given Q and h_L, the Moody diagram can be used only through a tedious trial-and-error method. This difficulty slowed the adoption of the scientific method for a whole generation of engineers. Surveys done 25 years after Moody's publication indicated that as many as 80 percent of engineers in the water supply practice still preferred to use old Hazen-Williams nomographs[2] instead of the Darcy-Weisbach equation. The need for this iterative solution was eliminated in the early 1970s when the information contained in the Moody diagram was replotted so that the unknown variables could be read off directly.

Wen-Hsiung Li replotted Moody's diagram as shown in Figure 4.4. In this graph the variables are arranged so that D appears only once. To use the graph, the term

$$\left(g\frac{S}{Q^2} \right)^{1/5} \tag{4.7}$$

must first be calculated. Again, S here represents h_L/L. All coordinates of the graph in Figure 4.4 are expressed as functions of this term. By substituting the known values of Q, the viscosity of the fluid ν, and the equivalent sand roughness e, the diameter D is simply obtained from the graph.

K. C. Asthana realized the oddity of using an equation along with a graph. He combined the Darcy-Weisbach equation and the Moody diagram into a single graph. His graph, shown in Figure 4.5, plots the Reynolds number and the relative roughness in the coordinates of (h_L/L), $(g \cdot e^3/\nu^2)$, and $Q/\nu e$. This method allows the direct determination of Q when all other variables are known.

To avoid confusion about the proper use of the three methods just reviewed, one must keep in mind the following:

1. To determine the energy loss, use Moody's diagram and the Darcy-Weisbach equation.
2. To determine the pipe diameter, use Li's diagram and the Darcy-Weisbach equation.
3. To determine the discharge, use Asthana's diagram directly.

All three methods are based on the Darcy-Weisbach equation and on the various experimental and theoretical attempts to define the friction coefficient. The only difference among the three procedures is the way in which the data and Equation

---

[2]The correct expression is *nomogram*. Western Union used to deliver telegrams, not telegraphs.

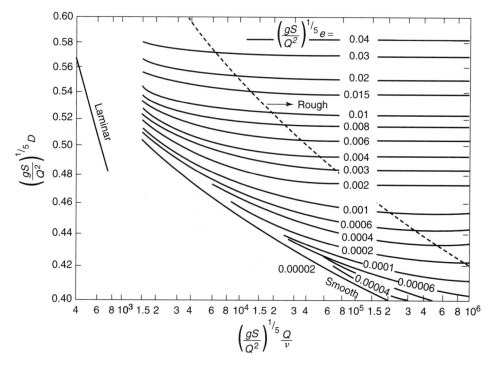

FIGURE 4.4 Li's pipe flow diagram. (From Wen-Hsiung Li, Direct Determination of Pipe Size, *Civil Engineering*, American Society of Civil Engineers, Vol. 44, No. 6, June 1974)

4.3 are plotted. The use of the Darcy-Weisbach equation can provide the best, most reliable solution for pipe flow problems. From the standpoint of practical work one criticism may be raised: The roughness of the pipe is still an unknown, and it also tends to change over time as deposits clog the pipe. This factor introduces a degree of uncertainty in the results, although not as much as with the Hazen-Williams formula.

4.4 THE CONVEYANCE METHOD

In most technical applications pipelines are designed for an efficient delivery of the design discharge. Thus, the discharge will surely be carried in the pipe at a fully turbulent condition. Accordingly, the laminar and transitional ranges of the Moody diagram—in which the experimental data appear to diverge from the analytic formulas—are of little value in practical design. In practice the friction factor will be controlled not by the Reynolds number but by the relative roughness. For this case, substituting Equation 4.6 into Equation 4.3 and rearranging, we obtain

$$Q = K \sqrt{\frac{h_L}{L}} = K \sqrt{S} \qquad (4.8)$$

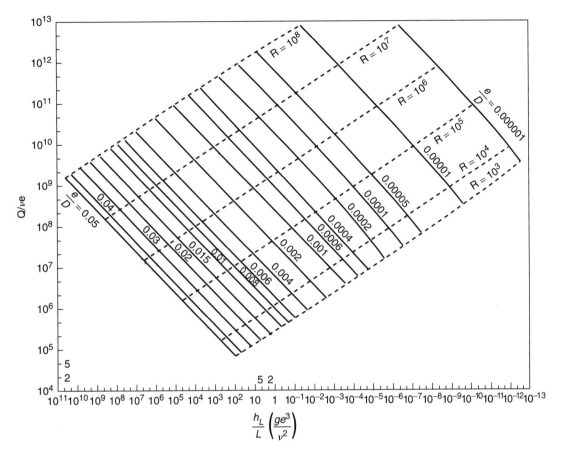

FIGURE 4.5 Asthana's pipe flow diagram. (Courtesy of K. C. Asthana, University of Addis Ababa, Addis Ababa, Ethiopia.)

in which the coefficient K is named *conveyance* and is defined as

$$K = \frac{\pi}{4}\sqrt{2g}\left(2\log\frac{D}{e} + 1.14\right)D^{2.5} \tag{4.9}$$

This formula contains only pipe-related variables in addition to the gravitational constant. Accordingly, for pipe flow design under fully turbulent conditions the pipe manufacturer could just as well stamp the conveyance value onto the pipe when it is made. Conveyance values for various pipe diameters and different pipe materials are compared in Figure 4.6. The graph indicates that the pipe roughness has a relatively small influence on the conveyance, compared with the pipe diameter. Table 4.3 gives K values for selected steel pipes.

The conveyance formula, Equation 4.8, allows us to solve quickly and simply most pipe flow problems, without the sometimes cumbersome procedure associated

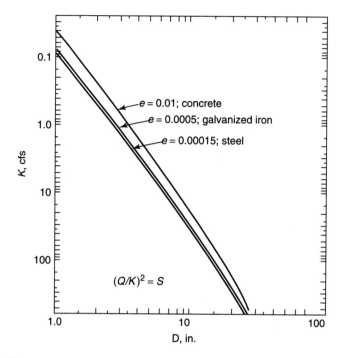

FIGURE 4.6   Pipe conveyance graph.

with the use of Moody's diagram. Its only restriction is the assumption that Kármán's equation, Equation 4.6, representing Nikuradse's experimental data, is correct. The conveyance method will not give reliable results for flows within the transitional region, but neither will Moody's diagram.

4.5   LOCAL LOSSES

In the hydraulic design of pipelines the energy lost through friction along the pipe is dominant for pipes of 100 ft (30 m) or longer. For shorter pipe lengths the aggregate of local energy losses at elbows, valves, inlet devices, and the like may be equal to or more than the frictional losses along the pipe. For this reason it was

TABLE 4.3   *K* for Steel Pipe

	K		
D in.	gpm	ft³/sec	liters/s
2	224	0.5	14
6	4086	9.1	258
8	8666	19.5	546
12	24875	55.4	1570
24	150415	335	9480

necessary to develop formulas to compute local energy losses. In particular, this matter became important in the suction pipe of pumps to prevent cavitation in the pump by careful design (Chapter 5).

Local losses in piping fixtures were found to be proportional to the amount of kinetic energy entering the fixture. The configuration of the fixture determines the constant of proportionality. Accordingly, local loss in a pipe fixture is computed from

$$h_v = k \frac{v^2}{2g} \qquad (4.10)$$

in which k is the so-called local loss coefficient, and v is the velocity in the pipe before the fixture, unless otherwise specified. Table 4.4 lists local loss coefficients for a variety of fixtures.

Converting Equation 4.10 to express the energy loss through a valve or other fixture in terms of the discharge leads to

$$h_v = \frac{8k}{g \pi^2 D^4} Q^2 \qquad (4.11)$$

Rearranging this formula and introducing the concept of conveyance for local losses, we get

$$h_v = \left(\frac{Q}{K_l}\right)^2 \qquad (4.12)$$

where K_l represents the conveyance of a hydraulic device creating a local loss in a piping system, where

$$K_l = \pi D^2 \sqrt{\frac{g}{8k}} \qquad (4.13)$$

Another way to deal with local losses is to convert their effects into an *equivalent length* of pipe, which is then added to the actual pipe length considered in the design. The equivalent length of a local loss is derived by equating the Darcy-Weisbach equation with Equation 4.10 and solving for L, which results in

$$L_{\text{equivalent}} = \frac{kD}{f} \qquad (4.14)$$

in which k is the local loss coefficient, D is the diameter, and f is the friction factor determined for the pipe flow.

4.6 DESIGN OF PIPE NETWORKS

In practical applications one often finds pipes of different sizes connected together in various ways. In the case of two pipes, they may be connected either in *series* or in *parallel*. When pipes are connected in series with each other, the discharge

TABLE 4.4   Local Loss Coefficients

Use the equation $h_v = kv^2/2g$ unless otherwise indicated. Energy loss E_L equals h_v head loss in feet.

①		Perpendicular square entrance: $k = 0.50$ if edge is sharp

②		Perpendicular rounded entrance: $\begin{array}{c\|c\|c\|c\|c\|c}R/d & = & 0.05 & 0.1 & 0.2 & 0.3 & 0.4 \\ \hline k & = & 0.25 & 0.17 & 0.08 & 0.05 & 0.04\end{array}$

③		Perpendicular reentrant entrance: $k = 0.8$

④		Additional loss due to skewed entrance: $k = 0.505 + 0.303 \sin\alpha + 0.226 \sin^2\alpha$

⑤		Suction pipe in sump with conical mouthpiece: $$E_L = D + \frac{5.6Q}{\sqrt{2g}D^{1.5}} - \frac{v^2}{2g}$$ Without mouthpiece: $$E_L = 0.53D + \frac{4Q}{\sqrt{2g}D^{1.5}} - \frac{v^2}{2g}$$ Width of sump shown: 3.5D <div align="right">(After I. Vágás)</div>

⑥		Strainer bucket:
		   $k = 10$ with foot valve
		   $k = 5.5$ without foot valve
		<div align="right">(By Agroskin)</div>

⑦		Standard Tee, entrance to minor line $k = 1.8$

in each will be the same, but the velocity may differ according to the diameter of each individual pipe. The total head loss in the system may be computed by summing the individual head losses in each length. In precise calculations the local losses occurring at the joints where the pipes are connected should also be considered.

TABLE 4.4 *continued*

⑧ 	**Sudden expansion:** $$E_L = (1 - \frac{v_2}{v_1})^2 \frac{v_1^2}{2g}$$ or $$E_L = (\frac{v_1}{v_2} - 1)^2 \frac{v_2^2}{2g}$$

⑨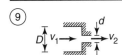

Sudden contraction:

$d/D =$	0.01	0.1	0.2	0.4	0.6	0.8
$k =$	0.5	0.5	0.42	0.33	0.25	0.15

use v_2 in Equation 4.10

⑩

Diffusor:

$$E_L = k(v_1^2 - v_2^2)/2g$$

$\alpha° =$	20	40	60	80
$k =$	0.20	0.28	0.32	0.35

⑪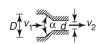

Confusor:

$$E_L = k(v_2^2 - v_1^2)/2g$$

$\alpha° =$	6	10	20	40	60	80	100	120	140
k for $D = 3d$	0.12	0.16	0.39	0.80	1.0	1.06	1.04	1.04	1.04
$D = 1.5d$	0.12	0.16	0.39	0.96	1.22	1.16	1.10	1.06	1.04

⑫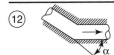

Sharp elbow:

$$k = 67.6 \times 10^{-6}(\alpha°)^{2.17}$$

(By Gibson)

⑬

Bends:

$$k = (0.13 + 1.85(r/R)^{3.5})\sqrt{\alpha°/180°}$$

(By Hinds)

⑭

Close return bend:

$$k = 2.2$$

Two pipes connecting two common end points in parallel deliver a total discharge that is the sum of the individual discharges of the two. Because the pressure at the common end points must be single-valued, the head loss through both pipes must be identical, regardless of the possible differences in lengths, diameters, or friction factors of the two pipes. The distribution of the total discharge between the two parallel pipes will depend on these differences. The total discharge of the system may be computed by writing the Darcy-Weisbach equation for each pipe to describe the common head loss. A third equation is written to express that the

TABLE 4.4 *continued*

⑮		Gate valve:

$e/D =$	0	1/4	3/8	1/2	5/8	3/4	7/8
k =	0.15	0.26	0.81	2.06	5.52	17.0	97.8

---

⑯   Globe valve:

                        $k = 10$ when fully open

---

⑰   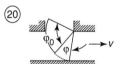   Rotary valve:

$\alpha° =$	5	10	20	30	40	50	60	70	80
k =	0.05	0.29	1.56	5.47	17.3	52.6	206	485	∞

(By Agroskin)

---

⑱   Check valves:

         Swing type $k = 2.5$ when fully open
         Ball type $k = 70.0$
         Lift type $k = 12.0$

---

⑲   Angle valve:

         $k = 5.0$ if fully open

---

⑳   Segment gate in rectangular conduit:

$$k = 0.8 + 1.3 \left(\frac{1}{n} - n \right)^2$$

where $n = \varphi/\varphi_0$ = the rate of opening with respect to the central angle.

(By Abelyev)

㉑   Sluice gate in rectangular conduit:

$$k = 0.3 + 1.9 \left[\left(\frac{1}{n} \right) - n \right]^2$$

where $n = h/H$.

(By Burkov)

---

sum of the two discharges is the total discharge flowing through the system. The same computational procedure applies for three or more pipes connected in parallel.

Complex pipe networks are generally composed of a number of branching pipes connected into common joints where three or more pipes meet. The simplest case of branching pipes is the *three-reservoir problem*. In this case three reservoirs

TABLE 4.4 *continued*

(22)		Measuring nozzle: $E_L = 0.3 \, \Delta p$ for $d = 0.8D$ $E_L = 0.95 \, \Delta p$ for $d = 0.2D$ where Δp is the measured pressure drop. (By A.S.M.E.)
(23)	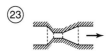	Venturi meter: $E_L = 0.1 \, \Delta p$ to $0.2\Delta p$ where Δp is the measured pressure drop.
(24)	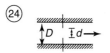	Measuring orifice, square edged: $E_L = \Delta p \left[1 - \left(\dfrac{d}{D} \right)^2 \right]$ where Δp is the measured pressure drop.

(25)  Confusor outlet:

$d/D =$	0.5	0.6	0.8	0.9
$k =$	5.5	4	2.55	1.1

(By Mostkov)

(26) 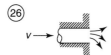 Exit from pipe into reservoir:

$k = 1.0$

(27) Diffusor outlet for $D/d > 2$:

$\alpha° =$	8	15	30	45
$k =$	0.05	0.18	0.5	0.6

(By Mostkov)

are connected together by three pipes joined at a common point, as shown in Figure 4.7. This problem is solved by writing the continuity equation for the branching point, where

$$\sum Q = Q_1 + Q_2 + Q_3 = 0 \tag{4.15}$$

The pressure at the branching point must be single-valued. There exists an H energy

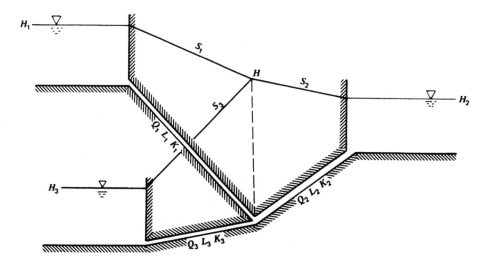

FIGURE 4.7 Branching pipes.

at the branching point. The energy losses through each pipe are then

$$S_i = \frac{H_i - H}{L_i} = \left(\frac{Q_i}{K_i}\right)^2 \tag{4.16}$$

where i represents pipes 1, 2, and 3, respectively, and the K values are the individual conveyances of the three pipes. Depending on the relative magnitude of H, the values of S for each pipe may be positive or negative, depending on whether the flow at the branching point is toward or away from the junction.

There could be a number of conditions for known or unknown parameters in such problems. The solution of any of these can be obtained by substituting the known terms into the preceding four formulas and solving them in a simultaneous manner. By the basic concepts of algebra, there may be four unknowns for the four equations. These problems are usually solved by trial-and-error iteration or by matrix methods.

The principles used in the solution of branching pipes are also valid for the solution of pipe networks where there are scores of internal branching points. In municipal and industrial piping networks many interconnected pipes are involved. The number of resulting nonlinear algebraic equations to be solved is large. To solve the multitude of interdependent equations in practical applications, digital computers are utilized. Several computer programs are available for the design and analysis of complex networks.

To solve smaller network problems a systematic trial-and-error procedure may be used. It is called the *relaxation method* in mathematics.[3] Its development for

---

[3]The method was first introduced by Oxford professor Richard Vynne Southwell, who later became the president of the Imperial College of Science and Technology at London (England, b. 1888).

pipe networks was presented in 1936 by Hardy Cross, who had developed a similar solution for statically indeterminate beams earlier. The first step in the application of the Cross method is to assume initial values for either all discharges in the individual pipes or all piezometric heads at the junction points. Once either set of values is assumed, the corresponding head losses or flows may be computed for each junction or pipe, respectively. If the computed results do not satisfy the requirements set by the appropriate law of conservation, an error is calculated by using the procedure proposed by Cross, and an adjustment is made on the initially assumed values. Both methods are described further.

In the procedure of *discharge balancing* the computation begins by assuming a reasonable discharge for each pipe of the network satisfying the continuity equation at each junction point. Next, the head loss through each of the pipes is calculated by using any of the pipe flow computational methods introduced previously. Then, with careful attention to the direction of the flow (hence, the slope of the energy grade line) in each pipe, the energy losses are summed through selected closed pipe loops that encompass the whole network. Because only one pressure value can exist at each junction, adding all head losses (or gains) in a consistent direction (say, clockwise) must result in zero if the discharge in each of the pipes was assumed correctly. If they were not, then the sum of the energy losses will be nonzero, indicating an error in the assumed flows that needs to be adjusted.

The adjustment procedure for discharges using any one of the flow formulas is based on the fact that they may be put in a general form of

$$h_L = N Q^n \tag{4.17}$$

In the conveyance method, for example, $N = L/K^2$, and n is 2. If the original discharge is Q_1, and the discharge correction is ΔQ, then the corrected discharge is

$$Q_2 = Q_1 - \Delta Q \tag{4.18}$$

Substituting this equation into Equation 4.17 and expanding the series leads to

$$h_L = N Q_2^n = N(Q_1 - \Delta Q)^n = N(Q_1^n - \Delta Q n Q_1^{n-1} \ldots) \tag{4.19}$$

If the discharge correction is relatively small, the remaining terms of the series may be neglected. Writing the sum of the head losses through a closed pipe loop results in

$$\sum h_L = \sum N (Q_1^n - \Delta Q n Q_1^{n-1}) \tag{4.20}$$

If ΔQ is the correct balancing flow at the given junction, Equation 4.20 will equal zero. With the assumption that the discharge correction will be the same in all pipes included in the closed pipe loop, we may rearrange Equation 4.20 to express the required discharge correction as

$$\Delta Q = \frac{\sum N Q_1^n}{\sum N n Q_1^{n-1}} \tag{4.21}$$

The denominator in this formula is taken as an absolute value, without regard to the direction of flow. Substituting the conveyance formula for the N and n values, for example, Equation 4.21 takes the form

$$\Delta Q = \frac{\sum (L/K^2)\, Q_1^2}{\sum |(2L/K^2)Q_1|} \tag{4.22}$$

Once the discharge correction is obtained for each of the closed pipe loops of the network, it is applied to adjust the originally assumed discharges. The whole procedure is repeated until the discharge error in each loop is below a preset level of acceptability.

Head balancing is another method used to solve the network problem iteratively. Recall from the introduction of Bernoulli's equation in Chapter 2 that by "head" we mean the energy term in feet at the pipe junctions. This solution process starts by making reasonable assumptions for the pressure head H at each one of the junctions of the network. For the assumed initial piezometric heads at all internal junction points the corresponding discharges can be calculated because the head loss h_L for each pipe is given by the difference in piezometric heads at the two ends. (Note that local losses at the junction points are neglected.) After calculating the discharges for each pipe we must write the continuity equation, Equation 4.15, for all junctions. For the initially assumed piezometric heads to be correct, which is a very remote possibility, all continuity equations must be satisfied. Otherwise, an error in the continuity equations, ΔQ, will appear at each junction in the form of

$$\sum Q = \Delta Q \tag{4.23}$$

This indicates that the piezometric heads were assumed incorrectly and must be adjusted. The amount of adjustment at each pipe junction, ΔH, is done in a systematic manner, along the same lines as in discharge balancing. The resultant equation, based on the concepts of the conveyance method, is

$$\Delta H = \frac{2\,\Delta Q}{\sum |Q/h_L|} \tag{4.24}$$

The denominator of this equation is a weighting factor, similar to that of Equation 4.21. It is the sum of the Q/h_L values of all pipes entering the junction, regardless of the direction of the flow. After all ΔQ values are calculated for the initially assumed piezometric heads, the required head corrections can be determined by Equation 4.24 at each junction point. For these new piezometric heads new head loss and discharge values may then be computed for each connecting pipe. The process is repeated until the errors in discharges are reduced to an acceptable level.

The correction formulas, Equations 4.22 and 4.24, are not sacrosanct. The experienced user may adjust the assumed discharge or head values in any arbitrary manner to obtain improved results. For example, faster convergence—and less excessive oscillation in the successive trial results—is found if the factor 2 is dropped from Equation 4.22 or 4.24.

Numerous computer programs are available commercially for solving pipe network problems. The most sophisticated ones are based on AutoCAD or similar software technology. All of them are based on the concepts introduced in this chapter. The discussion of the use of these programs is beyond the scope of this text.

Note that only internal pressure heads are adjusted in the trial-and-error procedure. Exit and inlet discharge points may have fixed pressure values throughout the computation depending on external controlling conditions.

Either the external pressures or the external (inflow and outflow) discharges are assumed to be known at the start in the two methods shown here for pipe network analysis. The methods provide for a systematic determination of the corresponding internal pressures and discharges, subject to the physical characteristics of the network and the initially assumed external values. Because virtually any assumption may be made for initial pressure or discharge values, the methods will provide corresponding internal solutions. This, however, leads to a forced solution. In nature, there can be only one solution for a pipe network flow problem: the one at which the most discharge flows with the least energy spent. If, for example, the outlet valves were opened in a network, the outflow discharges and pressures would correspond to a unique solution, with the flows seeking the path of least resistance. Analytically such a solution method would seek to optimize the sum of energy expended in each one of the pipe segments, measured in terms of the head loss divided by the discharge. Once this concept is understood one may question the value of forced solutions. The answer must be based on practical considerations. In a pipe network the exit discharges are rarely defined and are known to vary with time.

The design discharge for a network is the extreme flow under certain conditions, such as the maximum pressure and discharge requirement at certain locations during otherwise normal operating conditions. Extreme conditions in pipe networks are set by fire flow requirements. These, in turn, depend on the structures to be protected and are determined by the insurance companies. Better fire flows mean better insurance ratings and lower insurance premiums. Discharges for firefighting requirements range from less than 40 gpm for a small hose to 1000 gpm for a heavy turret gun. The suction hose of a pumper truck will collapse if the pressure in the fire hydrant is less than 20 psi at full flow. Hydrants are spaced between 200 to 500 ft apart. In a major fire two or more hydrants are operated at the same time, and the pipe network must be adequate to supply the required fire flow. In light of these considerations the usefulness of forced solutions in pipe network analysis is quite apparent.

In residential water supply the water pressure ranges from 50 to as much as 120 psi. In large supply systems, particularly in hilly areas, the network may need to be separated into pressure zones to maintain the water pressure within reasonable limits. Booster pumps and pressure reducers are therefore integral parts of larger systems. From the treatment plant the water is pumped through main feeder lines, and primary and secondary mains. In modern design practice secondary mains are rarely smaller than 6 in., and 8, 12, and 16 in. diameter pipes are common. Because the cost of the piping is a major part of the expense of water supply development,

TABLE 4.5   Celerity c^*, Bulk Modulus of Elasticity E, and Density ρ for Some Common Liquids at 60°F

Liquid	c^* ft/sec	E lb/ft²	ρ lb sec²/ft⁴
Water	4950	45×10^6	1.94
Seawater	4750	45×10^6	1.99
Benzene	3510	21×10^6	1.71
Crude oil	4600	35.9×10^6	1.70
Mercury	1460	56.2×10^6	26.3
Carbon tetrachloride	3060	28.9×10^6	3.095

in developing countries economic feasibility considerations often force the designers to use water mains with diameters as small as 2 in.

4.7   WATER HAMMER

Depending on the relative position of the pipe with respect to the datum plane, the energy at any point in the line will be composed of the kinetic energy $(v^2/2g)$ and pressure energy (p/γ). Water will move through a pipe regardless of its position, as long as it is below the energy grade line. The pressure on the pipe walls will be determined by the pressure energy. Wall thicknesses of standard pipes are designed to withstand ordinary service pressure. Sudden changes of discharge may, however, result in stresses of sufficient magnitude to exceed these design stresses.

Any change of discharge in a pipe (for example, valve closure, pipe fracture, or pump stoppage) results in a change of momentum of the flow. Controlled by the impulse-momentum equation, an impulse force is created, which is commonly called *water hammer*. This force should be checked because it may be detrimental to the system.

The theory of water hammer was developed by Nicolai Egorovich Zhukovsky[4] (Russia, 1847–1921) according to the following concepts: Consider a pipeline on which an open valve is located at a distance L downstream from a reservoir. Initially the fluid flows at a velocity v, and the fluid pressure at the valve is p_0. If the valve is closed instantaneously, the fluid rams into the closed gate, creating a pressure shock as it decelerates to zero velocity. According to Newton, pressure shocks in fluids of infinite extent travel at a velocity given by the formula

$$c^* = \sqrt{\frac{E}{\rho}} \tag{4.25}$$

which is called *celerity*. E in this equation is the bulk modulus of elasticity of the fluid, and ρ is its density. Table 4.5 lists values of c^*, E, and ρ for some common liquids.

---

[4]Spelled Joukowsky in older publications. It represents the German transliteration of the original Russian name.

TABLE 4.6   Modulus of Elasticity E_p of Various Pipe Materials

Pipe Material	E_p	
	million psi	lb/ft²
Lead	0.045	6.48×10^6
Lucite (at 73°F)	0.4	57.6×10^6
Rubber (vulcanized)	2	288×10^6
Aluminum	10	1440×10^6
Glass (silica)	10	1440×10^6
Brass, bronze	13	1872×10^6
Copper	14	2016×10^6
Cast iron, gray	16	2304×10^6
Cast iron, malleable	23	3312×10^6
Steel	28	4032×10^6

If we compress the fluid in an elastic pipe, the pipe will expand. The modulus of elasticity E_c of a system composed of elastic fluid and elastic pipe may be calculated from the equation

$$\frac{1}{E_c} = \frac{1}{E} + \frac{D}{wE_p} \tag{4.26}$$

where D is the pipe diameter, w is the thickness of the pipe wall, and E_p is the modulus of elasticity of the pipe material. Values of E_p for different pipe materials are listed in Table 4.6.

Equation 4.26 refers to circular pipes only. The value of E_c for noncircular conduits is considerably smaller because of their limited structural rigidity under outward pressure.

The celerity of a shock wave in a pipe, c, can then be computed from

$$\frac{c}{c^*} = \frac{1}{\sqrt{1 + ED/wE_p}} \tag{4.27}$$

which is plotted in Figure 4.8. The graph indicates that the pipe rigidity exerts a considerable influence on the velocity of the shock wave.

The shock waves that travel upstream and downstream from the valve under adjustment ultimately reach the ends of the pipe. There, the pressure is controlled by the stationary energy level—by reservoir depth, for example. The time t for a shock wave to reach such a point at a distance L from the valve is

$$t = \frac{L}{c} \tag{4.28}$$

at which time the shock disappears. At that instant the compressed, halted fluid in the pipe is not balanced from that end. Hence, to relieve the compression, the fluid starts to flow in the opposite direction. This creates a relief pressure shock that

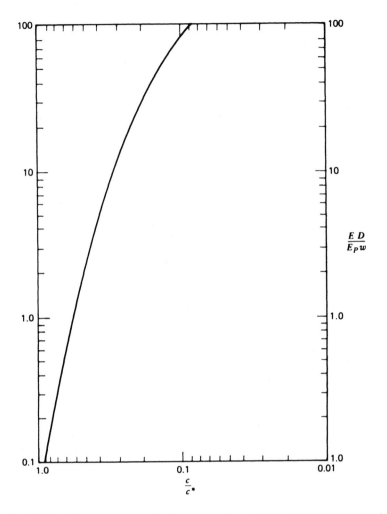

FIGURE 4.8 Celerity of pressure waves in pipes; c equals celerity in elastic pipe; c* equals celerity in a fluid of infinite extent.

travels back to the valve. The time period T during which the shock pressure acts on the valve is the time it takes for the pressure wave to travel from and back to the valve, that is,

$$T = 2t = \frac{2L}{c} \tag{4.29}$$

At this time all the fluid is moving backward at a velocity v. Because the valve is closed, there is no supply for this flow; hence, a negative pressure shock (suction) is created at the valve. This shock again travels to and back from the reservoir, reversing the flow. Such oscillations of pressure and periodic reversal of flow will

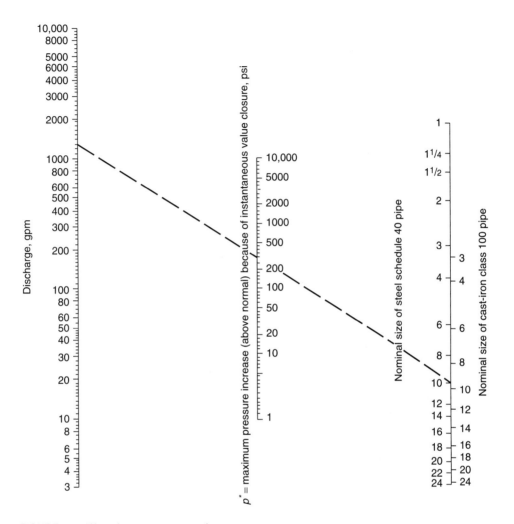

FIGURE 4.9   Water hammer nomograph.

persist until the kinetic energy is dissipated by friction. The process described will occur both upstream and downstream from the valve, the only difference being that the initial shock will be positive on the upstream side and negative on the downstream side. The magnitude of the pressure shock at instantaneous valve closure is

$$p^* = \rho c v \qquad (4.30)$$

and the pressure will oscillate in the pipe within the range

$$p = p_0 \pm p^* \qquad (4.31)$$

Either value can be detrimental to the pipeline. A nomographic solution of Equation 4.30 is shown in Figure 4.9.

The time of closure of a valve in reality is not zero but a certain finite period, say, T_c. The water hammer pressure increases gradually with the rate of closure of the valve. Whether T_c is smaller or larger than T of Equation 4.29 differentiates between quick and slow closure.

If T_c is smaller than T, the shock pressure will actually reach its maximum value, p^*. This means that quick closure is equivalent to instantaneous closure. If T_c exceeds T, p^* will not develop fully because reflected negative shock waves arriving at the valve after time T will counteract it. Lorenzo Allievi (Italy, 1856–1941) developed the theoretical fundamentals of slow closures. For slow valve closures the maximum pressure may be calculated by the Allievi formula

$$p = p_0 \left(\frac{N}{2} + \sqrt{\frac{N^2}{4} + N} \right) \tag{4.32}$$

where N is defined as

$$N = \left(\frac{Lv\rho}{p_0 T_c} \right)^2 \tag{4.33}$$

Checking pipelines for water hammer effects involves the following procedures:

First, we determine the energy grade line and the hydraulic grade line at the discharge flowing. On this basis we can locate the most critical points: where sudden change of discharge may occur (e.g., valves, pumps) and where the pipe may burst or collapse (e.g., at low and high points). Next we determine the parameters c^* and E for the fluid from Table 4.5 and note the pipe size (D and w) and its elastic modulus E_p (Table 4.6). From these values we calculate $ED/E_p w$. Using Figure 4.9 we obtain the system celerity, c. Next, we compute the wave travel times T from Equation 4.29 for both upstream and downstream directions. From operational considerations we assume the probable period of the time of closure, T_c. Quite often, closing mechanisms are designed so that they cannot be closed too fast. By use of Equation 4.30 we find the maximum and minimum pressures at any point along the pipe. Paying particular attention to selected critical points, we can check the pipe for bursting pressure. Similarly, we check for cavitation and collapse by comparing p with the vapor pressure. If $T < T_c$, the water hammer pressure will be reduced as computed by the Allievi formula.

Water hammer effects can lead to destructive results in hydraulic installations. Particularly in the large penstocks (inlet pipes) of turbines a sudden change of discharge, caused by rapid change in electric load, could cause a major failure. Figure 4.10 shows a typical penstock. Note the heavy concrete blocks holding the steel pipes. Water hammer may be controlled by using slow-closing valves, standpipes, pneumatic shock absorbers, pressure relief valves, or similar devices.

4.8 SLURRY FLOW

Solid–water mixtures are frequently pumped through pipes, as in dredging, in hydraulic transporting of soils for fills, and in moving fly ash or coal. The subject

FIGURE 4.10 Steel penstocks behind concrete dam. Large concrete block at bends counteract water hammer effects. (Courtesy of U.S. Army Corps of Engineers, Sacrámento, CA)

has been studied extensively by the chemical manufacturing industry under the name *two-phase flow*. Handling such design problems requires some extension of the basic pipe flow formulas.

To enable the relatively heavy solid particles to be carried by fluid, there must be a significant degree of turbulence in the water. The turbulent velocity components acting upward against gravity create shear forces on the solid particles. These forces counteract the gravitational force that pulls the particles downward and allow them to be kept in suspension.

To compute the required pipe diameter and the energy losses that will occur when pumping slurries, one must first know the physical characteristics of the slurry.

Specifically, the following three parameters must be determined:

1. The density of the solids carried: ρ_s (kg/m^3).
2. The statistical mean of the particle diameters: d (m).
3. The concentration or the percentage of solids in the liquid, by volume: c (percent).

If these parameters are known, the *submerged specific gravity* of the solids and the *slurry density* may be calculated by the following formulas:

Submerged specific gravity of solids:

$$S_{\text{submerged}} = \frac{\rho_s - \rho_{\text{water}}}{\rho_{\text{water}}} \tag{4.34}$$

Slurry density:

$$\rho_{\text{slurry}} = c\rho_s + (1 - c)\,\rho_{\text{water}} \tag{4.35}$$

Because the density of water is unity for normal operating temperatures, the two equations are rather simple. The density for sand may be taken as 2.6 g/cm^3 or 2600 kg/m^3.

Theoretical and experimental studies of two-phase flow in pipes have led to Spells' equation, which relates the slurry discharge to the pipe diameter. It gives the required average velocity for a slurry of known physical characteristics to be kept from settling out. A modified and simplified version, using SI terminology, and applicable to water-based slurries, is

$$v = m(d\,S_{\text{submerged}})^{0.816}\,(D\,\rho_{\text{slurry}})^{0.633} \tag{4.36}$$

where D is the pipe diameter in meters; d is the average grain size in meters, and v is the velocity required to keep the solids from settling out, in meters per second. The value m was determined experimentally to be 161 for the minimum velocity at which the slurry is not uniformly mixed across the pipe but is denser on the bottom, but at which the solids do not settle out. For the optimum velocity at which the slurry is homogeneously mixed throughout the pipe m should be 475.

To keep the solids in suspension requires pumping at higher velocities, which increases the energy loss in the pipe and leads to higher pumping costs. It also creates higher pressures in the pipe, perhaps requiring thicker pipe wall size, which means higher costs. Higher velocity also increases the erosion of the pipe walls by the moving solid particles. Such unwarranted expenses may be minimized by reducing the velocity of the flow.

Once the design velocity for the slurry is determined, a slurry pipeline is designed as though it will be carrying a common liquid. There are, however, two significant factors to be kept in mind: one is that the density of the slurry is greater than that of water. This influences the magnitude of the Reynolds number, which, in turn, causes a change in f. Another concern in hydraulic design of slurry pipelines is the pump performance. In determining the mass flow rate pumped, one must take into account the slurry density, not the density of the fluid alone.

EXAMPLE PROBLEMS

Example 4.1

A 3000 ft long pipe with a 10 in. diameter carries 300 gpm of water. The C coefficient is assumed to be 100. Calculate the loss of energy due to friction using the Hazen-Williams Equation 4.1 and the nomograph shown in Figure 4.1.

Solution. The variables are:

$$Q = 300 \text{ gpm}$$
$$D = 10 \text{ in.}$$
$$C = 100$$

Rearranging Equation 4.1, we can determine the slope of the energy grade line:

$$S = \left(\frac{Q}{0.285CD^{2.63}}\right)^{1/0.54} = \left(\frac{300}{0.285(100)(10)^{2.63}}\right)^{1.85} = 0.001 \text{ ft/ft}$$

This value may also be determined using Figure 4.1.

In the nomograph connect Q and D with a straight line and mark off the intersection on the turning line. Connect the marked point with C and extend the line to S. The resulting S is 0.001 ft per foot. For 3000 ft the head loss is $3000 \times 0.001 = 3$ ft. □

Example 4.2

A 12 km long pipe carries 0.1 m³/s of water. The pipe diameter is 30 cm; its relative roughness is 0.0001. Compute the change in head loss if the temperature of the water changes from 30 to 10°C.

Solution. The average velocity of the flow is

$$\frac{Q}{A} = \frac{0.1}{(\pi/4)(0.3)^2} = 1.4 \text{ m/s}$$

From Figure 1.5 the kinematic viscosities are

At 30°C $\nu = 8 \times 10^{-7} \text{ m}^2/\text{s}$
At 10°C $\nu = 1.3 \times 10^{-6} \text{ m}^2/\text{s}$

From Equation 2.23

$$R_{30°C} = \frac{1.4(0.3)}{8} \times 10^7 = 5.3 \times 10^5$$

$$R_{10°C} = \frac{1.4(0.3)}{1.3} \times 10^6 = 3.2 \times 10^5$$

Using Moody's diagram we find that

$$f_{30°C} = 0.014$$
$$f_{10°C} = 0.015$$

Consequently, the change is $\Delta f = 0.0010$
The change of head loss may be computed from Equation 4.3,

$$\Delta h_{\mathrm{L}} = \Delta f \frac{L}{D}\frac{v^2}{2g} = 0.0010 \frac{(12{,}000)}{0.3}\frac{(1.4)^2}{2 \times 9.81}$$

$$= 4\ \mathrm{m} \qquad \square$$

Example 4.3

An 8 in. cast-iron pipe carries water at 40°F from reservoir A and discharges it into reservoir B (Figure E4.3). The length of the pipe is 5000 ft. The elevation of the water surface is 3300 ft for A and 3100 ft for B. Calculate the discharge of the pipe.

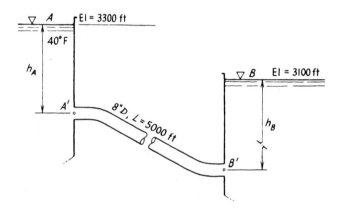

FIGURE E4.3

Solution. Because flow is unknown in this problem, we use Asthana's diagram. First, we obtain the roughness of cast-iron pipe from Table 4.2, $e = 0.00085$ ft, and the viscosity from Figure 1.5, $v = 1.6 \times 10^{-5}$. Then we solve for

$$\frac{h_{\mathrm{L}}}{L}\left(\frac{ge^3}{v^2}\right) = \left(\frac{200}{5000}\right)\left[\frac{32.2(0.00085)^3}{(1.6 \times 10^{-5})^2}\right] = 3.09$$

and

$$\epsilon = \frac{e}{D} = \frac{0.00085}{8/12} = 0.00128$$

Entering Figure 4.5 with these values gives

$$\frac{Q}{ve} = 3 \times 10^8$$

Therefore,

$$Q = ve(3 \times 10^8) = 1.6 \times 10^{-5}(0.00085)(3 \times 10^8) = 4\ \mathrm{cfs} \qquad \square$$

Example 4.4

Water at 65°F and moving at 900 gpm is to be transferred over 2000 ft to an elevation 60 ft lower than the supply source (Figure E4.4). Determine the required diameter of the concrete pipe such that the friction loss is equal to the difference in elevation.

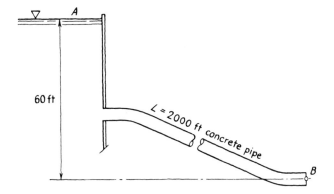

FIGURE E4.4

Solution. Because the diameter is sought, we use Li's diagram. From Figure 1.5,

$$\nu = 1.2 \times 10^{-5}\ \text{ft}^2/\text{sec}$$

The slope of the energy grade line is

$$\frac{h_L}{L} = S = \frac{60}{2000} = 0.03$$

From Table 4.2 the average concrete roughness is $e = 5 \times 10^{-3}$ ft.

$$Q = 900\ \text{gpm} \times 2.23 \times 10^{-3} = 2.0\ \text{cfs}$$

in which the conversion coefficient is obtained from the pertinent table in the appendix.
We use Li's graph and Equation 4.7:

$$\left(\frac{gS}{Q^2}\right)^{1/5} = \left(\frac{32.2 \times 0.03}{2.0^2}\right)^{1/5} = 0.24^{0.2} = 0.75$$

The following products are needed to use Figure 4.4:

$$\left(\frac{gS}{Q^2}\right)^{1/5}\frac{Q}{\nu} = 0.75\frac{2.0}{1.2 \times 10^{-5}} = 1.25 \times 10^5$$

$$\left(\frac{gS}{Q^2}\right)^{1/5} e = 0.75(5 \times 10^{-3}) = 3.75 \times 10^{-3}$$

From Figure 4.4 we find that

$$\left(\frac{gS}{Q^2}\right)^{1/5} D = 0.49$$

Hence, the required diameter is

$$D = \frac{0.49}{0.75} = 0.65\ \text{ft, or approximately 8 in.}$$

□

Example 4.5   Calculate the conveyance of a 3 ft diameter concrete pipe.

Solution. From Equation 4.9

$$K = \frac{\pi}{4}\sqrt{2g}\left(2\log\frac{D}{e} + 1.14\right)D^{2.5}$$

As a median value for concrete pipe $e = 0.005$ ft may be taken from Table 4.2. Then

$$K = \frac{\pi}{4}\sqrt{2g}\left(2\log\frac{3}{0.005} + 1.14\right)3^{2.5}$$

$$= [6.303(2 \times 2.778 + 1.14)15.59]$$

$$= 658$$

The conveyance of the 3 ft concrete pipe is approximately 660 cfs.   □

Example 4.6   Determine the discharge of a 12 in. galvanized pipe if the length of the pipe is 1200 ft and the head loss is 100 ft.

Solution. The given conditions are

$$h_L = 100 \text{ ft}$$
$$D = 1 \text{ ft (galvanized pipe)}$$
$$L = 1200 \text{ ft}$$

From Equation 4.8

$$Q = K\sqrt{S}$$

$$\sqrt{S} = \sqrt{\frac{h_L}{L}} = \left(\frac{100}{1200}\right)^{1/2} = 0.29$$

From Table 4.2 $e = 0.0005$ ft for the galvanized pipe.
   Then, from Figure 4.6

$$K = 49$$

Hence,

$$Q = 49 \times 0.29 = 14 \text{ cfs}$$ □

Example 4.7   By using the conveyance method, determine the required pipe diameter for the problem described in Example 4.4.

Solution. From Equation 4.8 $Q = K\sqrt{S}$

$$\sqrt{S} = \sqrt{\left(\frac{60}{2000}\right)} = 0.173$$

$$Q = 900 \text{ gpm} = 2 \text{ cfs}$$

Then,

$$K = \frac{2}{0.173} = 11.6 \text{ cfs}$$

From Table 4.3 we obtain $D = 8$ in. as the nearest standard pipe size.   □

Example 4.8

An 8 in. nominal diameter steel pipe is 1000 m long. How much is the discharge in the pipe if the head loss over the whole distance is 12 m?

Solution. From Table 4.3 we see that the conveyance K of the pipe is 0.546 m³/s. The slope of the energy line is

$$S = \frac{12}{1000} = 0.012 \text{ m/m}$$

From Equation 4.8

$$Q = K\sqrt{S} = 0.546\sqrt{0.012} = 0.06 \text{ m}^3/\text{s}$$ □

Example 4.9

Using the Darcy-Weisbach equation, determine the discharge of a 4 in. cast-iron pipe of 150 ft length, if the system includes a strainer bucket with a foot valve, a gate valve half-closed, a globe valve fully open, a square-edged measuring orifice that reduces the diameter to 2 in., and a 2 in. diameter confusor at the outlet. The total head loss in the system is 25 ft. The measured head loss in the orifice may be taken as 0.2 ft. Use the Moody diagram.
Assume the water temperature is 68°F (20°C).

Solution. For the given total head losses

$h_T = 25$ ft

$h_T =$ friction loss + loss due to strainer bucket with foot valve + gate valve half-closed + globe value fully open + square-edged measuring orifice + confusor outlet

From Table 4.4, k values are determined as follows:

$$k_1 = 10 \text{ for strainer bucket with foot valve}$$
$$k_2 = 2.06 \text{ for gate valve half-closed}$$
$$k_3 = 10 \text{ for globe valve fully open}$$
$$k_4 = 5.5 \text{ for the confusor with } \frac{d}{D} = 0.5$$
$$h = 0.2 \text{ ft for the square-edged measuring orifice loss}$$

For the cast-iron pipe $e = 0.00085$ ft.
Then,

$$\varepsilon = \frac{e}{D} = \frac{0.00085}{4/12} = 0.0026$$

From Tables 1.2 and 1.5

$$\rho = 1.94 \text{ slug/ft}^3$$
$$\mu = 2.09 \times 10^{-5} \text{ lb sec/ft}^2$$

Let v = average velocity of water in the pipe, yet unknown.

$$\mathbf{R} = \frac{\rho v D}{\mu}$$

$$= \frac{1.94 \times 4/12}{2.09 \times 10^{-5}} v$$

$$= 3.09 \times 10^4 v$$

for

$$h_T = 25 \text{ ft} = f\frac{L}{D}\frac{v^2}{2g} + (k_1 + k_2 + k_3 + k_4)\frac{v^2}{2g} + h$$

$$= f\frac{150}{(4/12)}\frac{v^2}{2g} + (10 + 2.06 + 10 + 5.5)\frac{v^2}{2g} + 0.2$$

$$24.8 = \frac{v^2}{2g}(450f + 27.56)$$

$$v = \frac{40}{\sqrt{450f + 27.56}}$$

Then,

$$\mathbf{R} = 3.09 \times 10^4 \frac{40}{\sqrt{450f + 27.56}}$$

$$= \frac{109.6 \times 10^4}{\sqrt{450f + 27.56}}$$

We try $f = 0.025$:

$$\mathbf{R} = \frac{123.6 \times 10^4}{\sqrt{450 \times 0.025 + 27.56}} = 1.98 \times 10^5$$

with $\varepsilon = e/D = 0.0026$.

Entering the Moody diagram, we get $f = 0.026$, which is close to our initial guess. Therefore, we use $f = 0.025$. Hence,

$$v = \frac{40}{\sqrt{450 \times 0.025 + 27.56}}$$

$$= 6.4 \text{ fps}$$

Hence, the discharge

$$Q = Av = \frac{\pi}{4}\left(\frac{4}{12}\right)^2 6.4 = 0.56 \text{ cfs}$$

□

Example 4.10   Determine the equivalent length of pipe for the piping system described in Example 4.9.

Solution. From Equation 4.14

$$L_{\text{equivalent}} = k\frac{D}{f}$$

Therefore, the equivalent length of pipe due to minor losses is

$$L_{equivalent} = \frac{27.56\,(4/12)}{0.025} = 367\ ft$$

and the total equivalent length for the pipe system is $150 + 367 = 517$ ft.   ☐

Example 4.11

Assume that the pipe described in Example 4.8 is discharging freely into the atmosphere through a rotary valve at the downstream end. How many degrees will the valve have to be turned to reduce the flow by 25 percent?

Solution. The discharge of the pipe was 0.06 m³/s, and the energy loss in the 1000 m long pipe was 12 m. The reduced discharge will be

$$0.06\,(0.75) = 0.045\ m^3/s$$

At that discharge the energy loss, by the conveyance formula, Equation 4.8, is

$$1000\left(\frac{0.045}{0.546}\right)^2 = 6.8\ m$$

Because the valve is discharging freely into the atmosphere, the remaining energy will have to be dissipated by the local loss due to the valve, h_v, or

$$h_v = 12 - 6.8 = 5.2\ m$$

We rearrange Equation 4.11 to

$$Q = \pi D^2 \sqrt{\frac{g}{8k}}\,\sqrt{h_v} \qquad or \qquad Q = C_v\sqrt{h_v}$$

where C_v represents the conveyance of a hydraulic device creating a local loss in a piping system. Thus,

$$C_v = \frac{Q}{\sqrt{h_v}} = \frac{0.045}{\sqrt{5.2}} = 0.02$$

and

$$C_v = 0.02 = \pi D^2 \sqrt{\frac{g}{8k}}\ m^3/s$$

in which $D = 8$ in. $= 0.203$ m and $g = 9.81$ m/s². Hence,

$$\frac{0.02}{3.14(0.203)^2}\sqrt{\frac{8}{9.81}} = \sqrt{\frac{1}{k}} = 0.140$$

and

$$k = 51.3$$

By Case 17 of Table 4.4 this requires a 50° closure of the rotary valve.   ☐

Example 4.12

From a water tower at point A, where the water elevation is 1000 ft above datum level, a pipe network is fed as shown in Figure E.4.12. The network is composed of steel pipes with an equivalent sand roughness of $e = 0.0005$ ft. The pipe lengths and diameters are tabulated in the solution. Assuming that at point E the exit pressure is maintained at 750 ft above datum level, calculate the discharge throughout the network using the Cross method.

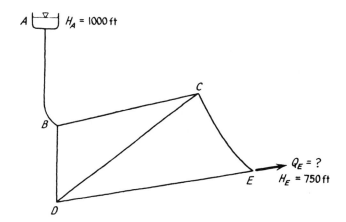

FIGURE E4.12

Solution. The first step in the procedure is to determine the conveyance value of each pipe. In this example the values were taken from Table 4.3. We calculate the reciprocals of the pipe resistance values, L/K^2.

The conveyance and the pipe resistance values for all pipes are shown in the first five columns of Table E4.12a. Next, we assume initial pressure head values for all internal junction points, as shown in the second column of Table E4.12b.

TABLE E4.12a

Line	L_{ft}	$D_{in.}$	K_{cfs}	K^2/L	h_L	Q	$\left\|\dfrac{Q}{h_L}\right\|$
AB	1000	12	55.4	3.07	100	17.5	0.18
BC	1500	8	19.5	0.254	50	3.56	0.07
BD	1200	12	55.4	2.56	100	16.0	0.16
CD	1500	6	9.1	0.055	50	1.66	0.03
CE	2000	12	55.4	1.54	100	12.39	0.12
DE	1000	8	19.5	0.38	50	4.36	0.09

Now, in the first table we show all head loss values for these trial pressure heads and then we calculate by Equation 4.8 the corresponding discharges, shown in the second to last column of Table E4.12a.

Next, we calculate the excess discharges by Equation 4.23. Then, we compute the head corrections by Equation 4.24, ignoring the factor of 2 to avoid excessive oscillation in the trial solutions. Note that the new H at junction D is now greater than the new H at junction C. Thus, flow is now from D to C. This change needs to be considered when summing the flows at each respective junction.

TABLE E4.12*b*

Junction	H	$\sum Q = \Delta Q$	$\sum \lvert Q/h_{\text{L}} \rvert$	ΔH	New H
A	1000	$17.5 - 17.5 = 0$	—	—	1000.0[a]
B	900	$17.5 - 3.56 - 16.0 = -2.06$	0.41	-5.0	895.0
C	850	$3.56 - 1.66 - 12.39 = -10.49$	0.22	-47.7	802.3
D	800	$16.0 + 1.66 - 4.36 = +13.30$	0.28	47.5	847.5
E	750	$4.36 + 12.39 - 17.5^b = -0.75$	—	—	750.0[b]

[a] Fixed elevations.
[b] By continuity external flows must equal.

Detailed computations for the successive trials are not included. However, the results of some iterations are shown in Table E4.12*c*.

TABLE E4.12*c*

Junction	H_1	H_2	H_5	H_{10}	H_{25}
A	1000.0	1000.0	1000.0	1000.0	1000.0
B	895.0	899.7	910.2	918.3	923.3
C	802.3	793.3	787.9	790.2	792.3
D	847.5	857.7	871.8	881.6	887.6
E	750.0	750.0	750.0	750.0	750.0

For the last set of H values the discharges are as shown in Table E4.12*d*.

TABLE E4.12*d*

Pipe	Discharge in cfs
AB	15.35
BC	5.76
BD	9.55
DC	2.29
CE	8.05
DE	7.23

As is evident in Table E4.12*d*, the solution is converging in a monotonous manner. By continuity Q_E exit flow must be equal to the flow in pipe AB and also equal to the sum of the flows in CE and DE. Computer spreadsheets are helpful for this type of problem. □

Example 4.13   When a 1000 ft long 12 in. diameter steel pipe with 0.375 in. wall thickness leads from a reservoir to a level 250 ft below the control gate, fully open, the water exits freely (Figure E4.13). Compute the discharge and the head loss of the system. What is the maximum pressure if the valve is closed in 0.4 sec, and how long will this pressure be sustained at the valve?

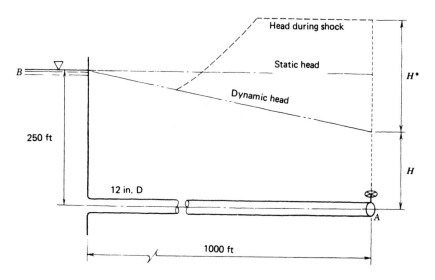

FIGURE E4.13

Solution.

$$D = 12 \text{ in.} = 1 \text{ ft}$$
$$w = 0.375 \text{ in.}$$
$$e = 1.5 \times 10^{-4} \text{ ft} \quad \text{(Table 4.2)}$$

and the kinematic viscosity is $\nu = 1.2 \times 10^{-5}$ ft²/sec (60°F, Figure 1.5).

Let v = average velocity of water in pipe at the steady state. Then, by writing Bernoulli's equation

$$\frac{p_A}{\gamma} + z_A + \frac{v_A^2}{2g} + f\frac{L}{D}\frac{v_A^2}{2g} + \frac{k_1 v_A^2}{2g} = \frac{p_B}{\gamma} + z_B + \frac{v_B^2}{2g}$$

in which $k_1 = 0.15$ (from Table 4.4, Case 15)

$$0 + 0 + \frac{v^2}{2g}(1 + f1000 + 0.15) = 0 + 250 + 0$$

$$\frac{v^2}{2g}(1.15 + 1000f) = 250$$

$$v = \frac{127}{\sqrt{1.15 + 1000f}}$$

$$R = \frac{vD}{\nu}$$

$$= \frac{127 \times 1}{\sqrt{1.15 + 1000f}\,(1.2 \times 10^{-5})}$$

$$= \frac{106 \times 10^5}{\sqrt{1.15 + 1000f}}$$

We assume that $f = 0.013$.

Then,

$$\mathbf{R} = 2.82 \times 10^6$$

From Equation 4.4

$$\varepsilon = \frac{e}{D} = 0.00015$$

Entering the Moody diagram, we find that $f \approx 0.013$. Therefore,

$$v = \frac{127}{\sqrt{1.15 + 1000 \times 0.013}} = 33.8 \text{ fps}$$

The head losses of the system are friction loss and exit loss:

$$h_L = f\frac{L}{D}\frac{v^2}{2g} + k_1\frac{v^2}{2g}$$

$$= \left(0.013 \times \frac{1000}{1} + 0.15\right)\frac{(33.8)^2}{2g}$$

$$h_L = 233 \text{ ft}$$

The steady-state discharge is

$$Q = \frac{\pi}{4}(1)^2(33.8)$$

$$= 27 \text{ cfs}$$

If the valve is suddenly closed, the water hammer will affect the pressure in the pipe.
From Equation 4.25,

$$c^* = \sqrt{\frac{E}{\rho}}$$

where

$$E = 45 \times 10^6 \text{ lbf/ft}^2$$

For water from Table 4.5

$$\rho = 1.94 \text{ lbf sec}^2/\text{ft}^4$$

Then,

$$c^* = \sqrt{\frac{45 \times 10^6}{1.94}}$$

$$= 4816 \text{ fps}$$

From the celerity of a shock wave in a pipe, c, in feet per second, we get from Equation 4.27

$$\frac{c}{c^*} = \frac{1}{\sqrt{1 + ED/E_p w}}$$

(which may be solved by Figure 4.8 also), where

$$E_p = \text{Young's modulus of elasticity for pipe material}$$
$$= 4.032 \times 10^9 \text{ lbf/ft}^2$$
$$D = \text{pipe diameter} = 1 \text{ ft}$$
$$w = \text{pipe wall thickness} = 0.375 \text{ in.}$$

Then

$$c = \cfrac{4816}{\sqrt{1 + \cfrac{45 \times 10^6}{4.032 \times 10^9} \times \left(\cfrac{12}{0.375}\right)}}$$

$$= 4134 \text{ fps}$$

From Equation 4.28,

$$t = \frac{L}{c}$$

$$= \frac{1000}{4134} = 0.24 \text{ sec}$$

and $T = 2t = 0.48$ sec, which is larger than the valve closure time of 0.4 sec. Accordingly, we have quick closure. By using Equation 4.30 or Figure 4.9 with $Q = 449(27) = 12,100$ gpm and a 12 in. steel pipe, we get $p^* = 1900$ psi $= 2.74 \times 10^5$ lb/ft^2 above the steady flow pressure level, sustained over 0.48 sec, which will certainly destroy the pipe unless the time of closure is significantly reduced.

Assuming $T_c = 3$ min from Equation 4.32 we obtain

$$p = p_0 \left(\frac{N}{2} + \sqrt{\frac{N^2}{4} + N}\right)$$

in which

$$N = \left(\frac{Lv\rho}{p_0 T_c}\right)^2 = \left(\frac{Lv\rho}{\gamma H_0 T_c}\right)^2$$

$$= \left(\frac{Lv\rho}{\rho g H_0 T_c}\right)^2 \left(\frac{L_v}{g H_0 T_c}\right)^2$$

where

$$L = 1000 \text{ ft}$$
$$v = 33.8 \text{ fps}$$
$$H_0 = 250 - 233 = 17 \text{ ft}$$
$$T_c = 180 \text{ sec}$$

so that

$$N = \left(\frac{1000 \times 33.8}{32.2 \times 17 \times 180}\right)^2$$

$$= 0.12$$

Hence,

$$p = p_0 \left[\frac{0.12}{2} + \sqrt{\frac{0.014}{4} + 0.12}\right]$$

$$\gamma H = \gamma H_0 (0.41)$$
$$H = 17 (0.41)$$
$$= 7 \text{ ft, or about 3 psi of pressure}$$

This is the pressure head rise at the valve if the valve is closed slowly over a 3 min period.   □

Example 4.14   A slurry is composed of a mixture of sand and water. The mean diameter of the sand was determined by sieve analysis to be 1.5 mm (0.06 in.). The density of the sand is 2.64 g/cm³. The slurry concentration is 15 percent solids by volume. Determine the density of the slurry, the submerged density of the sand, and the minimum and optimum velocities if the slurry is pumped through a 0.25 m diameter steel pipe. Also compute the head loss if the slurry pipe is 350 m long.

Solution. By Equation 4.34 the submerged specific gravity of the sand is

$$S_{submerged} = \frac{2.64 - 1.00}{1.00} = 1.64$$

The slurry density, with $c = 0.15$, and by Equation 4.35, is

$$\rho_{slurry} = 2.64(0.15) + 1.0(1 - 0.15) = 1.246 \text{ g/cm}^3 = 1246 \text{ kg/m}^3$$

Substituting into Equation 4.36, we obtain

$$v = m[1.64(0.0015)]^{0.816}[1.246(0.25)]^{0.633}$$
$$= m(0.00355)$$

For minimum velocity $m = 161$; therefore,

$$v_{minimum} = 0.57 \text{ m/s}$$

For optimum velocity $m = 475$; therefore,

$$v_{optimum} = 1.69 \text{ m/s}$$

The Reynolds number for the flow is

$$R = v \cdot D \cdot \rho/\mu = 0.57(0.25)1246/(10^{-3}) = 1.8 \times 10^5$$

The relative roughness is

$$e/D = 0.46 \times 10^{-4}/0.25 = 0.00018$$

Entering the Moody diagram, we find that

$$f = 0.017$$

The Darcy-Weisbach equation then gives

$$h_L = 0.017 \frac{350}{0.25} \frac{0.57^2}{2(9.81)} = 0.39 \text{ m}$$

□

PROBLEMS

4.1 The allowable head loss in a very smooth plastic pipe is 40 ft. The pipe is 2000 ft long. Determine the diameter of the pipe if the discharge is 600 gpm by using the Hazen-Williams formula.

4.2 A 700 m long, 8 in. diameter pipe discharges water at 0.025 m³/s. What is the approximate energy loss if the coefficient is estimated at 120?

4.3 The relative roughness of a pipe is 0.001. Determine the friction factor for a flow characterized by a Reynolds number of 2×10^5.

4.4 A 30 cm diameter pipe carries a flow of 0.030 m³/s. The length of the galvanized iron pipe is 300 m. Determine the head loss due to friction.

4.5 Check the result obtained for Problem 4.1 assuming that the water is at 70°F. Use Li's diagram.

4.6 In a 30 cm galvanized iron pipe that is 300 m long the head loss is 0.18 m. Determine the discharge using Asthana's diagram. Refer to Problem 4.4.

4.7 For a 6 in. diameter galvanized iron pipe calculate the slope of the hydraulic grade line if the discharge is 2.0 cfs. Use the conveyance formula (Equation 4.8).

4.8 Repeat Problem 4.7 for a steel pipe.

4.9 A 200 m long, 8 in. steel pipe carries a discharge of 30 liters/s from a reservoir. The pipe includes a rotary valve closed 40°, a gate valve closed 50 percent, and a perpendicular reentrant inlet. The pipe discharges freely into the atmosphere. Compute the total head loss. What is the equivalent length of pipe for the local losses?

4.10 A 6 in. diameter steel pipe branching network connects three reservoirs. Each of the three pipes is 1500 ft long. The water elevations in the three reservoirs are 1000, 900, and 850 ft above the datum level. Determine the flow in each pipe.

4.11 A 1000 ft long steel pipe carries water from a reservoir with a water elevation at 826 ft. The pipe has a diameter of 12 in. with a wall thickness of 0.5 in. A gate valve at the downstream end is slowly closed during a half-minute period. Determine the maximum water hammer pressure if the initial pressure at the valve was measured to be 742 ft above datum. The elevation of the valve is 545 ft.

4.12 Determine the amount of sand (in grams) in the slurry pipe in Example 4.14.

4.13 Repeat Example 4.14 using bituminous coal with a unit weight of 1.32 g/cm³ instead of sand.

MULTIPLE CHOICE QUESTIONS

4.14 The shortcoming of the Hazen-Williams formula is that
   A. It requires use of the Moody diagram.
   B. It applies for water only.
   C. Its C coefficient is a "fudge factor."
   D. Its range of validity is for laminar flow.
   E. Both B and C.

4.15 The C coefficient for a new smooth steel pipe in the Hazen-Williams formula is
   A. Less than 100.
   B. 100.
   C. About 120.
   D. About 150.
   E. About 40.

4.16 The Darcy-Weisbach formula states that the rate of energy loss in a pipe is
   A. Inversely proportional to the friction factor.
   B. Proportional to the kinetic energy.
   C. Linearly proportional to the Reynolds number.
   D. Proportional to the discharge.
   E. Inversely proportional to the pipe's area.

4.17 The friction factor f in the Darcy-Weisbach formula
   A. Increases proportionally to the Reynolds number.
   B. Is independent of the roughness in turbulent flow.
   C. May be described by Manning's equation.
   D. Is proportional to $\log(D/e)$.
   E. All of the above.

4.18 The first reliable experimental data for the determination of friction factors for pipe flow was provided by
   A. Prandtl.
   B. von Kármán.
   C. Manning.
   D. Nikuradse.
   E. Moody.

4.19 Ludwig Prandtl's contribution to pipe flow computations was the development of
   A. The plotting of $\log R$ versus f.
   B. The concept of sand grain roughness.
   C. Boundary layer theory.
   D. Nomography.
   E. Conveyance.

4.20 When the discharge is to be determined, Li's diagram is the most efficient to use.
   A. True.
   B. False.

4.21 Asthana's diagram is the best to use if the roughness e is unknown.
   A. True.
   B. False.

4.22 The conveyance formula is valid only for fully turbulent flow conditions.
   A. True.
   B. False.

4.23 The conveyance K for pipes depends on
   A. D, e, and g.
   B. D, Q, and e.
   C. D, f, and R.

D. S and D.

E. Q, $\log(D/e)$, and g.

4.24 With the conveyance method the rate of energy loss in a pipe is

A. Inversely proportional to (Q/K).

B. Inversely proportional to $(Q/K)^2$.

C. Directly proportional to Q/K.

D. Proportional to Q times K.

E. None of the above.

4.25 In computing local losses the velocity we use for kinetic energy is usually the

A. One before the fixture.

B. One after the fixture.

C. Average of inflow and outflow velocities.

D. One that is the smaller.

E. One that is the larger.

4.26 The procedure used by Hardy Cross to solve pipe network problems involves

A. Assumption of discharges in pipes and exits.

B. Assumption of pressures at all pipe junctions.

C. Systematic correction for continuity at junctions.

D. Adding and correcting head losses through closed loops.

E. All of the above.

4.27 Celerity in water hammer analysis depends

A. On the temperature, wall thickness, and fluid density.

B. Elasticity of pipe and fluid, and the pipe diameter.

C. On the type of fluid in the pipe.

D. All of the above.

E. None of the above.

4.28 The formula for celerity in a boundless fluid mass was first given by

A. Allievi.

B. Nikuradse.

C. Zhukovsky.

D. Bernoulli.

E. Newton.

4.29 In the mechanism of hydraulic shock upstream of a suddenly closed valve, identify the correct sequence of events:

A. Flow stops, shock wave travels up then back, flow reverses.

B. Shock wave travels up, flow reverses, shock wave returns.

C. Flow stops, shock wave travels up, flow reverses, shock wave returns.

D. Shock wave travels up and down, then flow stops and reverses.

E. None of the above.

4.30 An important purpose of standpipes in water conduits is

A. Storage of water.

B. Prevention of freezing.

C. Maintenance of pressure.

D. Reduction of shock pressure.

E. None of the above.

4.31 Assuming that the celerity in a given pipe is 1450 fps and a valve is closed instantaneously when the initial velocity of the flow is 24 fps, estimate the magnitude of the hydraulic shock.

A. 67,500 psf.

B. 34,800 psf.

C. 2813 psf.

D. 61 psf.

E. None of the above.

5

Hydraulic Machinery

The proper selection of pumps is an integral part of the design of piping systems. Pumps are also often used in drainage, irrigation, and other areas, wherever it is required to add hydraulic energy to the water. The basic hydraulic concepts of pumps are outlined to provide sufficient understanding of their design, selection, operation, and maintenance. A short review of water power generation closes the chapter.

5.1 PUMP CLASSIFICATIONS

Unless water is moved by gravity at an adequate discharge and pressure, it may be necessary to install pumps. The pumps add energy to water. Many kinds of pumps are used in the various technological fields. The three main classes are centrifugal, rotary, and reciprocating pumps. These divisions refer to the different ways pumps move the liquid. The different classes can be further subdivided into pumps of different types. For example, *centrifugal pumps* include the following:

> propeller (axial flow)
> mixed-flow
> vertical turbine
> regenerative turbine
> diffuser
> volute

Centrifugal pumps are classified according to the way the rotating component, the impeller, imparts energy to the water. In turbine pumps or radial flow pumps the impeller is shaped to force water outward at right angles to its axis. In mixed-flow pumps the impeller forces water in a radial as well as in an axial direction. In propeller pumps the impeller forces water in an axial direction only. Within the radial flow pump category we speak of a volute, diffuser, or circular type of casing, referring to the way in which water is collected and guided toward the exit pipe after it leaves the impeller.

Any one of these types of pumps can be single stage or multistage; stage refers to the number of impellers in a pump. Another distinguishing characteristic is the position of the shaft, which can be vertical or horizontal. Furthermore, there are single-suction and double-suction pumps. Pumps can also be grouped by their construction materials: bronze, stainless steel, iron, and their various possible combinations. Material becomes important when corrosive liquids are handled.

In the field of hydraulics, where we deal specifically with water, the most common pumps are centrifugal. Our discussion will therefore focus on these.

5.2 THE SPECIFIC SPEED OF PUMPS

The selection of the type of pump for a particular service is based on the relative quantity of discharge and energy needed. Lifting large quantities of water over a relatively small elevation—for example, removing irrigation water from a canal and putting it onto a field—requires a different kind of pump than when pumping a relatively small quantity of water to great heights, such as in furnishing water from a valley to a ski lodge at the top of a mountain. To make the proper selection for any application one needs to be familiar with the basic concepts of operation of the main types of centrifugal pumps.

The water enters the pump at the shaft that rotates the *impeller*. The impeller is a series of propellers, vanes, or blades that are arranged peripherally about the shaft and that may or may not be held together by one or two circular plates. As the motor rotates the shaft at high speed exerting a torque T, the water is whirled around as it enters the impeller. The angular velocity ω imparted to the water particles throws them outward onto the wall of the casing. The casing is built so that it leads the water toward the exit pipe either by vanes or by its gradually expanding spiral shape. In the casing the kinetic energy gradually changes into pressure energy.

Impellers of large radius and narrow flow passages transfer more kinetic energy per unit volume than smaller-radius impellers with large water passages. Pumps designed so that the water exits the impeller at a radial direction impart more centrifugal acceleration than those from which water exits axially or at an angle. Therefore, the relative shape of the impeller determines the general field of application of a centrifugal pump. We do not need to take a pump apart to see the shape of its impeller before we select one for a particular use. Few people design pumps; those who use them generally deal with them on a "black box" basis. To simplify things the discharge, head, and speed at optimum performance of various pumps are consolidated into one number called *specific speed*. The specific speed n_s of a pump is computed from

$$n_s = \frac{\text{rpm}\sqrt{\text{gpm}}}{H^{3/4}} \qquad (5.1)$$

where n_s is the specific speed in revolutions per minute, rpm is the rotational speed of the shaft in revolutions per minute, gpm is the discharge in gallons per minute, and H is the total dynamic head in feet that the pump is expected to develop, all at optimum efficiency. Pumps built with identical proportions but different sizes have the same specific speed.

Equation 5.1 contains terms in conventional units that are not dimensionally homogeneous. Although it is common to use the specific speed formula in this manner, with the increase in the use of SI a dimensionless form might be more appropriate. For this reason it has been proposed that the gravitational acceleration to the $\frac{3}{4}$ power be incorporated into the denominator.

The specific speed of a pump is not really a speed in the physical sense, although it can be used in the sense that a pump reduced in size to deliver 1 gpm to a height of 1 ft will run at its specific speed. In practice, specific speed is only a number well suited to characterize the various types of centrifugal pumps. Generally, pumps with low specific speeds (500 to 2000 rpm) are made to deliver small discharges at high pressures. Pumps characterized by high specific speeds (5000 to 15,000 rpm) deliver large discharges at low pressures. The approximate relationships between specific speed, impeller shape, discharge, and efficiency are shown in Figure 5.1. Pumps with large-diameter impeller shapes that have relatively small flow passages generate high pressures but deliver small discharges. They have low specific speeds, as shown in the graph.

5.3 HEAD AND POWER REQUIREMENTS

The dynamic head H to be developed by a pump is computed according to the methods described in Chapter 4. The pressure or head to be developed is the sum of the height to which the water is to be lifted from the level of the reservoir or sump from which the water is pumped. In addition to this height the friction losses occurring in the suction and discharge pipes must be taken into account. Losses of energy through inlet devices such as strainer and foot valve, elbows, valves, and other components must also be considered. The pressure or kinetic energy required at the end of the supply line is also a part of the *total dynamic head*.

Pumps of identical design are manufactured in various sizes. The pump size is expressed by the diameter of its exit pipe in inches. The approximate capacity in gallons per minute can be determined by multiplying the square of the exit diameter by 25. Selection of the *pipe diameter* in pipe design should ordinarily conform to this diameter. The pipe size should be selected to minimize the initial installation cost as well as the capitalized cost of operation and maintenance. Although a smaller-

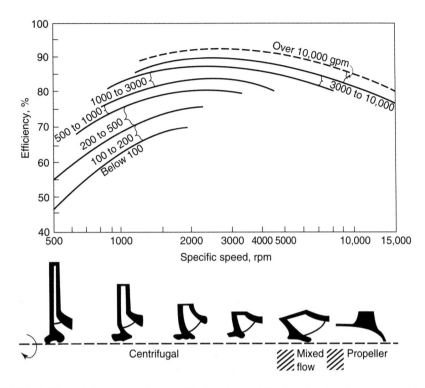

FIGURE 5.1 The relation of specific speed to the shape of the impeller, the discharge, and the efficiency of the pump. (Courtesy of Worthington Pump, Inc.)

diameter pipe is cheaper to install, the cost of the power required to overcome the larger friction losses throughout the life of the installation may be too high; reducing the cost by increasing the pipe size may be warranted. The designer should strive to attain a judicious economical balance between these expenditures.

The *speed of the impeller* is expressed in rpm, meaning rotations per minute. Although some pump motors are built such that the speed can be varied, this is a somewhat expensive and rare case. For most pumps standard electric motors are used. Standard speeds of AC synchronous induction motors at 60 cycles and 220 to 440 V are 3600, 1800, 1200, 900, 720, 600, and 514 rpm, depending on the number of poles. For a 50-cycle operation these speeds reduce to 3000, 1500, 1000 rpm, and so on. Therefore, the selection of the motor will determine the specific speed of the pump to a certain degree.

When selecting the *motor* for a pump and designing the wiring, fuses, and switching devices, it is important to know that pumps need more power for starting than during continuous operation. Because the full-load speed of a regular electric motor is reduced by about 3 to 5 percent, the requirement of power increases considerably.

The *power required* by a pump may be computed from the formula

$$P = \gamma QH \tag{5.2}$$

where Q is the discharge in cubic feet per second, γ is the unit weight of the fluid in pounds per cubic foot, and H is the total dynamic head in feet. The dimension of P power is then foot-pounds per second. From engineering mechanics we know that power may be expressed in terms of torque T and angular velocity ω

$$P = T\omega \tag{5.3}$$

where T is mass flow rate $\times$ radius $\times$ velocity. The angular velocity of the impeller can be computed from

$$\omega = \frac{2\pi}{60} \, \text{rpm} \tag{5.4}$$

in radians per second.

Power is commonly expressed in terms of horsepower, HP, which, in conventional American units, as in Equation 5.2, requires that we divide the right side of Equation 5.2 by 550, that is,

$$\text{water HP} = \frac{\gamma QH}{550} \tag{5.5}$$

which is called *water horsepower* because this is the power required by a pump to move water. (Recall from Chapter 2 that 1 HP = 550 ft-lb/sec.) Equation 2.19 may be used for metric units. Because pumps never perform at 100 percent efficiency, the water horsepower must be exceeded by a factor dependent on the *efficiency* of

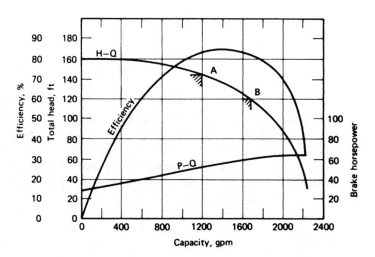

FIGURE 5.2 Typical pump performance graph. (From Hicks and Edwards, *Pump Application Engineering*. Copyright 1971, McGraw-Hill Book Company. Used with permission of McGraw-Hill Book Company)

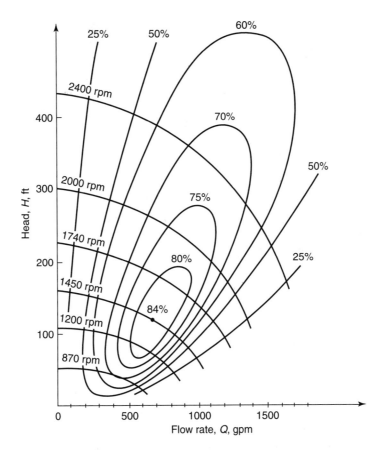

FIGURE 5.3 Typical characteristic curves of a centrifugal pump operated at different speeds. Contour lines show constant efficiencies.

the operation. The power needed by the motor is called *brake horsepower* and may be computed as follows:

$$\text{brake HP} = \frac{\text{water HP}}{\text{efficiency}} \qquad (5.6)$$

The efficiency of a pump varies with Q and H. The value is included in the manufacturer's performance graphs available in pump catalogs, an example of which is shown in Figure 5.2. Three graphs are shown here. One relates the total pumping head and the discharge of the pump. Points A and B compare two operating conditions: A at 145 ft pumping head and B at 120 ft. As the head is reduced the discharge increases from 1200 to 1680 gpm. The P-Q curve relates the brake horsepower to the discharge. The efficiency curve shows that this particular pump operates with a peak efficiency of 85 percent when the discharge is pumped at 1400 gpm.

FIGURE 5.4 Industrial pumping installation. Electric motor is on the right. Flexible coupling is covered with bent metal sheet. Note heavy concrete pedestal to reduce vibrations. (Courtesy of Ingersoll Rand, Inc.)

For each centrifugal pump there is a certain speed of revolution at which the pump operates with the highest efficiency. For all other turning speeds the efficiency is less. Figure 5.3 shows an example of the changing of the head, discharge, and efficiency characteristics as a function of rotating speeds. In the case shown, the peak efficiency of 84 percent occurs with a rotating speed of 1450 rpm. The pump at this speed would deliver 680 gpm against a pumping head of 120 ft.

The relative curvature of the pump characteristic lines—the relationship between head and discharge—is quite different for different types of pumps. Axial flow pumps, which have high specific speeds, generally have steep characteristic curves. For radial flow centrifugal pumps, which have low specific speeds, the characteristic curves are usually flat. With flat characteristic curves small changes of discharge result in small changes of head. With steep characteristic curves small changes of discharge may bring about large changes of head. This is a disadvantage of axial flow pumps when they are operated at less than peak efficiency. During start-up, for example, their power needs may be much larger than under efficient, normal operating conditions. To minimize vibration during pump operation, pumps are installed on heavy concrete pedestals. An example of a typical pump installation is shown in Figure 5.4.

Pump selection in terms of the power required for a given discharge and dynamic head may be made by using the chart shown in Figure 5.5. It shows the various

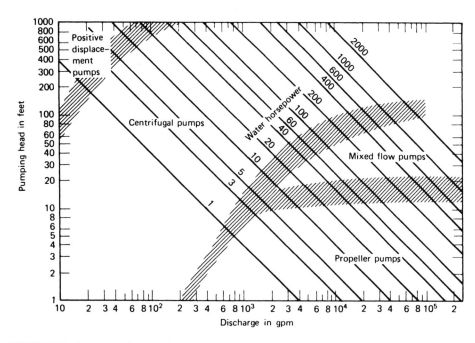

FIGURE 5.5   Power requirement for various discharges and heads of different types of pumps at optimum efficiency. (Courtesy of Fairbanks, Morse & Co.)

ranges in terms of discharge and head within which the different types of pumps are appropriate. The shaded areas are indeterminate zones.

5.4   PUMP SELECTION AND ALTERATION

In general hydraulic design work, the selection of the proper type and size of pump and motor is very often left to the pump manufacturer. Catalogs from major manufacturers are available on request. Although these catalogs include complete data on the hydraulic performance criteria of pumps such as the graph shown in Figure 5.2, most manufacturers supply a questionnaire on which the design requirements of a pump may be listed. From the data provided, the manufacturer will advise the designer about the most suitable models. Some pumps are often already available to the industrial designer in salvage yards or in other storage facilities. The work may involve changes in existing plants because of changing pumping needs, and in such cases the old pumps are at hand. All attempts should be made to convert existing pumps, since pumps are expensive.

Pump performance may be changed either by changing the impeller or by changing or rewiring the motor or both. To change the pump performance characteristics, certain basic laws, called *affinity laws*, valid for all centrifugal pumps can be applied.

Changing the impeller diameter D in the pump results in changes of Q, H, and P according to the relations

$$\frac{Q_1}{Q_2} = \left(\frac{D_1}{D_2}\right)$$

$$\frac{H_1}{H_2} = \left(\frac{D_1}{D_2}\right)^2 \qquad\qquad (5.7)$$

$$\frac{P_1}{P_2} = \left(\frac{D_1}{D_2}\right)^3$$

Subscripts 1 and 2 refer to values of the parameters before and after the change, respectively. In Equation 5.7 it is assumed that $N_1 = N_2$ and that the fluid remains the same. N_1 and N_2 refer to the rpm of the motor before and after the change.

Changing the speed of the impeller brings about similar changes in discharge, dynamic head, and power according to the following relationships:

$$\frac{Q_1}{Q_2} = \frac{N_1}{N_2}$$

$$\frac{H_1}{H_2} = \left(\frac{N_1}{N_2}\right)^2 \qquad\qquad (5.8)$$

$$\frac{P_1}{P_2} = \left(\frac{N_1}{N_2}\right)^3$$

It is assumed in these equations that the impeller diameter and the fluid remain the same. Equation 5.8 is valid only if the efficiency does not change markedly with the change of speed. If the ratio of the two speeds is less than 2 to 1, the error is usually small.

5.5   THE ALLOWABLE SUCTION HEAD

An important point in the design of pumping installations is the elevation of the pump over the water level in the sump or reservoir from which the water is taken. The water in the suction line is in tension. The pressure is therefore lower than the atmospheric pressure. Adding to this already lowered pressure is the energy loss between reservoir and pump because of local losses and pipeline friction. The pressure is reduced even more because part of the energy at the pump is used in the form of kinetic energy. This is particularly true if there are high velocities in the pump casing around the impellers. This latter effect is related to the specific speed of the pump. Adding the elevation of the pump, the kinetic energy head, and the friction losses in the suction pipe of a pump, one obtains the *total suction head* H_s. If this total suction head corresponds to a pressure reduction in the pump that is equal to or less than the vapor pressure, the water will change into vapor. This phenomenon, called *cavitation*, was introduced in Chapter 1. More than half

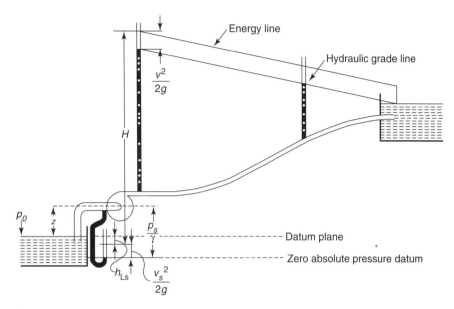

FIGURE 5.6   Notation for the derivation of the cavitation parameter.

of the troubles experienced with centrifugal pumps can be traced to the suction side, and many of these problems involve cavitation. If the water vaporizes in the pump, small vapor bubbles form at the suction passages and at the impeller inlet. These bubbles collapse when they reach the region of high pressure. These collapses may occur with such violence that damage to the metal can result. Successive bubbles break up with significant impact force, causing local high stresses on the metal surfaces that pit them along the grain boundaries of the metal of the casing and at the tips of the impeller. The presence of cavitation is easily recognized; the vibration and noise make the pump sound as if it were full of gravel. Cavitation brings about a significant drop in efficiency and a subsequent mechanical failure of the pump because of the cavitational erosion of the casing and the impeller and fatigue failure of the seals and shaft.

The allowable suction head in a given pump is the highest elevation above the downstream water level at which the pump will operate without a notable loss of efficiency due to cavitation. This height is expressed in terms of the total head H that the pump is required to deliver multiplied by a factor of proportionality σ called the *cavitation parameter*. The value of σ is determined by the manufacturer on the basis of the following simple test. The pump is set up on a pedestal of adjustable elevation between two reservoirs. The total head H delivered by the pump is composed of the elevation difference between the reservoir levels plus all energy losses, as shown in Figure 5.6. After the brake horsepower and the discharge of the pump at various pump inlet elevations z are measured, the efficiency of the pump may be calculated by Equations 5.2 and 5.6. As the pump's pedestal is raised beyond a certain elevation the efficiency will begin to drop, indicating the onset of

TABLE 5.1   Variation of pressure, temperature, and boiling point with elevation (U.S. standard atmosphere)

Elev. ft msl	Pressure			Air temp °F	Boiling point °F
	In. of mercury	mb	Ft of water		
−1,000	31.02	1050.5	35.12	62.6	213.8
0	29.92	1013.2	33.87	59.0	212.0
1,000	28.86	977.3	32.67	55.4	210.2
2,000	27.82	942.1	31.50	51.8	208.4
3,000	26.81	907.9	30.35	48.4	206.5
4,000	25.84	875.0	29.25	44.8	204.7
5,000	24.89	842.9	28.18	41.2	202.9
6,000	23.98	812.1	27.15	37.6	201.1
7,000	23.09	781.9	26.14	34.0	199.2
8,000	22.22	752.5	25.16	30.6	197.4
9,000	21.38	724.0	24.20	27.0	195.6
10,000	20.58	696.9	23.30	23.4	193.7
11,000	19.79	670.2	22.40	19.8	191.9
12,000	19.03	644.4	21.54	16.2	190.1
13,000	18.29	619.4	20.71	12.6	188.2
14,000	17.57	595.0	19.89	9.1	186.4

The data of this table are based on average conditions and must be adjusted to actual meteorological conditions for a specific problem.

cavitation. This means that p_0, the absolute pressure at the pump inlet, has been reduced to vapor pressure p_v as shown in Table 1.1. Table 5.1 shows the barometric pressure at various elevations over mean sea level. The table also lists the corresponding average air temperature and the ambient boiling pressure of water. Using the notations of Figure 5.6 and writing the energy equation between the surface of the downstream reservoir and the pump's intake, we may derive the formula for the cavitation parameter:

$$\sigma = \left(\frac{p_0 - p_v}{\gamma} - z - h_{LS} \right) \frac{1}{H}$$

$$= \left(\frac{p_s - p_v}{\gamma} + \frac{v_s^2}{2g} \right) \frac{1}{H}$$

(5.9)

where p_0 is the ambient atmospheric pressure at the water surface in terms of absolute pressure, p_v is the vapor pressure from Table 1.1, h_{LS} is the sum of the energy losses in the suction pipe, p_s is the gage pressure measured in absolute terms at the centerline of the intake of the pump, and v_s is the velocity at the intake.

The terms in parentheses in Equation 5.9 are the *net positive suction head*, or NPSH.[1] H is the total dynamic head. Values of the σ cavitation parameter range from 0.05 for a specific speed of 1000 to 1.0 for a specific speed of 8000. The value

---

[1]Such notations were quite characteristic of early American hydraulic formulas.

of σ is usually furnished by the manufacturer on the basis of tests. For pumps of high specific speed, that is, low heads with large discharge capacity, the allowable net positive suction head may be less than zero. This indicates that propeller pumps, for example, should be installed well below the water surface to eliminate possible cavitation. In these instances the pump needs to be of the vertical shaft type so that the motor can be installed at an elevation above any possible flood levels. Cavitation may also occur if the pump is not operating at or near the optimum point of its efficiency curve. An incorrectly selected pump will often show up by cavitating. Pumps will also cavitate if the atmospheric pressure was incorrectly assumed in design. At higher elevations pumps are known to cavitate when an oncoming warm weather front causes the vapor pressure to rise.

Before a centrifugal pump is operated, it should be first primed, unless the pump is self-priming. *Priming* means to fill the pump with water so that the impeller can create suction. Foot valves serve the purpose of keeping the water in the pump between periodic operations, but often they leak. Priming requires that there be valves before and after the pump. These are to be closed before priming and starting the pump. After the pump is started, the inlet valve is always opened first. After a short period the outlet valve may be slowly opened. A closed outlet valve does not harm a centrifugal pump. With the outlet valve closed the pump pressure is increased by about 15 to 30 percent. The power load on the motor when the pump outlet is closed is reduced by about 50 to 60 percent.

5.6 TROUBLESHOOTING

Pump manufacture has reached such a level of sophistication that a pump is expected to give troublefree service for a long period of time.

Troubles may arise from improper design or installation or from poor maintenance. Some common problems and their probable causes are listed.
If no or not enough water is delivered:

> Pump is not primed.
> Speed is too low; check wiring.
> Discharge head is too high.
> Suction head is higher than allowed.
> Impeller is plugged up.
> Impeller rotates in the wrong direction.
> Intake is clogged.
> Air leak is in intake pipe.
> Mechanical trouble (seals, impeller, etc.).
> Foot valve is too small.
> Suction pipe end is not submerged enough.

Not enough pressure:

> Air is getting into the water through leak in suction pipe.
> Impeller diameter is too small.

Speed is too low.
The valve setting is incorrect.
Impeller is damaged.
Packing in casing is defective.

Erratic action:

Leaks are in suction line.
The shaft is misaligned.
Air is in water.

Pump uses too much power:

Speed is too high.
Poorly selected pump.
Water is too cold.
Mechanical defects.

Vibration and noise:

Cavitation.
Motor is out of balance.
Bearings are worn.
Propeller is out of balance (blade damaged).
Suction pipe is picking up air through vortex action in sump.
Water hammer is in the piping system.

Noise due to cavitation can be eliminated by allowing some air to enter the suction pipe; however, this will not eliminate cavitation.

5.7 PUMPS IN PARALLEL OR SERIES

In pumping stations the discharge and head requirements may fluctuate considerably in time. At optimum efficiency the output of a single pump may fluctuate only within a rather narrow limit. Hence, for fluctuating needs it is advantageous to install several pumps in a pumping station. Pumps installed together may operate in series or in parallel. In either case the individual performance characteristics of the pumps must be carefully matched to attain the best overall efficiency. It is advantageous to install pumps of identical size because they will be matched best from a hydraulic standpoint. Also, from a practical standpoint the cost of stocking spare parts will be less, and interchangeability of the various components will facilitate repair and maintenance. To eliminate downtime for repair, an additional pump of similar size may also be added to the system. It is put on line when any one of the other pumps needs repair or maintenance.

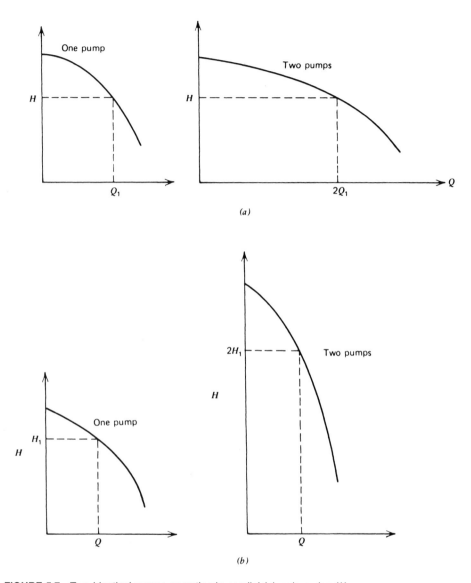

FIGURE 5.7   Two identical pumps operating in parallel (*a*) or in series (*b*).

When two identical pumps are installed together in series, as shown in Figure 5.7, the total output of the two is the same as the discharge of a single pump, but the output pressure is doubled. When connected in parallel, the total delivered discharge of two identical pumps is twice that of a single pump, but the output pressure remains the same as that of the single pump. These conditions are true only if the pumps operating together discharge into the atmosphere. When delivering water into a piping system that offers frictional resistances, two pumps op-

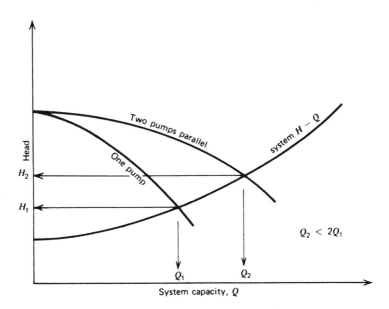

FIGURE 5.8   Operating characteristics of a piping system with one pump or with two pumps in parallel.

erating in parallel will encounter greater resistance to flow because their combined output results in an increase of velocity in the pipe. Hence, they will have a different operating point than when they operate alone in the same piping system. Figure 5.8 illustrates this concept. In this figure the head-versus-discharge curve is plotted for the piping system. The operating points for the single and parallel operation of two pumps are also shown. The intersection of these lines indicates the operating points. As indicated in the graph, the joint discharge of two pumps in parallel is less than twice the discharge of a single pump. Similarly, two pumps pumping in series will not double the head in the whole system.

The joint operation of any arrangement of pumps can be analyzed using these basic concepts if the frictional resistances of the piping system are known and the characteristic curves of the pumps are available. The output of the whole system under various operating conditions can then be determined by graphical analysis.

5.8 WATER POWER GENERATION

Water is an alternative source of electric power. In the United States only about 5 percent of the electric energy is produced from hydropower, and less than a quarter of the potential capacity is utilized. In contrast, in Switzerland and Norway, for example, the contribution of hydraulic energy to the total produced is about 99 percent. Many European countries generate over 50 percent of the power for their electric needs from water. Although hydropower facilities are expensive to

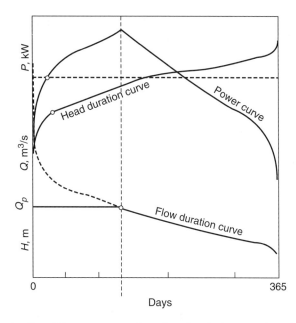

FIGURE 5.9 Construction of the power curve based on the discharge and head duration for an average year. Q_p is the rated capacity of the turbines installed.

build, once the plants are in operation, hydraulic power, which is a form of solar power, is free. Water power plants are 90 percent efficient, whereas power plants that use fossil fuel lose more than 60 percent of the energy as waste heat.

The power that can be generated from water is expressed by Equation 5.2. It shows that the same amount of power can be produced from a small quantity of water falling over a large distance as from a large discharge falling a small distance. The amount of power available at any site is determined by constructing a power curve, shown in Figure 5.9. By analyzing the daily flows over a period of time, one can construct daily discharge and head curves for an average year (see also Chapter 7). In this chart, flood discharges occurring during a small percentage of an average year's time are shown on the left, and the lowest discharge in an average year is at the right, of the *flow duration curve*. The maximum discharge capacity of the turbines, Q_p, limits the power utilization at the site, holding part of the flow duration curve constant at that level. Installing greater turbine capacity may allow the utilization of more power, but at a higher installation cost. Also shown is the *head duration curve* representing the available *drop* of head at the power plant during various parts of an average year. One must keep in mind that during floods the downstream water level is higher than under low flow conditions, so even though the discharge is high, the power generation capacity may be low. Toward the left of the head duration curve it drops, representing some days during the year when there is insufficient flow to maintain the water level necessary for power generation. This limit is marked on the graph with a dot. The product of the head, discharge values,

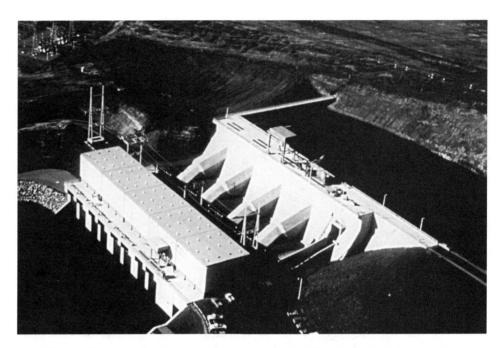

FIGURE 5.10 Thermolito (California) hydraulic power station. (Courtesy of California Department of Water Resources)

and γ gives the *power curve*. The utilizable hydropower is the portion above the dot on the power curve corresponding to the minimum utilizable head. During part of the year the utilizable power is greater than the power marked with the dotted line; during the rest of the year it is less. Power station design, particularly the selection of turbine capacity, is a matter of economy. One of the fundamental questions to be answered is whether the facility is to supply a minimum desired amount of power on a virtually permanent basis or whether it is to utilize the occasionally available power peaks.

The water from the reservoir enters the generating turbines either directly or through tunnels, steel pipes called *penstocks* at high dams, or power canals at low dams. Figure 5.10 shows a typical hydropower station behind a concrete dam, with the penstocks clearly shown.

Hydraulic power is converted into mechanical power by turbines. The turbines drive electric generators through a rather complex coupler mechanism to maintain constant turning speed in the generators. A typical set of hydropower generators, from Shasta Dam in California, is shown in Figure 5.11. There are three major designs: the Pelton, Francis, and Kaplan turbines. In high-head applications the appropriate turbine is the Pelton wheel,[2] shown in one application in Figure 5.12.

---

[2]Named after Lester Allen Pelton (United States, 1829–1908), who developed it in California for powering stamp mills and related machinery.

FIGURE 5.11   Hydroelectric generators of Shasta Dam. (Courtesy of U.S. Army Corps of Engineers, Sacramento, CA)

One or more high-velocity jets are aimed at small buckets on the wheel that are designed to efficiently convert the impulse force of the jet into momentum. In low-head applications, with so-called run of the river dams, and when the discharge is

FIGURE 5.12   A typical high-head hydropower installation using Pelton wheels. Note the jet control device to the right of the wheel.

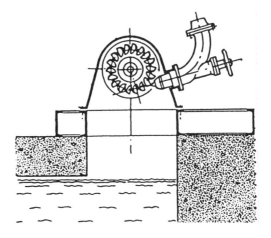

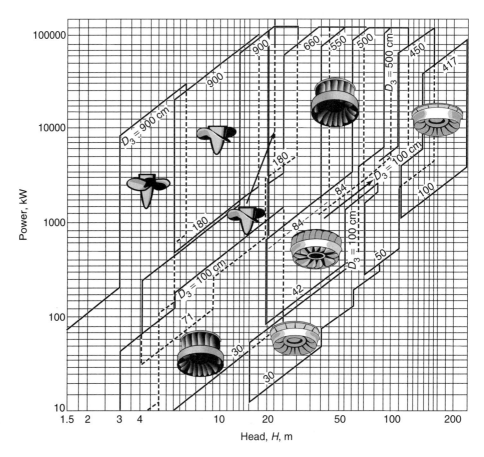

FIGURE 5.13 Chart to select the type and size of turbine for various power and head conditions. (After V. S. Kvatkovsky)

relatively small and constant, Francis (or reaction) turbines[3] are used. These are somewhat similar in design to mixed-flow centrifugal pumps. Their design resembles the shape of the fan in the common forced-air furnace. A few typical Francis wheels are shown on the right side of Figure 5.13. This chart aids in the selection of the appropriate turbine type for a given power and available head.

Kaplan turbines[4] are essentially propeller wheels with the pitch of the propellers adjustable through the shaft. A typical design is shown in Figure 5.14. The design details show the adjustable blades as well as the adjustable gates in the inlets. The

---

[3]James Bicheno Francis (England, 1815–1892) immigrated to the United States at an early age and worked as chief engineer for a group of manufacturers in Lowell, Massachusetts. He improved the design of a turbine originally conceived by others by optimizing the entrance and exit conditions.

[4]The adjustable-blade turbine was developed by Victor Kaplan (Germany, 1876–1934).

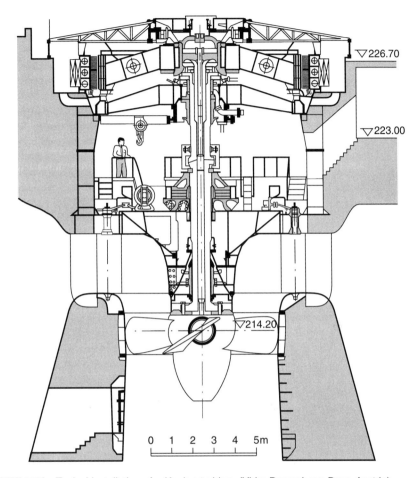

FIGURE 5.14   Typical installation of a Kaplan turbine. (Ybbs-Persenbeug Dam, Austria)

turning speed of the generator on the top can be made constant by controlling both the pitch of the turbine blades and the inlet gates. Figure 5.15 shows a cross section of Egypt's Aswan Dam through one of the turbines, indicating the various elements of a typical hydropower installation. Kaplan turbines are also placed with horizontal or slanted shafts in modern "bulb-type" installations. Figure 5.16 shows a bulb-type hydropower plant, on the Rhone River in France. This power plant utilizes tidal energy as well as the run of the river. The turbines in tidal plants operate in both directions, depending on the water levels inside and outside the inland basin.

Because turbines are connected to electric generators, they must maintain a constant rotational speed. Under varying power needs the amount of discharge allowed to flow through, or the pitch of the propellers, is changed. Just like pumps,

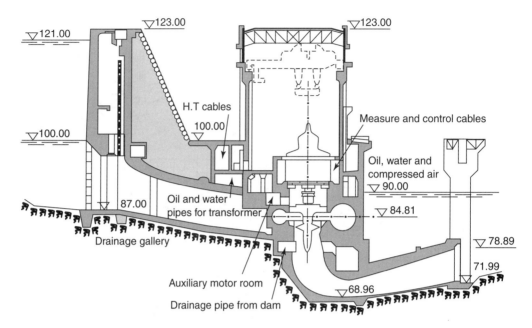

FIGURE 5.15   The powerhouse of the Aswan Dam, Egypt.

turbines of identical shape and different size have identical specific speeds. The specific speed formula for turbines is different from that for pumps. It is expressed as

$$n_s = \frac{NP^{1/2}}{H^{5/4}} \tag{5.10}$$

where N is the rotational speed at peak efficiency, in rpm, and P is the power generated. The generating power of a water turbine can be expressed by the formula

$$P = \frac{KD^2}{\sqrt{H}} \tag{5.11}$$

where K is a constant of proportionality. Because the power is proportional to the square of the turbine diameter, the latter is the most important variable in selecting turbines for various applications. With the outer diameter denoted by D_3, Figure 5.13 shows the recommended turbine types and sizes for different heads and power capacities. For the preliminary design of the size of inlet and outlet structures Emil Mosonyi[5] presents general guidelines, illustrated in Figure 5.17. The dimensions refer to the runner diameter of a Kaplan turbine. This is useful information for preliminary design of a hydraulic power station because turbines are designed and built by a few worldwide specialist companies, not by the designers of dams.

---

[5]In his 2200 page book *Water Power Development*, Vols. 1 and 2, Budapest: Hungarian Academy of Sciences Press, 1963.

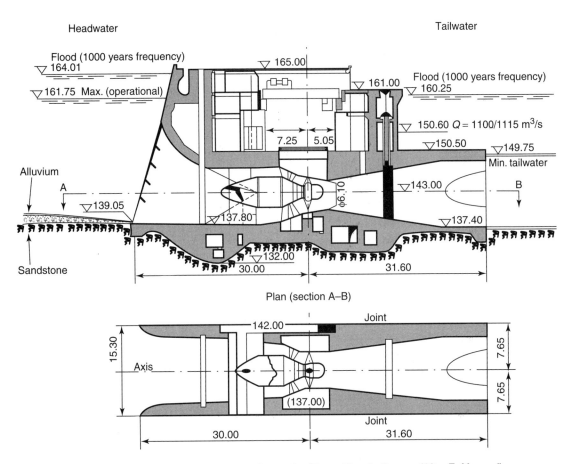

FIGURE 5.16 A modern bulb-type hydropower station on the Rhone River in France. (After E. Mosonyi)

In addition to conventional high-head and low-head power generation plants on rivers, there are various other types of hydropower plants. One, for instance, is the pumped storage plant, in which water is pumped to higher elevation during periods of low electric demand and released back through the turbines during high demand.[6] Another type is the tidal power plant placed in bays and river estuaries in regions of high tidal fluctuation. An example of such a plant is shown in Figure 5.16. The detailed review of design concepts for water power plants is outside the scope of this text.

EXAMPLE PROBLEMS

Example 5.1   Select the type of pump that is best suited for lifting 10,000 gpm over an 8 ft tall dike when the available motor to drive the pump operates at 500 rpm.

---

[6]One example is in operation at Niagara Falls, New York.

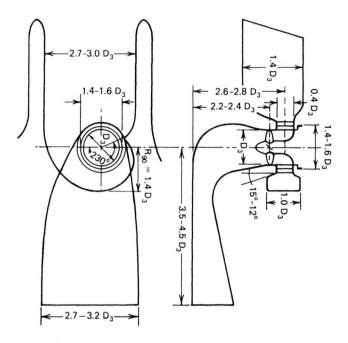

FIGURE 5.17 Approximate dimensions for preliminary design of a Kaplan turbine installation. (After E. Mosonyi: Water Power Development, Hungarian Academy of Sciences, Budapest, 1965)

Solution. To determine the optimum performance we first calculate the specific speed by substituting the data into Equation 5.1:

$$n_s = \frac{\text{rpm}\sqrt{\text{gpm}}}{H^{3/4}}$$

$$= \frac{500\sqrt{10,000}}{8^{3/4}} = 10,500$$

From Figure 5.1 we find that we need a propeller-type pump that has a probable efficiency of 81 percent. This pump type could also have been determined from Figure 5.5. □

Example 5.2

A flow of 500 gpm is to be pumped to a height of 60 ft, overcoming all losses through a 156 ft long cast-iron pipe that includes an open globe valve (Figure E5.2). Assuming that the motor runs at 3600 rpm, determine the pump size (Q, H, n_s) for a 6 and an 8 in. diameter pipeline.

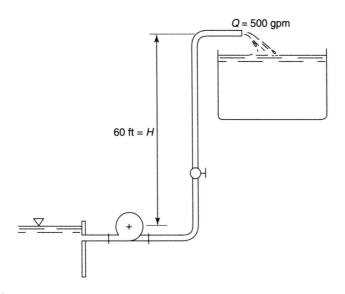

FIGURE E5.2

Solution.

$$500 \text{ gpm} = 1.1 \text{ cfs}$$

$$\text{average velocity} = \frac{Q}{A} = v$$

For a 6 in. pipe,

$$\text{average velocity} = \frac{1.1}{(\pi/4)(0.5)^2} = 5.6 \text{ fps}$$

For an 8 in. pipe,

$$\text{average velocity} = \frac{1.1}{(\pi/4)(8/12)^2} = 3.2 \text{ fps}$$

The total head loss equals friction loss in the pipe plus friction loss in the valve plus entrance loss plus exit loss.

We consider a 6 in. pipe and assume that the water temperature is 65°F. From the properties of water at 65°F

$$\mu = 2.2 \times 10^{-5} \text{ lbf sec/ft}^2$$

$$\rho = 1.94 \text{ slugs/ft}^3$$

$$\nu = \frac{\mu}{\rho}$$

$$= 1.1 \times 10^{-5} \frac{\text{ft}^2}{\text{sec}}$$

From Table 4.2 the roughness of the pipe (cast iron) = 0.00085 ft.

$$\frac{e}{D} = \frac{0.00085}{0.5} = 0.0017 \qquad \text{for a 6 in. pipe}$$

$$= \frac{0.00085}{8/12} = 0.00128 \qquad \text{for an 8 in. pipe}$$

Next, we calculate the Reynolds number from $\mathbf{R} = vD/\nu$.

$$\mathbf{R}_{6\,\text{in.}\,D} = \frac{5.6 \times 0.5}{1.1 \times 10^{-5}} = 2.5 \times 10^5 \qquad \text{for flow in a 6 in. pipe}$$

$$\mathbf{R}_{8\,\text{in.}\,D} = \frac{3.2 \times 8/12}{1.1 \times 10^{-5}} = 1.9 \times 10^5 \qquad \text{for flow in an 8 in. pipe}$$

By entering the Moody diagram, we get the following:

For a 6 in. D pipe condition $e/D = 0.0017$ and $\mathbf{R} = 2.5 \times 10^5$, we obtain $f = 0.023$.
For an 8 in. D pipe condition $e/D = 0.0013$ and $\mathbf{R} = 1.9 \times 10^5$, we obtain $f = 0.022$.
Next, we calculate the friction loss in the pipe, h_L.

$$h_L = f\frac{L}{D}\frac{v^2}{2g}$$

For a 6 in. pipe 156 ft long,

$$h_L = 0.023\left(\frac{156}{0.5}\right)\left(\frac{5.6^2}{2g}\right) = 3.5 \text{ ft}$$

and for an 8 in. pipe 156 ft long,

$$h_L = 0.022\left(\frac{156}{8/12}\right)\left(\frac{3.2^2}{2g}\right) = 0.8 \text{ ft}$$

The minor losses equal the loss in the valve plus entrance and exit losses, which equal

$$k_1\frac{v^2}{2g} + k_2\frac{v^2}{2g} + k_3\frac{v^2}{2g}$$

or

$$(k_1 + k_2 + k_3)\frac{v^2}{2g}$$

where

$$k_1 \text{ for the globe valve} = 10$$
$$k_2 \text{ for entrance loss} = 0.5$$
$$k_3 \text{ for exit loss} = 1$$

Then, minor losses are

$$(10 + 0.5 + 1)\left(\frac{5.6^2}{2g}\right) = 5.6 \text{ ft} \qquad \text{in the 6 in. pipe}$$

$$(10 + 0.5 + 1)\left(\frac{3.2^2}{2g}\right) = 1.8 \text{ ft} \qquad \text{in the 8 in. pipe}$$

Therefore, the total loss in a 6 in. pipe equals friction loss plus minor loss:

$$H_6 = 3.5 + 5.6 = 9.1 \text{ ft}$$

The total loss in an 8 in. pipe is

$$H_8 = 0.8 + 1.8 = 2.6 \text{ ft}$$

Then the total dynamic head is

$$H_d = 60 + 9.1 + \frac{5.6^2}{2g} = 69.6 \text{ ft} \qquad \text{for a 6 in. pipeline}$$

$$= 60 + 2.6 + \frac{3.2^2}{2g} = 62.8 \text{ ft} \qquad \text{for an 8 in. pipeline}$$

Let us now consider pump selection. From Equation 5.1 we determine the specific speed,

$$n_s = \frac{\text{rpm}\sqrt{\text{gpm}}}{H^{3/4}}$$

We get for a 6 in. pump,

$$n_s = \frac{3600\sqrt{500}}{(69.6)^{3/4}}$$

$$= 3340$$

and for an 8 in. pump,

$$n_s = \frac{3600\sqrt{500}}{(62.8)^{3/4}}$$

$$= 3610$$

From Figure 5.1 for n_s = 3340 and 3610 rpm, we note that the same kind of pump will suffice for both applications: a high specific speed radial flow pump with an approximate peak efficiency of 80 percent. Hence, we need a 6 in. pump having Q = 500 gpm, H = 69.6 ft, and n_s = 3340 rpm, or an 8 in. pump having Q = 500 gpm, H = 62.8 ft, and n_s = 3610 rpm. ☐

Example 5.3   A pump characterized by known performance data (see Figure 5.2) is operated by a 50 HP electric motor. Determine the discharge and dynamic head delivered and the efficiency of the operation.

Solution. From the P-Q curve shown in Figure 5.2 the discharge for a 50 HP motor is

$$Q = 1050 \text{ gpm}$$

Then from the efficiency curve at Q = 1050 gpm

$$e = 81\%$$

and from the H-Q curve at Q = 1050 gpm

$$H = 150 \text{ ft}$$ ☐

Example 5.4

Determine the electric power requirement of a pump pumping 5000 gpm with a total dynamic head of 30 ft if the overall efficiency is 80 percent.

Solution. The pump discharge is

$$5000 \times 2.227 \times 10^{-3}\ \text{cfs} = 11.14\ \text{cfs}$$

Thus, the power used by the pump from Equation 5.2 is

$$P = \gamma QH\ (\text{ft-lb/sec})$$

where

$$\gamma = 62.4\ \text{lb/ft}^3$$
$$Q = 11.14\ \text{cfs}$$
$$H = 30\ \text{ft}$$

Therefore,

$$P = 62.4(11.14)(30) = 20{,}854\ \text{ft-lb/sec}$$

Then water horsepower can be determined from Equation 5.5:

$$\text{water HP} = \frac{\gamma QH}{550} = \frac{20{,}854}{550} = 37.9$$

The overall efficiency of the pump is 80 percent; therefore, the brake horsepower can be computed from Equation 5.6:

$$\text{brake HP} = \frac{\text{water HP}}{\text{efficiency}}$$

$$= \frac{37.9}{0.8} = 47.4$$

Because 1 HP = 0.7457 kW,

$$47.4\ \text{HP} = (0.7457)(47.4) = 35.3\ \text{kW}$$

This is the electric power required to drive the motor of the pump and discharge 5000 gpm with a total dynamic head of 30 ft of water. □

Example 5.5

A pump delivers 500 gpm at 70 ft dynamic head when run at 3600 rpm. Determine the change in operating characteristics if the pump's impeller is reduced from 6 to 4 in.

Solution. By Equation 5.7, with $Q_1 = 500$ gpm and $H_1 = 70$ ft,

$$\frac{500}{Q_2} = \frac{6}{4}$$

$$Q_2 = 333\ \text{gpm}$$

$$\frac{70}{H_2} = \left(\frac{6}{4}\right)^2 = 2.25$$

$$H_2 = 31\ \text{ft}$$

The required power will change to

$$P_2 = \frac{P_1}{(D_1/D_2)^3} = 0.30\, P_1$$

From Equation 5.2

$$P_1 = \gamma Q_1 H_1 = 62.4(500)(2.23 \times 10^{-3})70 = 4.9 \times 10^3 \text{ ft-lb/sec}$$

Therefore,

$$P_2 = 0.30(4.9 \times 10^3) = 1.5 \times 10^3 \text{ ft-lb/sec} \qquad \square$$

Example 5.6   Determine the pressure in feet of water that must be available on the suction side of a pump if it operates at sea level, at 20°C, and with a total dynamic head of 140 ft. The cavitation parameter of the pump is 0.25, the velocity of flow in the inlet pipe is 4.0 fps, and the inlet has a strainer bucket with a foot valve. Consider the local loss due to the strainer bucket; other losses are negligible.

Solution. From the definition of the cavitation parameter σ, Equation 5.9,

$$\sigma = \frac{\dfrac{p_{atm}}{\gamma} - \dfrac{p_{vapor}}{\gamma} - z - h_{LS}}{H}$$

where

$$\gamma = \text{unit weight of water to be pumped}$$
$$= 62.4 \text{ lb/ft}^3$$
$$H = \text{total dynamic head} = 140 \text{ ft}$$
$$p_{atm} = 14.7 \times 144$$
$$= 2116.8 \,\frac{\text{lb}}{\text{ft}^2} \qquad \text{at sea level}$$
$$p_{vapor} = 0.34 \times 144 \qquad \text{(Table 1.1)}$$
$$= 49.0 \,\frac{\text{lb}}{\text{ft}^2}$$
$$h_{LS} = k_1 \frac{v^2}{2g} = 10\,\frac{4^2}{64.4} = 2.5 \text{ ft}$$
$$(k_1 = 10, \text{Table 4.4, Case 6})$$
$$\sigma = 0.25$$

Therefore,

$$0.25 = \frac{\dfrac{2116.8 - 49.0}{62.4} - z - 2.5}{140}$$

$$z = -4.4 \text{ ft}$$

Hence, the suction side of the pump must be at least 4.4 ft below water. $\square$

Example 5.7

Two identical pumps operating in parallel discharge into a pipeline where the head loss is $h_L = 2Q^2$. The performance characteristics of one pump are described in the following table. The system supplies water from a lower reservoir to another reservoir whose water level is 50 ft higher. Determine the discharge of the system with the two pumps. How much does the discharge increase by adding the second pump? Neglect local losses.

Table of Computations for Example 5.7

H ft	Q gpm	Q cfs	$2Q$ cfs	System Q cfs
100	—	—	—	5.00
90	—	—	—	4.47
80	500	1.11	2.23	3.87
70	990	2.20	4.41	3.16
60	1260	2.81	5.61	2.24
50	1460	3.25	6.50	0.00

Solution. The dynamic head of the system is $H = H_{elev} + h_L = 50 + 2Q^2$ (in ft, cfs). This function is tabulated in the table above. Two identical pumps operating in parallel deliver twice the discharge of one pump if losses due to pipe friction are neglected. The actual discharge for the two pumps is found at the intersection of the curve for two pumps in parallel without friction and the curve showing the system H-Q relationship. From Figure E5.7 this discharge is 3.56 cfs. With only one pump the discharge would be 2.65 cfs. Thus, adding the second pump increases the discharge by 34 percent.

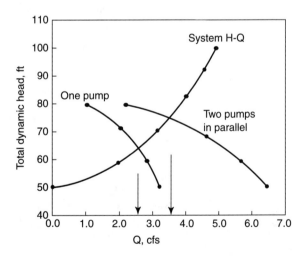

FIGURE E5.7 ☐

Example 5.8

The first three columns of data in the following table give the average discharges, their corresponding heads, and their duration for a hydroelectric plant at a reservoir.

Graph the power curve for the generators if the turbine capacity is 577.7 m³/s. What is the maximum power available? What power is available, on average, for half a year?

Table of Computations for Example 5.8

1	2	3	4	5
Q	Head		Usable Q	Usable P
m³/s	m	% Time	m³/s	MW
803.1	9.6	1.0	577.7	54.4
699.0	9.8	10.4	577.7	55.5
610.0	9.5	22.0	577.7	53.8
502.2	9.8	30.5	502.2	48.3
398.9	10.7	42.0	398.9	41.9
301.1	11.4	61.7	301.1	33.7
204.0	11.9	82.3	204.0	23.8
94.2	13.0	100.0	94.2	12.0

Solution. Because the turbine capacity is 577.7 m³/s, not all the discharge is available for power generation. Column 4 gives the usable portion of the discharge. Multiplying the product of columns 2 and 4 by 9810 N/m³ and dividing by 1×10^6 to change watts into MW (Equation 2.20), we find the usable power. These values are given in column 5. The power curve, Figure E5.8, is a graph of the values in column 5 versus those in column 3. From this figure we find that the maximum power available is 55.5 MW. The power available for half a year is 38.0 MW.

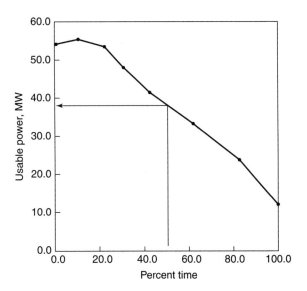

FIGURE E5.8   ☐

PROBLEMS

5.1 Calculate the specific speed of a pump that is operated by a 2000 rpm electric motor and delivers 150 gpm against a dynamic head of 45 ft. What type of pump would you select, and what is its estimated efficiency?

5.2   Calculate the torque required to operate the pump described in Problem 5.1. Neglect losses due to efficiency.

5.3   State the dynamic head, delivered discharge, efficiency, and brake horsepower for both operating conditions, A and B, shown in Figure 5.2.

5.4   A pump with characteristics shown in Figure 5.3 operates at 1450 rpm and delivers 685 gpm at a 120 ft head. If the rotational speed is changed to 1740 rpm and the discharge remains the same, determine the changes in head and in efficiency.

5.5   Determine the electric power requirement (in kW) of a pump operating at a discharge of 1000 gpm with a total dynamic head of 210 ft, a speed of 2000 rpm, and an efficiency of 78 percent.

5.6   For the pump with performance characteristics shown in Figure 5.2 delivering 1660 gpm (case B), determine the changes caused by changing the impeller from $D_1 = 6$ in. to $D_2 = 4.5$ in.

5.7   What will be the new torque requirement of the pump described in Problems 5.1 and 5.2 if the driving motor is changed to run at 3600 rpm? What will happen to the discharge and head?

5.8   The cavitation parameter of a pump is 0.85 according to the manufacturer's data. Determine the allowable suction head for the conditions defined in Problem 1.2 if the total dynamic head is 56 ft and the temperature is 20.0°C.

5.9   The suction head of a pump having a specific speed of 3000 rpm is 6 ft. The total dynamic head is 40 ft, which includes 15 ft due to frictional losses in the piping system. It is found that at optimum operational conditions cavitation occurs. Recommend various remedial actions.

5.10 Select a pump for a discharge of 6.0 cfs in an 8 in. diameter pipe with a total energy input equivalent to 60.6 ft of head. State the specific speed, pump type, pump efficiency, and brake horsepower for the pump with a speed of 1800 rpm.

5.11 Each of two pumps, identical to the one for which the characteristic curves are given in Figure 5.2, can deliver a pressure equivalent to 100 ft of water. Assuming that the pumps discharge into the atmosphere, what will be their combined discharge and head if they are connected in parallel? in series?

5.12 The two pumps described in Problem 5.11 are connected in series and deliver water into a horizontal 6 in. diameter cast-iron pipe that is 2000 ft long. At the end of the pipe the water discharges into a tank where the water level is 20 ft above the pipe. Determine the discharge.

5.13 Repeat Problem 5.12 for the pumps connected in parallel to a 12-in. pipe.

5.14 Repeat Example 5.8 if the turbine capacity is increased to 650 m³/s. In Example 5.8 the maximum power available was 55.5 MW. With the increased turbine capacity, for how much of the year can this level of power be supplied?

MULTIPLE CHOICE QUESTIONS

5.15 Pelton wheels are used in power generation when high heads and low discharges are encountered.
   A. True.
   B. False.

5.16 The "specific speed" of pumps involves three variables:
   A. Specific weight, rotating speed, pump elevation.
   B. Fluid density, discharge, vapor pressure.
   C. Rotating speed, dynamic head, cavitation parameter.
   D. Dynamic head, discharge, fluid density.
   E. Dynamic head, rotating speed, discharge.

5.17 A centrifugal pump has a relatively thin impeller with a large radius. What set of conditions best characterizes it?
   A. Large specific speed and discharge, small head.
   B. Large specific speed and head, small discharge.
   C. Small specific speed, discharge, and head.
   D. Small head and specific speed, large discharge.
   E. Small discharge and specific speed, large head.

5.18 What kind of pump is typically used for a flood control or irrigation application in terms of specific speed?
   A. Medium speed (2000–4000 rpm).
   B. High speed (10,000–15,000 rpm).
   C. Low speed (500–2000 rpm).

5.19 The exit diameter of a centrifugal pump is 4 in. Estimate its discharge capacity.
   A. 25 gpm.
   B. 50 gpm.
   C. 150 gpm.
   D. 400 gpm.
   E. 750 gpm.

5.20 A centrifugal pump has 75 percent efficiency when pumping water at 1.5 cfs against a 150 ft dynamic head. Determine the horsepower required.
   A. 34 HP.
   B. 5 HP.
   C. 50 HP.
   D. 45 HP.
   E. 150 HP.

5.21 A pump is turned at 1200 rpm. What is its angular speed?
   A. 60 rad/sec.
   B. 160 rad/sec.
   C. 125 rad/sec.
   D. 650 rad/sec.
   E. None of the above.

5.22 The size of the impeller in a centrifugal pump is reduced by half. By what ratio will the power requirement change?
   A. 1/2.
   B. 1/4.
   C. 2.
   D. 4.
   E. 1/8.

5.23 The diameter of the impeller in a centrifugal pump is reduced by 10 percent. By what ratio will the discharge be changed?
   A. 0.75.
   B. 1.23.

C. 1.45.

D. 2.5.

E. None of the above.

5.24 If the turning speed of a pump is increased by 25 percent, how will the discharge be affected, assuming no change in efficiency?

A. It will not be affected.

B. It will increase by the same proportion.

C. It will decrease by the same proportion.

D. It will decrease by a factor of 0.64.

E. It will grow by a factor of 1.56.

5.25 The cavitation coefficient for centrifugal pumps is

A. The ratio of the suction head to the head loss.

B. Larger at Seattle than at Denver.

C. In reference to the elevation where the pump's efficiency begins to drop.

D. Between 1 and 2.0.

E. None of the above.

5.26 If a pump does not deliver enough pressure,

A. There may be a leak in the pressure pipe.

B. The suction head may be too high.

C. The impeller is damaged.

D. The speed may be too low.

E. Both C and D.

5.27 When two pumps are installed in a system in parallel

A. The pressure will increase.

B. The system head-discharge curve will be changed.

C. Cavitation may occur.

D. The discharge will be less than twice that of one pump.

E. None of the above.

5.28 Affinity laws for centrifugal pumps

A. Refer to changes in vapor pressure.

B. Depend on the specific speed.

C. Relate to impeller or speed variations.

D. Consider their cavitation parameters.

E. None of the above.

5.29 Mixed-flow pumps are not true centrifugal pumps.

A. True.

B. False.

5.30 If the brake power requirement of a pump is 20,000 ft-lb/sec, and its turning speed is 125 rad/sec, how much torque is carried by the pump shaft?

A. 160 ft-lb.

B. 25 ft-lb.

C. 2500 ft-lb.

D. 125,000 ft-lb.

E. None of the above.

6

Seepage

The movement of water through the pores of soils often controls the safe and proper performance of earth works and hydraulic structures. Fundamental concepts of seepage are explained in this chapter. Methods of solution are presented for steady flow in saturated two-dimensional, confined flow fields as well as those with free surfaces. The hydraulic stability of soils subjected to seepage is explained. Fundamental formulas of well hydraulics are introduced. Examples of seepage in partially saturated and unsteady conditions are discussed in the closing of this chapter.

6.1 INTRODUCTION

The flow of water in soil pores is usually an imperceptible process. For this reason the analysis of seepage is often relegated to secondary importance in the field of hydraulics. Because the subject is inherently related to the study of soils, the concepts of groundwater flow are also considered to a certain extent in soil mechanics and hydrogeology. As is usual in such cases of common domain, seepage is rather inadequately covered in both hydraulics and soil mechanics books—not that the problems arising from seepage are unimportant. In fact, the opposite is true. Often the effects of seepage determine whether a structure will stand or fail. Failures caused by seepage are often the costliest in terms of life and property. A collapsing trench during excavation or a failing dam or levee during floods all too frequently dominates the headlines of newspapers. Many of these problems as well as the less spectacular ones—landslides on slopes and flooding basements, for example—could

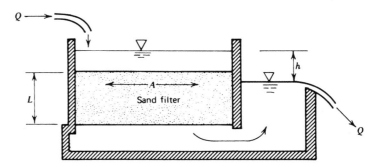

FIGURE 6.1 Darcy's sand filter experiment.

have been avoided by a little attention to the concepts of seepage. This, of course could come only with a fundamental understanding of these concepts.

6.2 PERMEABILITY AND SEEPAGE PRESSURE

In 1856 Henri Philibert Gaspard Darcy (France, 1803–1858), a water supply expert, explained the hydraulic behavior of sand filters used in the treatment of water. He postulated that the discharge Q of a sand filter of A surface and L thickness is proportional to the energy loss Δh that takes place across the filter. Figure 6.1 shows the arrangement considered by Darcy. The resulting formula is

$$Q = K \frac{\Delta h}{L} A \tag{6.1}$$

in which K, a constant of proportionality, was subsequently generalized for all cases of flow through porous materials in the form of

$$v = K \frac{\Delta h}{\Delta l} \tag{6.2}$$

which is referred to as Darcy's law of seepage.[1] This formula says that the velocity v at any point in the permeable material is proportional to the rate of loss in hydraulic energy at that point. Actually, this is not really a velocity but a *flux*. The true seepage velocity, as will be explained later in connection with Equation 6.23, depends on the effective porosity n_e of the soil.

---

[1]Interestingly, Equation 6.1 was hidden as a footnote in Darcy's voluminous book on the design and construction of municipal water supply systems.

TABLE 6.1   Darcy's Coefficient of Permeability

Soil Type	Average Grain Size mm	Range of K Coefficient cm/s	Order of Magnitude of K cm/s
Clean gravel	4–7	2.5–4.0	10
Fine gravel	2–4	1.0–3.5	1
Coarse, clean sand	0.5	0.01–1.0	10^{-1}
Mixed sand	0.1–0.3	0.005–0.01	10^{-2}
Fine sand	0.1	0.001–0.05	10^{-3}
Silty sand	0.02–0.1	0.0001–0.002	10^{-4}
Silt	0.002–0.02	0.00001–0.0005	10^{-5}
Clay	0.002	10^{-9}–10^{-6}	10^{-7}

Darcy's constant of proportionality, K, is called the *coefficient of permeability*[2] and has the dimension of velocity. Later it was determined that K represents several physical parameters relating to both the fluid and the porous solid. In fact,

$$K = k \frac{\gamma}{\mu} \qquad (6.3)$$

where k, called *intrinsic permeability*, depends only on the sizes, shapes, and other geometric properties of the pores and minute channels in the soil, with units of length squared. The variables γ and μ are the unit weight and dynamic viscosity of the fluid, respectively. Because the value of k for soils is difficult to measure and because in hydraulics we are almost always dealing with water of relatively uniform temperature, Equation 6.3 is used only to consider the effect of temperature variations on otherwise known conditions of seepage. It happens in rare instances in hydraulics, for instance, in the case of declining yield from shallow wells during cold winter months. In petroleum engineering, in contrast, the importance of temperature is paramount.

Typical values of Darcy's coefficient of permeability in different types of soils are shown in Table 6.1. The right column of this table shows that for different types of soils the order of magnitude of K varies by factors of 10. This means that the determination of the proper value of the permeability of a soil is at best an uncertain business. Laboratory evaluation of the K coefficients by using samples obtained from bore holes is standard practice. In spite of its obvious scientific appearance, this is a largely worthless operation. Seepage in most cases occurs in a horizontal direction; soil samples are usually tested for vertical flow. For geologic reasons most soil layers are far more permeable horizontally than vertically. Drilling and sampling of soils result in serious disturbances of the soil. Sand and silt are greatly loosened in the process, and clay is compacted. Furthermore, since seepage occurs in huge masses of soil, samples are only minute parts of the whole mass. Claims

---

[2]Another name often used in the literature for K is *hydraulic conductivity*. The term is much more common in hydrogeology, less common in soil mechanics.

that these samples are representative of the whole are hardly convincing. By the nature of the formation of sands and gravels during the geologic processes one can expect considerable variations in permeability from point to point. Even pumping tests, another common way to arrive at a design value for permeability, give uncertain results. One thin gravel seam undetected during drilling can supply many times the discharge of the rest of the layers. The conclusion is that although all efforts must be made to obtain knowledge about the soils through which seepage will take place, the average value and variations of the permeability coefficient must be used with considerable suspicion until the project demonstrates its actual magnitude. For preliminary calculations, values such as those in Table 6.1 may be sufficient, unless the magnitude of the project is such that the selection of permeability coefficients as well as the responsibility for such decisions can be placed on the shoulders of a reputable testing laboratory.

In the field of hydrogeology one often finds the expression *transmissivity* (T) in place of permeability. It is the product of the permeability and the thickness of the aquifer and represents the capability of the aquifer to discharge water. The dimension of T is length squared per time, such as m²/day.

Although the topic of seepage immediately conjures up applications where the question of discharge is important—in problems of dewatering, wells, and seepage losses from canals or reservoirs, for example—there are a far greater number of cases in practice where it is the least important. Seepage is neglected almost always when the flow of water is not obvious; in these cases damage results at a later time. For example, consider clay saturated with flowing water. Because of clay's very low permeability, flow of water to the surface is less than the rate of evaporation there; the clay surface may appear dry, and the existence of seepage is disregarded. To look at the problem more closely, let us express the pressure gradient I by rewriting Equation 6.2 in the form

$$I = \frac{\Delta h}{\Delta l} = \frac{v}{K} \tag{6.4}$$

in which the right side appears to represent the dimensionless ratio of two velocities. The v in this equation is the Darcy velocity, and K is in units of velocity but it expresses the coefficient of permeability. For clay the flow velocity is small because the permeability is also small. The ratio of two small numbers could be as large as or greater than that of two large numbers, even though for clays v, the actual velocity of the groundwater, is so small that it may go unobserved. On the exposed soil surface the rate of evaporation could well exceed the rate of flow, making the surface appear dry.

The left side of Equation 6.4 represents the rate of loss of hydraulic energy along the paths of the seeping water. In a trench cut into saturated clay (Figure 6.2), the flow can be assumed to be nearly horizontal.

Taking a small cubic element of $(\Delta l)^3$ size in the clay mass shown on the right of the trench in Figure 6.2, we can recognize that the hydraulic energy lost from the seepage of the water through the element is Δh. This hydraulic energy is in fact transferred by viscous shear from the fluid to the soil. It manifests itself by an

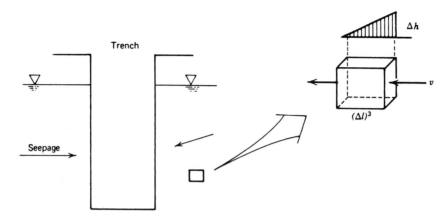

FIGURE 6.2   Seepage force in the case of flow into a trench.

amount of pressure $\Delta p = \gamma \, \Delta h$ acting in the direction of the flow, distributed evenly in the element along the walls of its pores. The *seepage pressure* just described exerts a force on the soil at all points within the flow field. At the boundary of the flow field, along the sides of the trench shown in Figure 6.2, the seepage force acting on the soil tends to force the soil into the trench. This force is resisted by the internal strength of the soil, which for clay is its cohesion. Should the force of seepage be large enough to exceed the value of cohesion, the trench may fail from the effect of the seepage, although perhaps no water is observed in the trench.

The *seepage force*, from the viewpoint of the soil mass, is a body force, similar to its weight. Its magnitude for an elementary cube of soil is obtained by multiplying the energy gradient $\Delta h / \Delta l$ acting on the cube by the unit weight of water and by the volume of the elementary cube. The direction of the seepage force is the same as the velocity of the water.

6.3 SEEPAGE ANALYSIS

In the previous section the hydraulic energy term Δh was dealt with in its elemental form. To solve actual problems encountered in practice such elemental notation is insufficient. One has to know the actual values to perform required computations. Herein lies the difficulty that makes the hydraulics of groundwater flow a science in itself. To solve seepage problems one has to be able to define the physical boundaries within which seepage takes place. Once these boundaries are well defined, the values of the hydraulic energy must be known where water enters and exits the flow field.

In the analysis of seepage, problems may be grouped as steady or unsteady flows. Because the velocity encountered in seepage is often imperceptibly slow, most solution methods assume that the flow is steady. Furthermore, problems are defined according to the flow boundaries. If all boundaries are fixed, we speak of

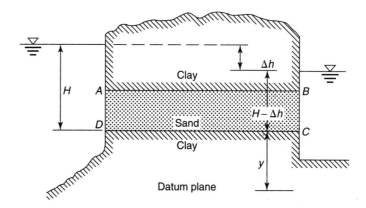

FIGURE 6.3 Seepage through a rectangular flow field.

confined seepage. If part of the boundary is a free surface, we have an *unconfined seepage* problem. Free surface problems present an additional degree of complexity. These will be discussed in the next section of this chapter. In the present section we deal with steady seepage within confined, fixed boundaries.

By the law of conservation of energy, we know that energy at inflow minus energy at outflow equals the energy transferred to the soil mass in the flow field. By Bernoulli's equation the hydraulic energy in seepage is composed of only two parts: pressure energy (p/γ) and elevation energy (y). The third term, the kinetic energy ($v^2/2g$), is not considered because the velocity values encountered in seepage are usually very small. Therefore, for seepage the hydraulic energy at any point is

$$H = \frac{p}{\gamma} + y \qquad (6.5)$$

where p is the pressure in the water, γ is the unit weight, and y is the elevation above a conveniently selected reference level. Let us consider the two-dimensional flow field shown in Figure 6.3. It represents the seepage through a horizontal, confined layer of sand. The layer connects two independent bodies of water. If there is a difference of elevation between the two bodies of water, there will be seepage through the sand layer. The flow will be subject to Darcy's law, embodied in Equation 6.1. Rearranging to separate the terms relating to the geometric properties leads to

$$\frac{Q}{K\,\Delta h} = \frac{A}{L} \qquad (6.6)$$

where the left side is called *relative discharge*, and the right side represents the particular geometry of the flow field. Obviously the larger the cross-sectional area, the greater the flow, and the longer the flow path, the less the flow. The right side of Equation 6.6 can be generalized into a new term called the *form factor*, Φ. In a two-dimensional case, Equation 6.6 may be written as

$$\frac{Q}{K \, \Delta h} = \frac{T}{L} = \Phi \qquad (6.7)$$

where T is the width of the flow channel. In the case shown in Figure 6.3 the magnitude of T equals the distance between points A and D. The discharge Q in this two-dimensional case refers to that of the layer of unit depth.

We now return to the consideration of potential values on the flow field shown in Figure 6.3. Assuming that the water level at the left side is higher than on the right, the seepage will occur from left to right. Interpreting Equation 6.5 along the entrance, between points A and D we find that it is constant. As we go from D to A, the pressure becomes less as the elevation grows. This potential energy can be taken arbitrarily as being equal to H, as shown in Figure 6.3. The potential energy along the surface B to C will also be constant, with a value of $H - \Delta h$. We may use Bernoulli's equation, with subscript 1 representing the conditions at the entrance, and subscript 2 those along the exit:

$$\frac{p_1}{\gamma} + y_1 = \frac{p_2}{\gamma} + y_2 + h_L \qquad (6.8)$$

$$H = (H - \Delta h) + h_L$$

In the second line we substituted the potential values defined previously. Thus, we find that the energy loss, h_L, equals Δh. This means that in seepage all available external energy is fully utilized.

The shortest path among all possible paths of flow from point A to point B will be the straight line along the boundary. The same is true for the flow between D and C. Using the shortest available paths will require the least energy utilization. From the minimum energy theorem it also follows that the water particles will travel in pure translation, without rotation. From the standpoint of practical work, this may be an innocuous concept. However, it allows the utilization of a wide field of higher mathematics in the solution of seepage problems.[3] Because the energy available will be fully utilized, this means that the discharge delivered through the flow field will be optimized. The form factor Φ in the flow field shown in Figure 6.3 will be the ratio of the distances between points A and D, and B and C. In two-dimensional cases this will be a dimensionless term.

The concept of the form factor provides distinct advantages for the solution of seepage problems, particularly for cases when two-dimensional representation is possible. Two-dimensional representation is valid in all cases where the flow field does not change regardless of where it may be considered. Seepage through levees,

---

[3]The conditions of continuity and irrotationality give rise to the Laplace equation, the fundamental partial differential equation for seepage as well as electrical conduction, magnetic, and thermal conduction field problems.

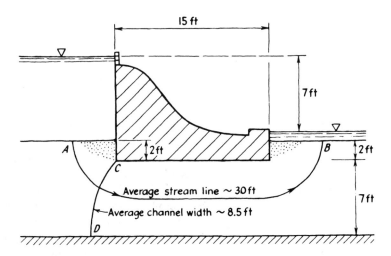

FIGURE 6.4   Example for the approximate determination of the form factor.

under long dams, and into channels are the most typical of these problems. In two-dimensional cases more complicated than the simple rectangle considered in Figure 6.3 the determination of the form factor gets more complicated. For simple solutions one may consider T as the "average" depth and L as the "average" length of the flow field. Reasonably good approximate solutions for such seepage problems can be made very quickly by approximating these average measures of the flow field and dividing them to obtain the form factor. One example is shown in Figure 6.4. The form factor here is $\Phi = 8.5/30$, or 0.28.

Another interpretation of the form factor may be made by simple scaling. Using a selected scale, meter or foot, we may determine the magnitude of T in terms of L as

$$T = n \cdot L \tag{6.9}$$

where n represents the ratio of the two terms. In this format the discharge flowing can be expressed as

$$Q = \frac{K\Delta h}{n} \tag{6.10}$$

Generally speaking, complete solution of a seepage problem involves the determination of the hydraulic potential values at all points of the flow field. The lines connecting all points having equal potential are called *equipotential lines*. The family of equipotential lines that are usually shown and labeled represent each 10 percent or 25, 50, and 75 percent of the total energy, $H_{in} - H_{out}$, that causes the flow. Seepage cannot take place without using up energy, which means that the velocity vector cannot have a component parallel with the equipotential line. This means that the velocity vectors must be perpendicular to equipotential lines everywhere. Perpendicular to these equipotential lines are the *stream lines* of the flow. Stream lines are defined as lines that are tangential to the direction of the velocity of seepage at all points of the flow field. A plot showing equipotential and stream lines of a flow field is called a *flow net*. The boundaries of flow fields are either stream lines or potential lines.[4]

The family of stream lines that are usually shown and labeled usually divide the total seepage discharge into percentages, for example, each 10 percent, or 25, 50, and 75 percent of Q, the total discharge passing through the flow field. Similarly, equipotential lines shown are percentages of the total energy used within the field, the difference between inlet and outlet energies.

Flow nets are graphically constructed such that pairs of adjacent equipotential lines and stream lines form a curvilinear square. If the flow is curved, these squares are distorted but retain their nature in the sense that at their corners the lines still cross perpendicularly. In each such square a circle may be drawn such that it touches on each of the four sides. Once a flow net is drawn correctly, the form factor may be computed by dividing the number of its stream lines by the number of its equipotential lines.

One of the simplest methods for solving two-dimensional seepage problems is the drawing of flow nets. The graphical construction of flow nets usually starts with the drawing of a few main stream lines (or potential lines) within the boundaries of the flow field. These, of course, are only rough approximations. Following this, the other family of lines are drawn, forming a curvilinear square net. One must remember that the boundary lines are also parts of the net. Invariably, the first trials result in incorrectly shaped "squares" or imperfect (nonperpendicular) intersections of lines. Hence, the graphical process involves frequent erasures in parts of the flow net, until an approximately correct plot is produced. One soon learns that even small local changes may cause significant variations in the whole plot.

One of the most frequent mistakes made by beginners is to attempt to make the net too fine, drawing too many small squares. Starting out with more than two or three stream lines in the beginning leads to this problem. A convenient method is to draw the fixed flow boundaries on an $8\frac{1}{2} \times 11$ in. paper, then to insert it in a transparent plastic cover (like Copco PS-5, for example). The trial flow net is then drawn on the plastic sheet, where it can be erased and corrected with ease. Once the flow net is accepted as correct, the stream lines and potential lines are numbered, each starting with zero at a boundary. The form factor of the flow field is then calculated by dividing the number of stream lines by the number of potential lines. With Φ at hand, the discharge flowing through the flow field can be computed by Equation 6.7. The flow field also allows the determination of pressures along the boundaries or inside the flow field by Equation 6.5.

There are many other ways to determine the form factor in a more precise manner for flow fields of different shapes, particularly for two-dimensional conditions. The most notable methods include *finite element and finite difference methods*

---

[4]There is one exception. In the case of free surface flow, discussed next, under the downstream end of the free surface the boundary is neither. It is called the *surface of seepage.*

of numerical analysis, and *analytic methods* that use conformal transformation.[5] In the past *analog solutions* involving electric[6] and viscous flow analogies were also used. Most of these methods are very cumbersome, complex, and theoretical, and far beyond the scope of this book. Texts listed in the bibliography will provide ample guidance to those who are interested in these particular topics.

The so-called relaxation method,[7] a type of finite difference method, is based on the fact that in a potential field the magnitude of the value of the potential or stream function at one point is the average of the values of equidistant nearby points. If the flow field is described on a rectangular grid, the potential or stream values can first be arbitrarily assumed at each internal grid point. At the boundaries of the flow field there are two possible cases: Either the potential or stream function is known, or it is an unknown depending on the flow field. If it is known, then it is inserted as a constant. If it is unknown, then the value of the function can be described as the average of the four adjacent points inside the flow field.

Once the values of the function are initially prescribed at all gridpoints, the procedure is to recompute one by one each value by replacing it with the average of the adjacent values. Fixed boundary values remain fixed during the computation. The process is repeated until the difference between previous and recalculated values become small. Once the function is determined at all grid points, equipotential or stream lines may be drawn. Before computers, this was a tedious, longhand calculation.

For relatively simple two-dimensional flow fields classical mathematics offers solution methods using *conformal transformation*. This is a complicated field of advanced mathematics involving complex numbers.[8] Fortunately, one does not have to be familiar with the mathematics because, for most practical cases, solutions already exist in the extensive literature on seepage.[9] For example, in the case of a dam with a sheet pile underneath, the general solution, based on conformal transformation, is shown in Figure 6.5. This figure provides solutions for two extreme cases also. One is when the sheet pile stands alone, in which case *b* in Figure 6.5 is set to zero. The other case is when there is no sheet pile; that is, *s* is set to zero.

The *stability* of a hydraulic structure encountering seepage underneath may be endangered by *uplift* as the water exerts hydrostatic pressure at its base. The magnitude of this uplift pressure may be computed at any point of the base of the structure by Equation 6.5, in which *y* is the elevation of the point above the reference level, and *H* is the hydraulic potential as determined by the potential values of the

---

[5]First introduced into the field of fluid mechanics in 1919 by the same Riabouchinsky who invented dimensional analysis, presented in Chapter 3. In 1919 he escaped the Russian Revolution and later became the associate director of the fluid mechanics laboratory at the University of Paris.

[6]Based on the similarity between Darcy's law and Ohm's law of electric resistance.

[7]This is akin to the Hardy Cross method introduced for pipe network analysis.

[8]The mathematical procedure is called Schwarz-Christoffel mapping.

[9]An outstanding collection of such solutions was developed and published by Pelagela Jakovlevna Polubarinova-Kochina (Russia, 1899–1980) which was translated into English.

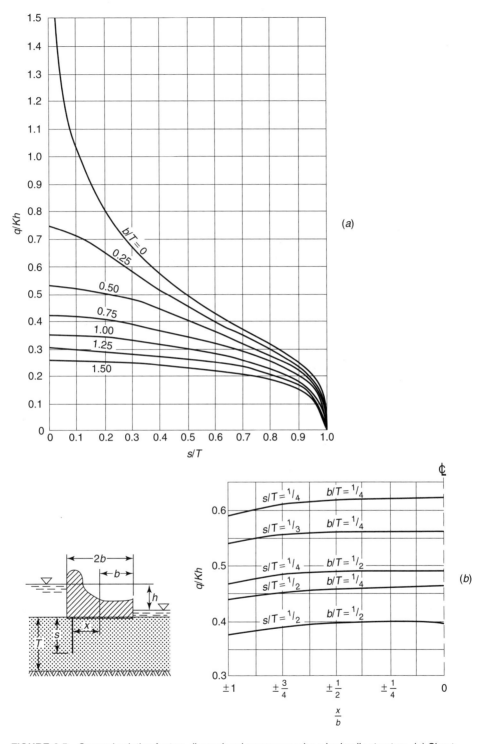

FIGURE 6.5 General solution for two-dimensional seepage under a hydraulic structure. (*a*) Sheet pile at center, $x = 0$. (*b*) Correction for an offset sheet pile, $x \neq 0$. (From M. E. Harr, *Groundwater and Seepage*. McGraw-Hill Book Company © 1962. Used with permission of McGraw-Hill Book Co.)

flow net boundary along the structure. The sum of all these hydraulic uplift pressures acting on the base of a structure is the *uplift force*. The magnitude and location of this force is important information: It determines the weight of a hydraulic structure required to counteract it. The uplift force decreases along the stream line representing the structure's boundary as the water seeping downstream loses energy due to loss of head. By placing sheet piles protruding below the structure, the boundary stream line is lengthened. This reduces the rate of energy loss along this critical stream line, reduces the uplift force, and decreases the potential of erosion under the structure by the flowing groundwater.

A concern regarding the stability of a dam is the potential of erosion at the downstream side caused by seepage. The most critical location for this is along the exit boundary of the flow field. If the seepage force is large enough to pick up and carry away particles of the soil, erosion may occur. Often, the critical location is marked by the smallest square of the flow net at the boundary, where the rate of change of the seepage pressure is the greatest. Soil erosion is often encountered in cohesionless granular soils subjected to seepage forces. When erosion occurs it is called *piping*.[10] The photograph in Figure 6.6 shows such boils at the downstream side of a dam. If erosion of the grains does not occur but all strength of the soil is lost because of seepage forces present, it is referred to as a *quick condition* (e.g., quicksand). In case of upward flow along a horizontal boundary of the field of seepage, the seepage force is resisted by the weight of the soil particles within the soil mass. Because the soil particles are submerged in water, their actual weight is their net buoyant weight, which is the dry unit weight of the soil less the unit weight of the water. For most granular soils the dry unit weight is about twice that of water. Hence, their net buoyant weight is about the same as water's. On this basis we now derive the hydraulic stability of a horizontal soil surface through which seepage takes place in an upward direction. Writing the difference of the net buoyant weight of the soil and the seepage force for a small cube of Δl size on the surface as

$$\text{net stabilizing weight} = \text{net buoyant weight} - \text{seepage force}$$

$$= (\gamma_{\text{submerged soil}} - \gamma_{\text{water}})\, \Delta l^3 \frac{\Delta h}{\Delta l} \tag{6.11}$$

where the last term is the energy gradient, which expresses the difference of potential values across the cube. By rearranging and substituting Equation 6.4 and replacing the dry unit weight of the soil with twice the unit weight of water, we obtain

$$\frac{v}{K} - 1 = \frac{\text{net stabilizing weight}}{\gamma_{\text{water}}(\Delta l)^3} \tag{6.12}$$

The critical condition is reached when the soil on the surface loses its net stabilizing weight—that is, the soil particles subjected to the upward seepage force appear to

---

[10]The term *suffosion* is also used for this phenomenon. It refers to the washing away of smaller particles from the soil matrix.

FIGURE 6.6   Piping behind leaky earthen dam. Notice 'boiling' caused by fine-grained sediment carried by the escaping water. (Courtesy of Ohio Department of Natural Resources, Division of Water)

be weightless in the water. When the net stabilizing weight is zero, the discharge velocity v should be equal to the permeability of the soil, that is

$$v_{\text{critical}} = K \tag{6.13}$$

Under this condition Equation 6.4 tells us that if the *exit gradient* (the rate of change of the potential value at the exit boundary of the flow field) equals unity, then the condition becomes critical. A relevant problem, often found in construction practice, is the case of an excavation in a river or lake, surrounded by sheet piles. The water is removed from the excavation area by pumps. As a result, the water flows around the sheet pile through the soil. The shortest path for the water is along the sheet pile. Because of this, the critical point for the exit gradient is at point E in Figure 6.7. For a two-dimensional case, such as a long trench protected by parallel sheet piles, mathematical solutions exist, as shown in Figure 6.7. This plot gives the exit gradient, I_E, for the flow under various conditions of excavation depths, sheet pile depths, and other geometric parameters.

In construction operations the presence of groundwater often requires pumping. Although this is sometimes considered a nuisance, the flow of groundwater provides a certain safety factor. Figure 6.8 shows two cases of identical geometry. In Figure

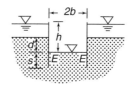

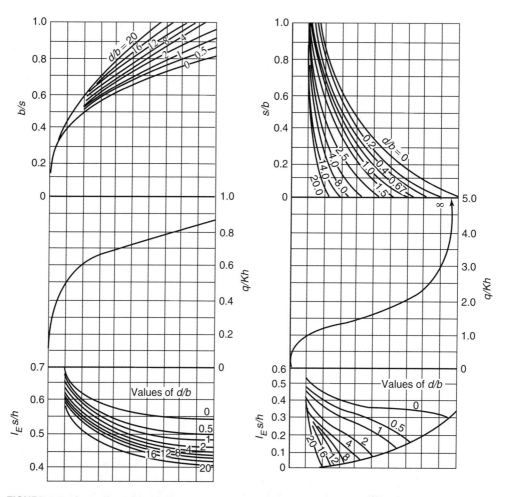

FIGURE 6.7 General solution for discharge and hydraulic stability for two-dimensional seepage into an excavation protected by sheet piles. (From M. E. Harr, *Groundwater and Seepage.* Copyright 1962, McGraw-Hill Book Company. Used with permission of McGraw-Hill Book Company)

6.8*a* the impervious shale layer is left untouched. Because there is no way for the water to flow upward into the excavation, there is no loss of pressure. As a result, the water pressure is allowed to build up to the level of the outside water elevation. This condition has caused catastrophic failures in excavation.

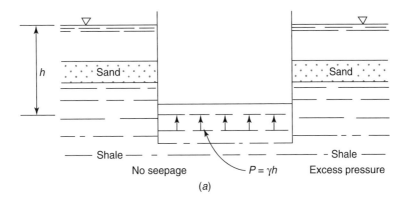

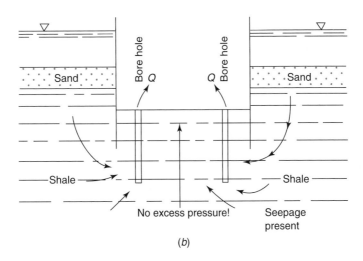

FIGURE 6.8 Hydraulic stability with an undrained (*a*) and drained (*b*) impervious layer at the bottom of an excavation.

In Figure 6.8*b* deep holes are drilled into the shale, allowing the water to flow out of the layer. Although this condition necessitates continuous pumping, the water pressure is reduced through seepage losses, a factor that prevents excessive pressure from building up under the excavation.

6.4 FREE SURFACE PROBLEMS

For all cases considered so far the flow field was predetermined because the uppermost flow line was always bounded by an impervious boundary; therefore, all parts of the flow moved at pressures larger than atmospheric pressure. Such problems are called confined flow problems. But in many practical problems there exists a free surface bounding the flow field from the top. These are the unconfined flow

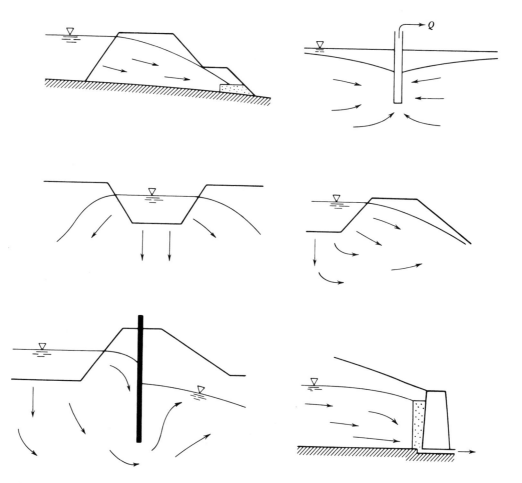

FIGURE 6.9 Examples of unconfined seepage.

cases. Some examples of unconfined flow are shown in Figure 6.9. Unconfined flow problems are even more difficult to solve than those involving confined flow. The problem is that the location of the free surface is unknown; therefore, the shape of the flow field, the form factor, is uncertain. In essence, there are two problems to be solved that are mutually dependent on each other. The location of the free surface depends on the shape of the flow field and vice versa. To solve such problems, first one has to define what conditions must exist along the free surface. There are two fundamental boundary conditions to be satisfied: First, the free surface is a stream line; hence, it must be perpendicular to the equipotential lines at all points. Second, the free surface is exposed to the atmosphere; hence, the potential value, Equation 6.5, must equal the elevation, y. The inclusion of these two conditions in analytic, numerical, or graphical solution methods allows the determination of free surface bounded flow fields in much the same way as in confined flow problems.

Graphical construction of flow nets for two-dimensional free surface problems is an iterative process. It starts with the drawing of a trial free surface, which then defines the boundaries of a trial flow field. A flow net is then constructed as already described. Checking whether the trial free surface was drawn correctly is based on the second free surface boundary condition, because the trial surface was initially selected as a stream line. The pressure must be zero on the free surface, that is, values of the potential lines must be equal to the elevation of the free surface. If this condition is not satisfied by the trial free surface, a new one must be drawn for which the process is repeated.

There are many analytic solutions for free surface problems to be found in the scientific literature. The hydraulic designer should use these as a general guide. For instance, solutions of a few very commonly occurring problems of flow through levees, earth dams, porous walls, and the like may be obtained by using the graphs shown in Figure 6.10. The graphs show the solution for flow through a porous vertical wall. This solution applies well in the case of a rock-fill dam or earth dam resting on an impermeable base and built with a core of low permeability. The core in this case can be considered as a vertical wall, and the remainder of the structure can be disregarded because of its relatively high permeability.

Figure 6.10 indicates that the exit point of the free surface at the downstream side is largely independent of the location of the downstream water level. This is logical because for all seepage flows the discharge will be the optimum under the given energy conditions. If the free surface is lower than necessary, the form factor becomes smaller because of the smaller average thickness of the flow field. Because nature tends to deliver the most for the least, the flow field will take the shape of whatever will provide the optimal discharge for the given energy. Therefore, the position of the exit point of the free surface S' will be somewhere above the downstream water level.

This principle is important in practice when the hydraulic performance of drains behind retaining walls is considered. Figure 6.11 shows a general solution of horizontal and vertical retaining wall drains developed by conformal transformation. In all cases it is assumed that the drains allow no downstream water level outside of the soil; that is, there is perfect drain operation. Hence, the drawdown is complete: $S = H$. Figure 6.11 shows that, regardless of the design of the drains, the water table cannot be lowered by much more than 50 percent of the still water table. The graphs also show that horizontal drains that are longer than $0.3H$ perform better than vertical drains behind completely drained retaining walls.

6.5   DISCHARGE OF WELLS

Wells extract water from permeable water-bearing layers called *aquifers*. One may distinguish between shallow wells that usually extract water from layers characterized by a free surface and deep wells taking water from a buried aquifer in which

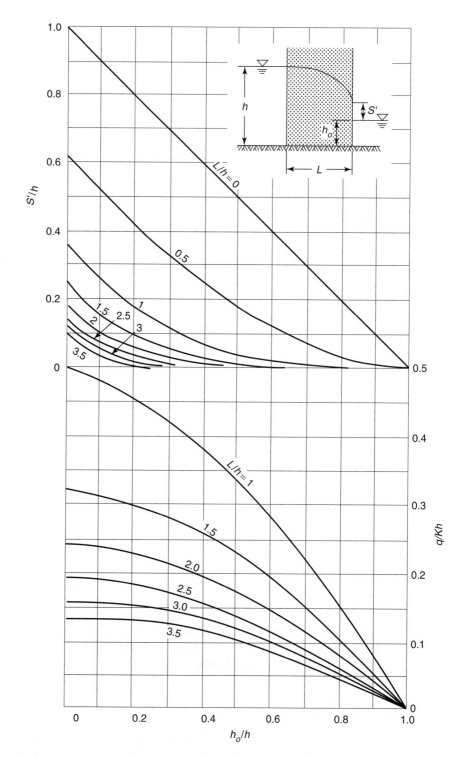

FIGURE 6.10 General solution for a two-dimensional free surface flow through porous walls. (From M. E. Harr, *Groundwater and Seepage*. Copyright 1962, McGraw-Hill Book Company. Used with permission of McGraw-Hill Book Company)

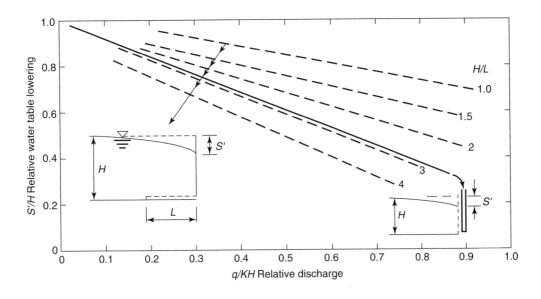

FIGURE 6.11   General solution for the elevation of the drained water surface behind retaining walls.

water is under hydrostatic pressure. These water-bearing layers are called *artesian aquifers.*

Shallow wells may be vertical and dug, driven, or drilled; or horizontal, such as *galleries* or *collector pipes* that are radially arranged around a vertical concrete shaft. The latter are called radial collector wells (Ranney wells), a typical design of which is shown in Figure 6.12. These will be discussed later.

Formulas to determine the discharge of wells were first developed by J. Dupuit[11] (France, 1804–1866) more than a century ago. With certain theoretical simplifications[12] Dupuit derived the discharge formula of a *shallow vertical well* under steady flow conditions, extracting water from an unconfined horizontal water-bearing layer of known depth and infinite horizontal extent. His formula is commonly written as

$$Q = \frac{K\pi(H^2 - h^2)}{\ln(R/r)} \qquad (6.14)$$

in which the variables are as shown in Figure 6.13. The term R in Equation 6.14 is called the *radius of influence,* which represents the distance from the well at which the free surface height remains unchanged *(H)* while the well is being pumped.

---

[11]His full name was Arsene Jules Emile Juvenal Dupuit. Although his name is most often mentioned in connection with the well formula, his most noteworthy contribution to hydraulics was in the area of surface profiles in gradually varied open channel flow.

[12]The so-called Dupuit-Forchheimer condition assumes that the velocity toward the well is horizontal throughout the flow field.

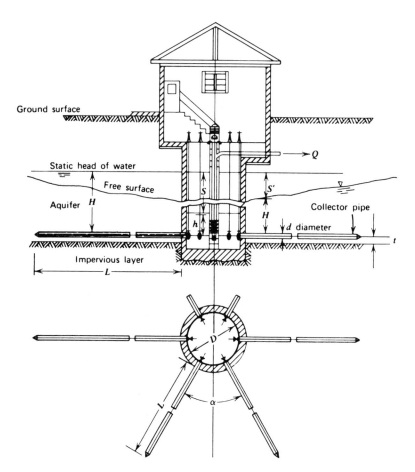

FIGURE 6.12 Horizontal (radial) collector well.

Clearly, this is a theoretical abstraction that would make the direct solution of Equation 6.14 impractical. In the past, empirical equations were proposed to overcome this difficulty. A typical one, shown here more as a historic curiosity than for practical use, is the Sichardt formula, which expresses R as a function of the drawdown in the well, S, and the permeability K of the soil in the form of

$$R = 3000 \, S \, \sqrt{K} \tag{6.15}$$

where K is to be substituted in meters per second and S in meters. This formula is dimensionally incorrect and conceptually questionable, but it may be used for rough estimates. In modern practice R is defined as the *effective* radius of the flow system over which the hydraulic energy represented by S is dissipated.

The difficulty of determining R is eliminated if there are other wells nearby—*observation wells*—where the height of the water level can be measured during pumping from the main well. In the case of *well fields* nearby wells may serve as

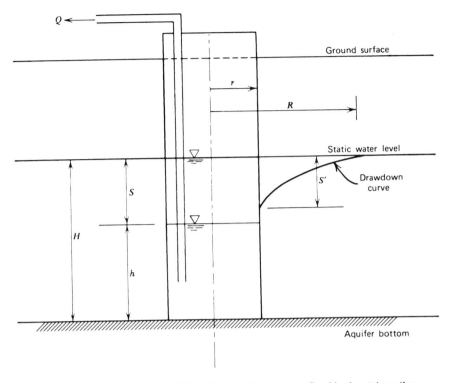

FIGURE 6.13   Interpretation of the variables for a well in an unconfined horizontal aquifer.

observation wells. For unconfined aquifers the discharge may be determined by Dupuit's equation, and by assuming horizontal flow throughout, as

$$Q = \frac{\pi K(h_2^2 - h_1^2)}{\ln(r_2/r_1)} \qquad (6.16)$$

where the subscripts refer to the two observation wells, r is their distance from the main well, and h is the depth of water in them measured from the bottom of the aquifer.

Equations 6.14 and 6.16 assume steady flow. It has been long recognized that wells are usually operated in a *transient* state. Transient well flow equations[13] generally assume that part of the water pumped comes from storage from the aquifer. These methods are quite involved.

The drawdown S is commonly measured inside the well. It is generally larger than the drawdown of the free surface level just outside the well casing or screen. This situation is analogous to the one shown in Figure 6.11 for retaining wall drains, except that here the well flow field has radial symmetry instead of being two-

---

[13]Also called *nonequilibrium formulas.*

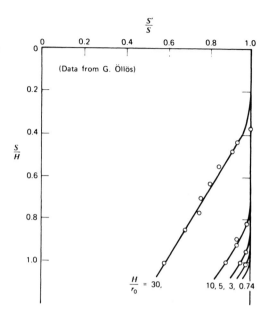

FIGURE 6.14 The drawdown lag in a shallow well. Experimentally measured drawdown characteristics of vertical wells with a free surface.

dimensional as before. Laboratory experiments using sand models have shown that the lag between S and S'—the drawdown in the soil—depends on the relative drawdown, S/H, and also on the relative size of the well radius. Figure 6.14 shows the results of these laboratory studies. These are quite important because to use a well formula correctly one would have to substitute the drawdown in the water-bearing layer rather than where it is commonly measured, inside the well. It is indeed fortunate that for small drawdowns in larger-diameter wells the lag between S and S' is not significant. But with small drilled wells, on the other hand, the error introduced can be considerable. The drawdown lag shown in Figure 6.14 (as well as in Figures 6.10 and 6.11) again shows nature's optimization principle in action. The available energy—hence, the velocity and discharge—grows with the drawdown. But as the drawdown increases the available flow area decreases. This, in turn, increases the velocity, which brings about an increase in energy losses. For a given discharge, S' represents the most economical energy utilization in the flow field.

It is important to remember that the permanent discharge of a well can be no greater than the discharge capacity of the aquifer. The *permanent yield* that may be obtained from a well is usually obtained by *pumping tests*. Standard pumping test methods in the literature are described. Most of these are time-consuming and tedious procedures. From a practical standpoint two methods may be mentioned. One is to pump the well at a gradually stepped up discharge over an extended period of time, during which the changes in drawdown are recorded. The maximum permanent yield of the well is the largest discharge that can be pumped indefinitely without any appreciable increase in drawdown and without a permanent inflow of fine sand and silt particles from the soil matrix. Another practical purpose of

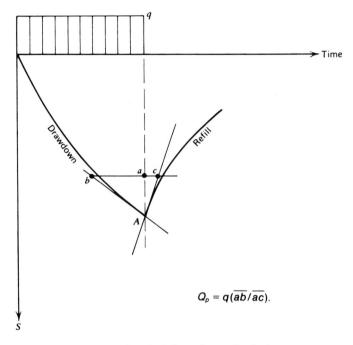

$$Q_p = q(\overline{ab}/\overline{ac}).$$

FIGURE 6.15   The Porchet method of analysis for well-pumping tests.

pumping tests is to remove fine soil particles from the vicinity of the well. This process is called "developing the well." A portion of the finer particles are not required to maintain the structural stability of the soil but actually reduce the permeability if left in place. For this reason the discharge levels during pumping are maintained until the water flows clean of silt.

Another simple procedure that may be used to determine the permanent yield in an indirect manner is the Porchet method, first proposed in France. This calls for pumping the well at a constant discharge q for a short period of time, during which the gradual increase of the drawdown is carefully measured. This drawdown measurement is then continued after the pumping is ceased, by which the rate of refill is established. The data are then plotted on a graph (Figure 6.15) in the form of time versus drawdown. The permanent yield, Q_p, is obtained by drawing tangents to the drawdown and refill lines at the peak point of the graph and expressing the ratio of the slopes of the two tangents from where

$$Q_p = q \frac{\overline{ab}}{\overline{ac}} \tag{6.17}$$

in which q is the constant pumping discharge applied, and ab and ac are as shown in Figure 6.15.

In using the Porchet method, care should be taken to obtain the slope data in a truly dynamic condition. After the pump is started, the drawdown rapidly increases in the well until an equilibrium is reached between the aquifer yield and the pump's capacity. Near this equilibrium the drawdown-versus-time curve tends to flatten out, making the slope measurements there invalid.

There are several standard methods for determining the aquifer permeability. Often, these are based on dynamic testing of wells: measuring the rate of rise upon a sudden removal of water by pumping or bailing. The *slug test, auger hole,* and *piezometric methods* may be mentioned among these. Because of the underlying mathematical complexity of the analysis these tests are not covered here. Aquifer tests generally assume that the aquifer is homogeneous and isotropic, that is, the permeability is constant throughout. Other basic assumptions include that the aquifer is infinite in extent in the horizontal direction and has a constant thickness; the well screen completely penetrates the aquifer; the flow is laminar; and the initial static water level is horizontal—that is, the water is at rest. Few of these basic assumptions can be affirmed in practice.

Horizontal or *radial collector wells* consist of a central concrete shaft from which a number of drain pipes are jacked out radially to a distance L. Valved stubs of pipes are placed into the shaft wall before the shaft is sunk into the ground. The collector pipes are then pushed into the soil horizontally by using hydraulic jacks. Construction of a horizontal collector well is considerably more expensive than drilling vertical wells, but its significantly greater yield more than offsets the difference in construction costs, particularly if the aquifer is relatively shallow. The yield of a single horizontal collector well is often greater than that of 7 to 15 conventional wells in a well field at the same location and could be as much as 5000 gpm or more. The discharge formula for horizontal collector wells was determined by laboratory experiments using sand models to be

$$\frac{Q}{KLS} = 1.2 + \frac{0.9}{n} \tag{6.18}$$

where n is the number of symmetrically placed collectors each of length L. Other variables are shown in Figure 6.12. In the experiments on which Equation 6.18 is based the relative head H/L ranged between 0.5 and 1.4. Field data collected in Europe and in the United States indicate that considerable lateral variation of permeability is encountered with horizontal collector wells. This, in turn, results in significant deviations from the design discharge obtained by Equation 6.18. The Q/KLS value was found to range between 0.08 and 2.5 in actual practice, whereas in laboratory experience with a sand model, where uniform permeability was maintained, it ranged between 0.7 and 1.2. The number of collector pipes in actual installations range between 2 and 12. To obtain the expected discharge from a collector well, we can increase the number of collector pipes almost at will.

Artesian[14] *wells* obtain their water from one or more water-bearing layers located at greater depths in which the water is under hydrostatic pressure. For the discharge

---

[14]The name comes from Artois County of France, where Dupuit first identified the concept.

of a horizontal artesian layer of T thickness, Dupuit's derivation yields

$$Q = KTS \frac{2\pi}{\ln(R/r)} \tag{6.19}$$

where R, again, is the radius of influence. With two observation wells available at distances r_1 and r_2 away from the main well, the discharge can be derived from Dupuit's equation as

$$Q = \frac{2\pi \, KT(h_2 - h_1)}{\ln(r_2/r_1)} \tag{6.20}$$

The well discharge formulas reviewed herein are similar in one aspect: They contain a single value for aquifer permeability. This is usually a gross idealization of the actual field conditions.

6.6   UNSATURATED AND UNSTEADY FLOWS

The physical concepts used in common applications of seepage theory are not necessarily valid in all cases encountered in nature. For example, the problems involving free surface flow are generally analyzed under the assumption that the free surface is the upper limit of the flow field. In soils containing silt and clay a *capillary layer* develops. This capillary layer can sometimes be thicker than the flow field proper. Table 6.2 shows the capillary rise that is encountered in some soils. The actual flow field in soils conducive to capillary action is much larger than that suggested by the classical theorems of groundwater flow. Although in the capillary fringe the water pressure is negative, it is no barrier to the flow. These problems are treated by recognizing that the capillary layer, or at least most parts of it, is a part of the whole flow field. When a capillary layer is present on the top of a gravity flow field, the free surface does not act as a boundary to the flow. It is not a limiting stream surface. Rather, it is only a piezometric surface, representing the location of atmospheric pressure in the flow field. Hence, it is the elevation to which water will rise in an exploratory bore hole. Because the capillary layer thickness depends on the surface tension, adhesion, and pore size, this layer will be nearly parallel to

TABLE 6.2 Capillary Rise in Various Soils			
	Soil Type	Average Grain Size mm	Capillary Rise m
	Sand	2–0.5	0.03–0.1
		0.5–0.2	0.1–0.3
		0.2–0.1	0.3–1.0
	Silt	0.1–0.05	0.3–1.0
		0.05–0.02	1.0–3.0
	Silt clay	0.02–0.006	3–10
		0.006–0.002	10–30
	Clay	<0.002	30–300

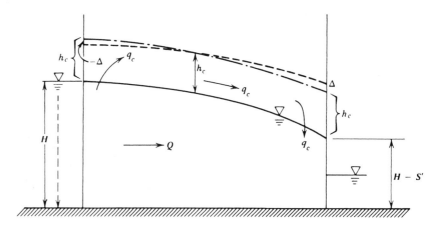

FIGURE 6.16 The mechanism of capillary flow over a piezometric surface bounding a gravity flow field.

the piezometric surface and in case of gravity flow in the flow field proper, the water will move within the capillary layer as well. Because the outside water level is below the capillary layer on the upstream side of the flow field, the water flowing in the capillary layer must enter through the piezometric surface along its upper portion, as shown in Figure 6.16. To facilitate this movement there must be an energy gradient upward through the upper entry portion of the piezometric surface. This necessitates the lowering of the capillary height within the entry portion of the capillary layer. From here the water in the capillary layer will flow along with the rest of the moving water toward the downstream portion of the capillary layer. Because of the negative pressures within the capillary layer, the water cannot exit from this layer directly to the atmosphere but must leave the capillary layer through the downstream portion of the piezometric surface. Since this requires excess pressure along the exit portion, the capillary height must increase there. Therefore, the top of the capillary layer over a gravity flow field will always have a flatter slope than the piezometric surface of the flow field.

Experimental investigations of the flow within a capillary layer on the top of a gravity flow field have shown that the velocity varies with a well-defined regularity from zero at the top of the capillary layer to an approximately uniform velocity near the piezometric surface. Considering Darcy's equation, one must conclude that because the horizontal energy gradient is not zero at the top of the capillary fringe the permeability must be zero. Indeed, this conclusion agrees with the well-known fact that the degree of saturation within the capillary layer decreases with height, and only the lower portion is fully saturated. The layer of partial saturation is a layer of varying permeability. The reason for this is the great variation in the size of pores in natural soils. Larger pores allow lower capillary heights. Water will rise higher through the smaller pores. Hence, the capillary rise in natural soils follows the random function of the pore size distribution. This fact prevents the application of the classical theoretical methods of seepage analysis. But based on several independent studies with sand models, an empirical formula may be used for estimating the discharge q_c of a capillary layer. This may be written as

$$\frac{q_c}{Q} = C \frac{h_c}{H} \left(\frac{S'}{H} \right)^{2.5} \tag{6.21}$$

in which Q is the discharge of the gravity flow field, H is the static head upstream, h_c is the thickness of the capillary layer, and S' is the height of the seepage face. The coefficient C in this equation was found to be 0.545 in the sand model experiments.

The preceding problem has been manifested in earth dams where the impermeable clay core was no higher than the water surface upstream. Seepage over the clay core through the capillary layer can be conspicuous.

Another case where the formal seepage theory breaks down occurs in connection with *seepage losses* from canals and rivers. Some of these problems can be analyzed by regular means, particularly when the groundwater level is only a little lower than the level of the water in the channel; but this is not true in many places. When the water level in a porous ground is well below the bottom of the channel, the water in the channel is not connected in a continuous manner with the groundwater. Particularly when the channel carries suspended sediments, silt, and clay, it plugs the pores of the channel bottom, creating a relatively less permeable layer there. This occurrence is called *colmatation*. The water that seeps through this layer because of the pressure on the channel bottom enters a more permeable soil. As the permeability increases downward the escaping water flows faster. When this is the case, part of the pores in the soil contain air, and as a result, the soil is unsaturated. Because the unsaturated soil pores are in contact with the atmosphere, the pressure there is atmospheric. The water seeping downward in such cases behaves somewhat like water flowing from a shower. Just as with the elevation of the bathtub, the location of the permanent groundwater table below the river is irrelevant. In these cases the rate of seepage loss from the channel is strictly dependent on the average depth of water, the permeability of the channel bottom layer, and its thickness. The permeability of soil itself, as long as it is larger than that of the layer at the bottom of the channel, does not influence the seepage losses.

An analytic solution for this problem was provided by V. V. Vedernikov, a prominent Russian scientist. According to this solution, the seepage loss per unit length of channel is

$$q = K(b + hF) \tag{6.22}$$

where b is the surface width in meters, h is the depth in meters, and F is a function of the channel shape as plotted in Figure 6.17. The coefficient of permeability is in meters per second, and q is in cubic meters per second per meter of channel length.

Darcy's equation essentially states that the velocity is directly proportional to the hydraulic gradient. The factor of proportionality is the permeability, K. Because water moves only though the pores and cracks in the soil and not through the solid material, the macroscopic, actual velocity is greater than Darcy's

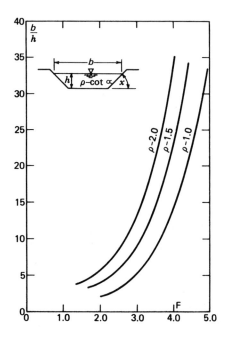

FIGURE 6.17 Vedernikov's function for seepage losses from open channels.

velocity. This *velocity of seepage* is often an important parameter in hydraulics. Pressure changes may be communicated instantaneously through a saturated soil mass if we assume complete incompressibility. In contrast, to relieve pore pressures in a soil mass, movement of water is required. Because of the slowness of the movement of fluids in a soil mass, there may be a very significant time lag between the external pressure (e.g., drawdown in a well) and internal pressures in the aquifer. The velocity of seepage is dependent on the porosity of the soil and on Darcy's equation, that is,

$$v_{\text{seepage}} = \frac{K}{n_e} \frac{\Delta h}{\Delta l} \tag{6.23}$$

where n_e is the *effective porosity*.[15] Table 6.3 lists some typical values of porosity of natural soils. The gradient $\Delta h/\Delta l$ may be replaced by the slope of the steepest descent of the groundwater table, denoted by i. Seepage velocities for a 1 percent (1 ft drop over 100 ft distance) free surface slope are shown for various soils in Table 6.4. The tabulated values show the time it takes for the water to travel 1 ft. In these calculations a porosity of 0.25 was used.

All the foregoing considerations of seepage pertain to steady flow. For steady flow conditions a constant supply of groundwater discharge is needed. Although

---

[15]Not all porosity necessarily contributes to the flow, as there can be "dead end" pore space.

TABLE 6.3 Porosity of Some
Natural Soils

Uniform sand, loose	0.46
Uniform sand, dense	0.34
Glacial till, very mixed grained	0.20
Soft glacial clay	0.55
Stiff glacial clay	0.37
Soft very organic clay	0.75
Soft bentonite	0.84

Source: C. Terzaghi and R. B. Peck, *Soil Mechanics in Engineering Practice.* New York: Wiley 1948.

we know that most of the groundwater comes from rainfall, part of which infiltrates into the ground, the seasonal fluctuations of the groundwater inflow affect our assumptions of steady seepage in a negligible manner. In water supply from wells, however, the groundwater inflow into the layers over long periods of time may be significantly less than the amount pumped out.[16] In such cases the well yields decrease because more is withdrawn than is replaced.

In the inflow and outflow boundaries of flow fields, the sudden change in water levels can create very large seepage forces in the soil. These seepage forces often exceed the internal strength of the soil, causing *slope failures* along the embankments. Figure 6.18 shows the upper end of the slope failure in an earth dam in Ohio. Shore lines of creeks and rivers often exhibit this phenomenon after floods.

Flow through *levees* that become slowly saturated during floods is another example of unsteady seepage. Compared with seepage velocities, the rise of flood water is relatively sudden. Levees are made of local, alluvial soil that is usually relatively permeable. They are not like dams, constructed to hold water permanently. Hence, they will retard the flow for only a short period of time, until the advancing free surface reaches the protected side. The width of levees, therefore, is selected so that they hold the water as long as the flood is expected to last.

TABLE 6.4 Typical Seepage Velocity Values for Various Soils

Soil Type	Time Required for Water to Move 1 Ft
Gravel	3.6 min
Coarse clean sand	36 min
Fine sand	2.5 days
Silty sand	25 days
Silt	250 days (0.685 year)
Clay	6.85 years (or more)

Note: The assumptions are that the slope of the groundwater table is 1 percent, and the porosity is 25 percent.

---

[16]In Texas, Arizona, and Hungary, heavy reliance on drilled wells has caused the sinking of aquifer water levels by dozens of feet.

FIGURE 6.18 Upper end of the slide surface in a failed earthen dam. Slope failure was caused by excessive seepage in poorly compacted fill. (Courtesy of Ohio Department of Natural Resources)

Unusually prolonged floods—those that exceed the duration of the design flood—may cause levees to leak. The danger of failure increases with time as leakage, piping, and increasing seepage forces develop. The Mississippi River floods of the summer of 1993 showed many such examples.

EXAMPLE PROBLEMS

Example 6.1 A levee built of clay is placed on top of a 6 in. sand layer. The bottom width of the levee is 60 ft. During a flood the water elevation difference between the inside and the outside of the levee is estimated to be 8 ft (Figure E6.1). If we assume that the permeability of the sand layer is 0.01 cm/s, calculate the expected seepage discharge for a 200 ft long portion of the levee.

FIGURE E6.1

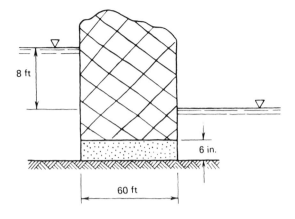

Solution. From Equation 6.1,

$$Q = AK\frac{\Delta h}{L}$$

$$A = \frac{1}{2} \times 200 = 100 \text{ ft}^2$$

$$K = 0.01 \frac{\text{cm}}{\text{s}} \times \frac{1}{2.54} \times \frac{1}{12} = 3.28 \times 10^{-4} \text{ ft/sec}$$

$$\Delta h = 8 \text{ ft}$$

$$L = 60 \text{ ft}$$

Thus,

$$Q = 100 \times 3.28 \times 10^{-4} \times \frac{8}{60} = 4.4 \times 10^{-3} \text{ cfs} \times 449 = 2.0 \text{ gpm} \qquad \square$$

Example 6.2

A sand filter of the configuration in Figure E6.2 is 1.5 m deep, and its surface area is 10 m². If the hydraulic conductivity is 0.05 cm/s and the effective porosity is 0.3, determine the discharge across the filter for a head difference of 1.0 m. Also, determine the average seepage velocity and its associated kinetic head.

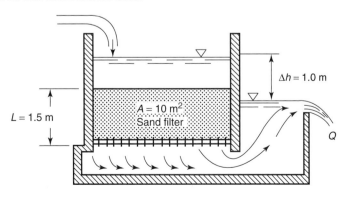

FIGURE E6.2

Solution. From Equation 6.1,

$$Q = K\frac{\Delta h}{L} A = 0.05 \times \frac{1}{100} \times \frac{1}{1.5} \times 10 = 3.3 \times 10^{-3}\,\text{m}^3/\text{s}$$

From Equation 6.23,

$$v_{\text{seepage}} = \frac{K}{n}\frac{\Delta h}{\Delta l} = \frac{(0.05/100)}{0.3}\frac{1.0}{1.5} = 1.1 \times 10^{-3}\,\text{m/s}$$

The kinetic head is

$$\frac{v^2}{2g} = \frac{(1.1 \times 10^{-3})^2}{2 \times 9.81} = 6.2 \times 10^{-8}\,\text{m}$$

Compared with $\Delta h = 1.0$ m, the kinetic head is insignificant. $\square$

Example 6.3

A concrete dam with a 15 ft base width creates a 7 ft drop in the water surface. The dam is embedded 2 ft into a 9 ft thick horizontal permeable layer that has a permeability coefficient of 0.1 cm/s. Estimate the seepage discharge under the dam by using the form factor method.

Solution. Draw the dam and the permeable layer to scale. Establish the approximate location of the average stream line length by roughly locating the position of the stream line representing 50 percent of the seepage flow above and below it. Find the distance A and B as shown in Figure E6.3. To obtain the average width of the flow field, we first recognize that it should be somewhat more than the depth of the layer below the dam, since the part of the flow field before and after the dam is considerably wider, although portions far from the dam carry almost no discharge. The average width selected was between points C and D, the length of which was measured to be 8.5 ft. The distance between A and B is about 30 ft.

Accordingly, the two-dimensional form factor is

$$\Phi = \frac{8.5}{30} = 0.28$$

By Equation 6.7 the relative discharge

$$\frac{Q}{K\,\Delta h} = \Phi = 0.28$$

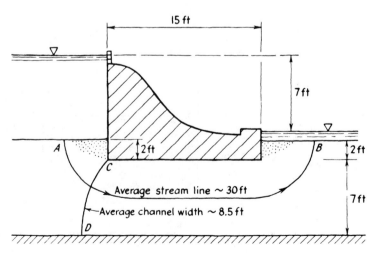

FIGURE E6.3

If

$$K = 0.1\,\frac{cm}{s} = 0.1 \times \frac{1}{2.54} \times \frac{1}{12}$$

$$K = 0.00328 \text{ ft/sec}$$

$$\Delta h = 7 \text{ ft}$$

Hence,

$$Q = \Phi \cdot K \cdot \Delta h = 0.28 \times 0.00328 \times 7$$

$$Q = 0.0064 \text{ cfs per unit length of the dam}$$

A well-drawn flow net should improve the accuracy of this estimate.   □

Example 6.4   A 100 ft wide concrete dam with a sheet pile at the center is built over a horizontal silt layer 100 ft deep. The permeability of the silt is 10^{-5} cm/s. The head loss across the dam is 18 ft (Figure E6.4). Determine the seepage discharge if the sheet pile extends vertically 50 ft into the silt. Compare this value with the seepage discharge if no sheet pile is used. Determine the seepage discharge if the sheet pile is placed at the front edge of the dam.

Solution. We need the following parameters to use Figure 6.5a:

$$\frac{s}{T} = 0.5 \qquad \text{and} \qquad \frac{b}{T} = 0.5$$

Then from Figure 6.5a

$$\frac{q}{Kh} = 0.41$$

$$q = 0.41\left(\frac{10^{-5}}{2.54} \times \frac{1}{12}\right) \times 18 = 2.4 \times 10^{-6} \text{ cfs/ft}$$

If the preceding procedure is repeated for $s/T = 0$, the discharge without a sheet pile is

$$q = 0.53 \times \left(\frac{10^{-5}}{2.54} \times \frac{1}{12}\right) \times 18 = 3.1 \times 10^{-6} \text{ cfs/ft}$$

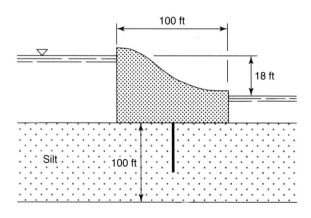

FIGURE E6.4

Thus, adding the sheet pile in the center of the dam reduces the discharge by 23 percent. If the sheet pile is placed at the front edge of the dam, then

$$\frac{s}{T} = 0.5 \qquad \frac{b}{T} = 0.5 \qquad \text{and} \qquad \frac{x}{b} = 1$$

From Figure 6.5b

$$\frac{q}{Kh} = 0.38$$

$$q = 0.38 \times \left(\frac{10^{-5}}{2.54} \times \frac{1}{12}\right) \times 18 = 2.2 \times 10^{-6}\,\text{cfs/ft}$$

Placing the sheet pile under the front edge of the dam slightly decreases the seepage discharge. This is also the best location for a sheet pile to minimize the uplift seepage pressure on the base of the dam. □

Example 6.5

For the conditions described in Example 6.4 with the sheet pile under the center of the dam, determine which method reduces the seepage discharge more: increasing the sheet pile depth by 25 ft or increasing the width of the dam by 25 ft.

Solution. From Example 6.4 and $s = 75$ ft we have

$$\frac{b}{T} = 0.5 \qquad \text{and} \qquad \frac{s}{T} = 0.75$$

Then from Figure 6.5a

$$\frac{q}{Kh} = 0.29$$

If s remains at 50 ft and $2b$ is increased to 125 ft, we have

$$\frac{b}{T} = 0.625 \qquad \text{and} \qquad \frac{s}{T} = 0.5$$

Then from Figure 6.5a

$$\frac{q}{Kh} = 0.37$$

Hence, increasing the sheet pile depth by 25 ft reduced the seepage discharge more than increasing the dam width by 25 ft. Which method would probably be cheaper to implement? □

Example 6.6

Determine the uplift pressure distribution along the horizontal base of the dam shown in Figure E6.6a. The width of the dam is 55 m, and the head difference across the dam is 15 m. What is the total uplift force per unit length of the dam?

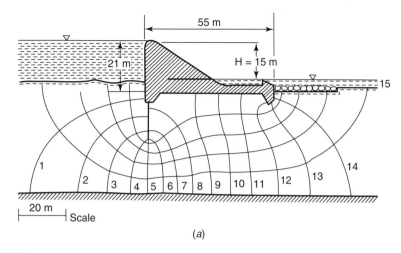

(a)

FIGURE E6.6a

Solution. Use the flow net to estimate the total hydraulic heads at the locations on the base of the dam given in column 3 of the following table. It is necessary to use the given dimensions on Figure E6.6*a* for scale. Use the broad horizontal base of the dam as the elevation datum. For example, the depth of water above the base of the dam is about 21 m. Because the head difference across the dam is 15 m and because there are 15 equipotential lines, each equipotential line encountered when flowing under the dam represents a loss of 1 m of head. The first location in column 1 is the point on the base of the dam just to the left of the sheet pile. Water moving from the upstream reservoir to this point loses about 1.3 m of head. Thus, the total hydraulic head at this point is 19.7 m (21 − 1.3 = 19.7). Subtracting the elevation head (−4.0 m) gives the pressure head at this point as 23.7 m. Repeating this procedure for each location gives the pressure heads given in column 5.

Table of Computations for Example 6.6

Location	x m	Hydraulic head m	Elevation head m	Pressure head m
Left of pile	1	19.7	−4.0	23.7
Right of pile	2	12.9	−4.0	16.9
Left angle	6	12.7	0.0	12.7
9	22	12.0	0.0	12.0
10	32	11.0	0.0	11.0
11	42	10.0	0.0	10.0
Right angle	51	9.4	0.0	9.4
12	53	9.0	−4.0	13.0
13	55	8.0	−4.0	12.0

Graph the pressure head distribution along the base of the dam. This is shown in Figure E6.6*b*. To calculate the total uplift force, sum the total area in the curve in Figure E6.6*b* by dividing it into quadrilateral sections and adding each contribution. Multiply the total area by the unit weight of water (9810 N/m³) and divide by 1000 to determine that the total uplift force on the base of the dam is 6430 kN per meter of length.

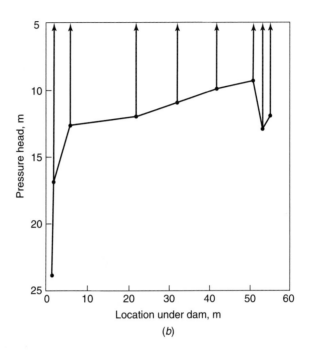

FIGURE E6.6b

Example 6.7

Determine the critical gradient for the excavation in a river under the protection of two parallel sets of sheet piles 200 ft apart, if the excavation is performed in 12 ft of water to a depth of 10 ft below water and if the sheet piles are driven to a depth of 30 ft below the depth of the river (Figure E6.7). The soil has a permeability of 10^{-4} cm/s. Determine the required pump capacity for each 100 ft length of the excavation, and determine the factor of safety with respect to piping or quick condition.

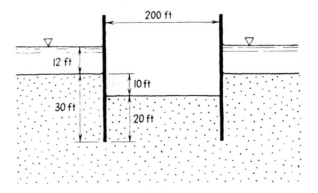

FIGURE E6.7

Solution. The variables are

$$h = 12\,\text{ft} + 10\,\text{ft} = 22\,\text{ft}$$
$$s = 20\,\text{ft}$$
$$d = 10\,\text{ft}$$
$$b = 100\,\text{ft}$$

Hence,

$$\frac{s}{b} = 0.20 \qquad \frac{d}{b} = 0.1$$

From Figure 6.7 $q/Kh = 1.8$, where q is for 1 ft of length:

$$K = 10^{-4}\frac{\text{cm}}{\text{s}} = 3.28 \times 10^{-6}\,\text{ft/sec}$$

For 100 ft of length:
$$Q = 100q = 180Kh = 180 \times 3.28 \times 10^{-6} \times 22$$
$$Q = 1.3 \times 10^{-2}\,\text{cfs}$$

Again, from Figure 6.7, we obtain

$$I_{\text{exit}}\frac{s}{h} = 0.33$$

Hence,

$$I_{\text{exit}} = 0.33 \times \frac{h}{s}$$

$$= 0.33 \times \frac{22}{20}$$

$$I_{\text{exit}} = 0.36 = \frac{v}{K}$$

Since by Equation 6.13 $\dfrac{v_{\text{critical}}}{K} = I_{\text{critical}} = 1$, the factor of safety is

$$\frac{I_{\text{critical}}}{I_{\text{exit}}} = \frac{1}{0.36} = 2.8 \qquad\qquad \square$$

Example 6.8   A 30 m wide porous wall has a permeability of 10^{-3} cm/s. On one side the water level is 20 m; on the other side it is 1 m (Figure E6.8). Determine the discharge for a 50 m long section of the wall, and locate the free surface elevation at the downstream side of the wall.

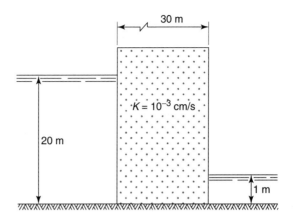

FIGURE E6.8

Solution. For the porous wall,

$$L = 30 \text{ m} \qquad h = 20 \text{ m}$$
$$K = 10^{-3} \text{ cm/s} \qquad h_0 = 1 \text{ m}$$

We have

$$\frac{L}{h} = \frac{30}{20} = 1.5$$

$$\frac{h_0}{h} = \frac{1}{20} = 0.05$$

From Figure 6.10 with L/h and h_0/h obtained above, $S'/h = 0.19$, $q/Kh = 0.32$; therefore, the free surface elevation $S' = 0.19 \times 20 = 3.8$ m over the downstream level, and $q = 0.32 \times 10^{-3} \times 10^{-2} \times 20 = 6.4 \times 10^{-5}$ m³/s for 1 m of wall. For a 50 m long wall,

$$Q = 6.4 \times 10^{-5} \times 50$$
$$Q = 3.2 \text{ liters/s} \qquad \square$$

Example 6.9

A vertical retaining wall supports a completely saturated horizontal fill with a thickness of 3 m. The permeability of the fill is 5×10^{-3} cm/s. Underneath the fill there is an impervious horizontal layer. If a completely drained vertical fill is discharging 0.1 liter/s for each meter length of the wall, determine the height of the saturation behind the drain.

Solution. The relative discharge is

$$\frac{9}{KH} = 10^{-4}/(5 \times 10^{-5} \times 3) = 0.67$$

Entering this value into Figure 6.11 and turning at the solid line representing vertical drains, we find that

$$S'/H = 0.48 \qquad \text{and} \qquad S' = 0.48 \times 3 = 1.44 \text{ m}$$

Therefore, the height of the saturated layer is

$$3 - 1.44 = 1.56 \, \text{m} \qquad \square$$

Example 6.10   A well is to be developed in an unconfined sandy aquifer to lower the water table for a small construction project. The steady-state drawdown must be at least 5 ft within a distance of 100 ft from the well and 10 ft within a distance of 10 ft from the well. The hydraulic conductivity of the sand is 10^{-2} cm/s. A relatively impermeable clay with a hydraulic conductivity of 10^{-8} cm/s forms the base of the aquifer. From the top of the clay layer the depth of water in the aquifer before pumping is 27 ft. Calculate the required discharge from the well.

Solution. The hydraulic conductivity of the silty sand is 10^{-2} (0.0328) = 3.28 × 10^{-4} ft/sec. From Equation 6.16

$$Q = \frac{\pi K (h_2^2 - h_1^2)}{\ln(r_2/r_1)} = \frac{\pi (3.28 \times 10^{-4})(22^2 - 17^2)}{\ln(100/10)} = 0.0873 \, \text{cfs}$$

Multiplying by 449 to convert cfs to gpm gives the discharge to be about 39 gpm. $\qquad \square$

Example 6.11   A radial collector well is built with 8 collector pipes each 50 ft long. The static head of the aquifer is 30 ft. The assumed average permeability of the aquifer, composed of coarse, clean sand, is 10^{-1} cm/s. Estimate the probable discharge of the well when the drawdown in the shaft is 8 ft.

Solution. By using Equation 6.18 with

$$K = 10^{-1} \frac{\text{cm}}{\text{s}} = 3.28 \times 10^{-3} \, \text{ft/sec}$$
$$L = 50 \, \text{ft}$$

and

$$S = 8 \, \text{ft}$$

we get

$$Q = \left(1.2 + \frac{0.9}{n}\right) K \cdot L \cdot S$$

$$= \left(1.2 + \frac{0.9}{8}\right) 3.28 \times 10^{-3} \times 50 \times 8$$

$$Q = 1.72 \, \text{cfs} = 770 \, \text{gpm}$$

Because $H/L = 30/50 = 0.6$, the problem is within the range of validity of Equation 6.18; hence, it furnishes a good estimate. $\qquad \square$

Example 6.12   For the problem discussed in Example 6.8 estimate the probable capillary discharge.

Solution. The upstream head in the problem was 20 m, and the exit point of the free surface measured from the downstream water level was 3.8 m. In Equation 6.21 used to determine the capillary discharge, the variable S' is to be measured from the upstream water level.

Hence, $S' = 20 - (3.8 + 1) = 15.2$ m. The discharge Q for unit length of wall was 6.4×10^{-5} m^3/s.

The permeability of the soil was 10^{-3} cm/s. From Table 6.1 this indicates a fine sand with an average grain size of 0.1 mm. In Table 6.2 the capillary rise h_c for such soils ranges between 0.3 and 1.0 m. Taking the maximum value we may substitute into Equation 6.21 as

$$q_c = 0.545 \left(\frac{1}{20}\right)\left(\frac{15.2}{20}\right)^{2.5} \times 6.4 \times 10^{-5}$$

$$= 8.8 \times 10^{-7} \text{ m}^3\text{/s for 1 m of wall}$$

For a wall section 50 m long, the total capillary discharge is

$$50 \times 8.8 \times 10^{-7} = 4.4 \times 10^{-5} \text{ m}^3\text{/s} = 4.4 \times 10^{-2} \text{ liters/s}$$

This is 1.4 percent of the seepage discharge calculated in Example 6.8. □

PROBLEMS

6.1 Water seeps horizontally into a trench from a layer of clay that has a permeability of 10^{-7} cm/s. The discharge velocity is estimated to be 3×10^{-8} cm/s. Express the magnitude of the horizontal seepage force acting on a unit volume of the soil.

6.2 A 2 m thick horizontal sand filter is 3 m wide and 8 m long. Determine the discharge through the filter if the hydraulic conductivity is 10^{-2} cm/s, and the head loss through the filter is 30 cm.

6.3 Determine the intrinsic permeability of the filter in Problem 6.2 if the water temperature averages 10°C.

6.4 For a permeable soil layer under an earth dam built of clay the form factor was determined to be $A/L = 0.08$ m, in which A is the average cross-sectional area of the permeable soil seam measured perpendicular to the flow, and L is the average length of the flow lines. Estimate the seepage discharge across the permeable layer if the head loss is 6 m and the assumed permeability is 10^{-3} cm/s.

6.5 For a 50 ft wide concrete dam built on top of a 50 ft deep horizontal uniform fine sand layer, a sheet pile should be driven at the center line to cut the seepage discharge by 25 percent. How deep should the sheet pile be? What will be the additional reduction of discharge if the sheet pile is driven at the front of the dam?

6.6 A trench built in a 6 ft deep lake involves an 18 ft wide excavation. The soil is silty sand with an average grain size of 0.02 mm. The depth of the excavation is to be 10 ft below the lake bottom. To what depth should sheet piles be driven to result in a maximum exit pressure gradient of 0.5? At this depth, what is the seepage discharge per unit length of trench?

6.7 In Problem 6.6 what will be the hydrostatic pressure on the bottom of the excavation if a 1 ft thick concrete mat is poured to eliminate water flow (unit weight of concrete is 155 lb/ft^3)?

6.8 A rock-fill dam built with a 10 m thick vertical clay core is resting on impermeable bedrock. The permeability of the clay is 10^{-6} cm/s. Determine the seepage discharge for a 100 m section of the dam (direction perpendicular to the plane of the page) for the case when the upstream water is 10 m above the base and the downstream water level is zero. How high will the zone of saturation (surface of seepage) be on the protected side, neglecting capillary effects?

6.9 For Example 6.6 determine the discharge for a 100 m width of the dam if the permeability of the fill under the dam is 5×10^{-7} m/s.

6.10 A retaining wall built to support a saturated 5 m high sand layer ($K = 10^{-2}$ cm/s) is provided with a vertical drain of well-graded gravel that results in complete drainage along the length of the wall. How high will the saturated thickness of the sand be behind the drain if the discharge per unit length is 0.15 liter/s?

6.11 Assuming that the permeability of a 10 ft thick shallow horizontal aquifer is 5×10^{-4} ft/s, determine the discharge of a radial collector well built with 6 symmetrically placed 45 ft long collector pipes if the drawdown in the shaft is 6 ft.

6.12 An artesian aquifer 6 m thick is pumped at a rate of 0.01 m³/s. Observation well 1 is 15 m away from the pumping well and has a water level 15.3 m above the base of the aquifer. Similarly, observation well 2 is 100 m away and has a water level at 16.8 m. What is the average horizontal hydraulic conductivity in the aquifer?

6.13 Determine the capillary discharge in Problem 6.8 if the height of the capillary layer is 3 m.

6.14 For an effective porosity of 0.25, estimate the seepage velocity in the aquifer in Problem 6.12 at a distance of 10 m from the pumping well.

MULTIPLE CHOICE QUESTIONS

6.15 Seepage velocity may be obtained from discharge velocity by
  A. Multiplying the latter by the permeability.
  B. Dividing it by the permeability.
  C. Multiplying the latter by the porosity.
  D. Dividing it by the porosity.
  E. Multiplying the latter by the void ratio.

6.16 Darcy's permeability coefficient depends on
  A. Geometric properties of the porous medium.
  B. Gravity, density, and viscosity of the fluid.
  C. Both A and B.
  D. Temperature.
  E. All of the above.

6.17 A permeability coefficient is quoted to be 10^{-5} cm/s. How would you describe the soil?
  A. Gravel.
  B. Coarse sand.
  C. Fine sand.
  D. Silt.
  E. Clay.

6.18 Which of these formulas gives the seepage pressure dp on a properly oriented unit cube of soil if the unit weight of water is denoted by w?
  A. $w \cdot v/K$.
  B. $I \cdot w/K$.
  C. $K \cdot w \cdot I$.

D. $I \cdot dh \cdot w/K$.

E. $K \cdot dh/dL$.

6.19 The correct definition of the equipotential line is the line
A. Along which the hydrostatic pressure is constant.
B. Along which all velocity vectors are constant.
C. To which all velocity vectors are parallel.
D. That is parallel to velocity at all points.
E. On which the Bernoulli equation is undefined.

6.20 An equipotential line is defined as the line
A. That is perpendicular to all stream lines.
B. To which the velocity vectors are perpendicular.
C. For which Bernoulli's equation gives the same value.
D. That separates the flow field into equal proportions of Q.
E. A, B, and C.

6.21 Stream lines are defined as lines
A. That are perpendicular to all equipotential lines.
B. On which the velocity vectors are tangential.
C. For which dS/dL is constant.
D. For which Bernoulli's equation gives a zero value.
E. A and B.

6.22 Basic assumptions when drawing a flow net are a
A. Homogeneous, isotropic media with Darcy's equation valid.
B. Answer A, plus a two-dimensional flow.
C. Answer B, plus an incompressible fluid.
D. Answer C, plus no capillary effects.
E. Answer D, plus a nonviscous fluid.

6.23 Quick conditions occur when
A. $dh/dL = 1$.
B. $I = 1$.
C. $v = K$.
D. All of the above.
E. None of the above.

6.24 Free surface in a flow field is defined where
A. v is tangential to the line.
B. Pressure is atmospheric.
C. Both A and B.
D. Bernoulli's equation gives the elevation.
E. Both C and D.

6.25 Unconfined flow is defined as
A. The flow field that may change its shape with changes of pressure.
B. The case when none of the flow field is exposed to atmospheric pressure.
C. The flow that cannot be solved by drawing a flow net.
D. A flow field bounded by a free surface.
E. A flow field partially bounded by an impervious surface.

6.26 Dupuit is known for
A. The development of theoretical solutions in seepage.
B. The analytic solution of free surface flow.

C. His well discharge formulas.

D. His analysis of horizontal collector wells.

E. The discharge-refill analysis of wells.

6.27 What type of soil is characterized by a capillary rise of 1 to 10 ft?

A. Gravel.

B. Sand.

C. Silt.

D. Clay.

E. Shale.

6.28 Capillary discharge means

A. The flow of water in the capillary layer.

B. The flow of water against gravity.

C. Flow through capillary tubes.

D. All of the above.

E. None of the above.

6.29 Sheet piles below dams are used to

A. Increase the resistance against sliding.

B. Reduce the seepage pressure in front of the dam.

C. Increase the stability against overturning.

D. Reduce the chance of piping downstream.

E. Cut off seepage.

7

Elements of Hydrology

In the design and analysis of open channels and hydraulic structures the most important variable is the magnitude of the flows that may be expected. Yet this variable is the least certain of the variables, since it depends on many interrelated topographic, geologic, and other elements of nature, in addition to its dependence on random meteorological factors. Fundamental principles and methods used in the determination of design discharges are reviewed in this chapter.

7.1 INTRODUCTION

Hydrology is the study of the circulation of water in nature. It covers the physics of what is called the *hydrologic cycle*, from the evaporation of the seas through the movement of the atmospheric moisture that makes up the weather. Hydrology includes the statistics of precipitation and the collection of runoff and rainwater through creeks and rivers. Our particular interest in this subject is focused on the magnitude and occurrence of floods in order to determine reasonable values for design discharges pertaining to open channels and hydraulic structures. Few areas of hydraulics have benefited as much as hydrology from recent advances in technology. Satellites, microwave technology (e.g., Doppler radar), and powerful computers analyzing huge quantities of hydrologic data have contributed immensely to the development of new hydrologic knowledge.

7.2 THE DESIGN DISCHARGE

In hydraulic design few things are more important than the selection of the design discharge. This is the maximum capacity a channel or structure is designed to handle. Basically the matter is a question of economics. The smaller the design discharge of a structure, the cheaper it is to build. The most important question is, what will happen if the design discharge is exceeded? In the case of a highway culvert we may be willing to put up with the occasional flooding of a few acres of agricultural land. Even the threat of periodic payment of damages would not deter the designer from selecting a design discharge corresponding to a rainfall expected once in 5 years or so. Designing culverts to the size corresponding to a flood that occurs once in 100 years would result in far too big an investment. On the other hand, an undersized spillway of a dam presents a great deal more danger. A dam may fail once its crest is overtopped, causing considerable losses of property and in even more critical cases, of life as well.

The period within which we cannot tolerate the design discharge to be exceeded by the random flows due to precipitation is called the *return period* of our design flood. The expressions "recurrence interval" and also "flood frequency" are used. The return period of a flood of a certain magnitude is a statistical concept. The larger the flood, the less frequently it will be encountered. As with all matters of random nature analyzed by statistical methods, flood frequencies represent a degree of uncertainty. The hapless designer selecting, for example, a 10-year design discharge could find the structure flooded out twice in the first 2 years, once by an 11-year flood, then by a 500-year one. Selecting a return period that exceeds the expected life of a structure is no assurance that larger flows will not be encountered within that period. The occurrence of floods is a matter of chance, which we try to counter with our educated guesses. Just how educated our guesses are may later be determined in the courtroom. However, designers will not stay in business long if their designs are consistently more expensive than those of their competitors.

The reciprocal of a return period T of a certain occurrence, such as a devastating flood or an unusually severe drought, is the probability of the occurrence,

$$P = \frac{1}{T} \tag{7.1}$$

In design, the planned or expected life of a structure is often stated. A navigation lock may be outmoded, or a reservoir may silt up in a foreseeable period of time. The probability P_n that a certain event with a return period T may occur in the first n years of the life of a structure is given by

$$P_n = 1 - \left(1 - \frac{1}{T}\right)^n \tag{7.2}$$

according to mathematical statistics. If one decides that the planned life of a structure is to be equal to the return period of the event, then the probability of the event's occurrence *within* the lifetime of the structure may be computed from Equation 7.2 by substituting T for n. With paper and pencil the reader can easily prove that with T equal to or greater than 10 years this probability ranges between 0.65 and 0.63. In other words, one takes a 65 percent chance that the event will, in fact, occur during the lifetime of the structure. If this is an unacceptable choice, one should decide what risk would be considered acceptable. For each hydraulic structure the acceptable risk may be different. A dam that endangers the life of thousands must be built with a much smaller risk factor than a minor structure. For design it is better to select the acceptable probability, or risk, that the designer is willing to take during the life of the structure and to set the cost of the design against the probable loss of life and property in case of failure due to the event considered. Often such decisions are not left to the designer but are incorporated into state codes that set the acceptable level of risk according to the potential danger downstream. Denoting the acceptable risk as U and the life of the structure as N, substituting into Equation 7.2, and rearranging gives

$$T = \frac{1}{1 - (1 - U)^{1/N}} \tag{7.3}$$

For instance, if the design life of a structure is 50 years, and the acceptable risk is selected to be 1 chance in 100, or $U = 0.01$, then Equation 7.3 will result in a return period of 5000 years, which is an extremely rare event. Unfortunately, aside from the floods on the Nile, there are no hydrologic data series available that will allow us to determine such extreme events. For this reason the design of major hydraulic structures in the United States is based on the concept of *probable maximum flood*, which comes from a meteorological estimate of the physical limit of rainfall over a watershed. Structures of lesser importance are designed for a percentage of the probable maximum flood, which is determined and published for specific regions of the country.[1] The U.S. Army Corps of Engineers uses the concept of *standard project storm* (SPS).[2] It is based on statistical analysis of the storms at a given drainage area, from which the *standard project flood* (SPF) is derived. In this model the infiltration capacity of the soil, seasonal effects, snow melt, general topographic effects on runoff—the most severe combination of meteorologic and hydrologic conditions that may occur, including rare effects—are considered. Both PMP and SPS methods will be described later in this chapter.

7.3 RAINFALL STATISTICS

Floods are generally caused by rains. Those caused by melting snow are only a more complicated variety. Rainfalls are measured by meteorological stations located at most airports and at other sites scattered about the country.

Data on rainfalls—sometimes going back over 100 years or more at some locations—are analyzed by statistical means. Such data are represented by three

---

[1] Probable Maximum Precipitation (PMP) was developed by the National Oceanic and Atmospheric Administration's (NOAA) National Weather Service (NWS) and described in the NOAA Hydrometeorological Report (HMS) series applicable to various areas of the United States.

[2] SPS design criteria are described in the Army Corps of Engineers' Civil Engineering Bulletin No. 52-8: *Standard Project Flood Determinations*, E. M. 11102–1411; Rev. March 1965.

important variables, namely, the intensity, the duration, and the frequency of rainfall.

Intensity is computed by dividing the rainfall depth of a particular rain by the period of time during which it fell. It is expressed in inches per hour or in millimeters per hour. Statistical findings have proved that generally the shorter the time of rainfall, the higher its intensity. *Duration* is the period of time of the rainfall. It is expressed in minutes, hours, or days.

Frequency of a rain refers to the return period of a particular rainfall characterized by a given duration, intensity, or both. It is expressed in years. Because of the relatively short record of rainfall at most places, statistical data are extrapolated on the basis of probability theory. This allows us to estimate rainfall and flood occurrences up to frequencies of 100 to 500 years even though data are available for a much shorter period. This assumes that the climate does not change at the particular place with time. Thus, a period of 30 years of measurement could easily contain rainfalls corresponding to frequencies much greater than 30 years. In 30 years of rainfall measurements several rains of 50-, 100-, or even 500-year frequencies may well be encountered. These extreme events become obvious only after the data have been subjected to statistical analysis.

Statistical analysis allows us to combine rainfall or flood data measured at several points in the geographic region in question. The method used for lumping these various sources of information is called *multicorrelation.* Because of the very large amount of data to be processed, the analysis is done by computers. Figure 7.1 shows rainfall depth, duration, and frequency curves developed for Columbus, Ohio, using multicorrelation.

Rainfall data for a particular geographic area are generally available to the designer of hydraulic projects from federal, state, or local sources. The U.S. National Weather Service of the NOAA, state departments of natural resources, state highway departments, nearby universities, and research organizations are all sources of hydrologic data. In developing countries, sometimes the best information is available from pertinent organizations of the former colonial powers.

Experience with storms has shown that heavy rains are associated with localized storm patterns of approximately elliptical shape that are aligned with the weather fronts. The concentration of rains over such storm patterns means that the smaller the area, the larger the maximum possible precipitation. Rainstorms creating the largest floods in a drainage basin are those whose storm patterns best overlap the drainage basin. If the topographic shape of a drainage basin extends in the general direction of the weather fronts commonly encountered in that region, large floods occur more frequently than in drainage basins that are perpendicular to the frontal patterns. Maximum possible 24 hr precipitations over 200 mi^2 areas within the United States east of the Rocky Mountains are shown in Figure 7.2. To convert these data to areas of other sizes and for rains of larger durations, Figure 7.3 may be used. The index rainfall ratio shown in this graph refers to the maximum possible rainfall as a percentage of the 200 mi^2, 24 hr rainfall obtained from Figure 7.2. Projects are rarely designed to handle the maximum possible precipitation unless the maximum possible safety is required, as in the design of spillways for high dams.

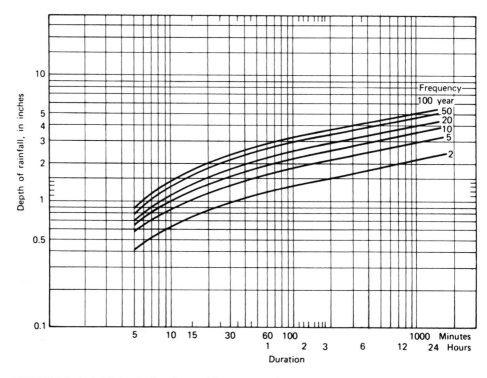

FIGURE 7.1 Rainfall depth, duration, and frequency at Columbus, Ohio.

U.S. Army Corps of Engineers projects indicate that about 40 to 60 percent of the maximum possible precipitation is usually selected as a basis of design for flood control projects.

7.4 THE MAGNITUDE OF FLOODS

Rainfall is only one of several parameters that contribute to the magnitude of floods encountered at a particular site. Other important parameters include the size of the land area on which the rain falls and the shape and average slope of this area along the main channel through which the rainwater is led to the site in consideration. The type of soil, its permeability, land use, and similar factors all contribute to the relative magnitude of a flood. Of all these variables the most important one for flood magnitude is the land area drained by a stream network. This is called the *watershed*[3] of the particular outflow point. The size of this drainage area is measured from topographic maps by the designer. Distinction is made by hydrology

---

[3]It is interesting to note that in technical parlance watershed means the area from which rainfall collects to a point of outflow. In liberal arts, watershed refers to the boundary line.

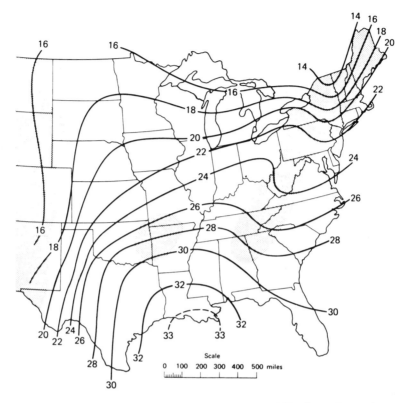

FIGURE 7.2 Maximum possible precipitation in inches over a 200 mi² area in one day.

experts between small watersheds and large watersheds. Drainage areas of less than 30 m² (about 80 km²) are considered small watersheds. For small watersheds the most important factor after the size of drainage area is the rainfall intensity. Second to this is the slope of the main channel. For large watersheds the most important variable after the drainage area size is the main channel slope, followed by the rainfall intensity.

Once the rain starts to fall, it begins its pattern of flow along the steepest descent of the land. The rate of collection depends on the relative density of available flow channels. Water flows faster in creeks and in other channels than on land. The velocity of the flow depends on the slope of the channels. Often in hydrologic computations the term *main channel slope, S*, is used. It is the slope between two points defined by excluding 10 percent of the downstream end and 15 percent of the upstream end along the main drainage path extending from the studied site to the watershed boundary. Thus, the steepest 15 percent and the flattest 10 percent of the land contours are excluded in this consideration.[4] The main

---

[4]Note that this is not so in the following formulas for time of concentration computations.

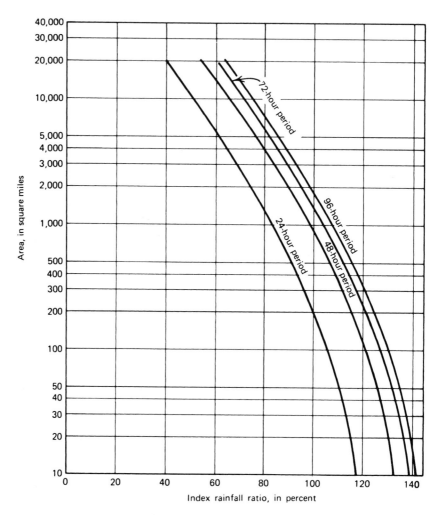

FIGURE 7.3 Relationships of depth, area, and duration from storm studies by the U.S. Army Corps of Engineers.

channel slope is determined by the design on the basis of available topographic maps.

The time it takes for the first raindrop that fell at the most distant point of the drainage area to reach the outlet of the watershed is called the *time of concentration,* T_c. Experience has proved that the most critical floods are from rains whose duration at least equals the time of concentration. For small drainage basins (less than 200 acres) of agricultural land the time of concentration in minutes may be estimated from the Kirpich[5] formula, which in SI terms is written as

[5]Z. P. Kirpich, *Civil Engineering*, Vol. 10, 1940.

$$T_c = \frac{0.0663\, L^{0.77}}{S^{0.385}} \qquad (7.4)$$

where T_c is in hours, L is the length of the principal water course from the outlet to the edge of the watershed in kilometers, and S is the slope between the maximum and minimum elevations. In customary American units T_c is in minutes, L is in feet, and the coefficient is 0.0078. For drainage of airports the Hathaway[6] formula is used, which in SI units is written as

$$T_c = \frac{0.606(Ln)^{0.467}}{S^{0.234}} \qquad (7.5)$$

The variable n is the roughness factor, a somewhat uncertain parameter that ranges from 0.02 for a smooth, impervious surface to 0.8 in the case of timberland.

The effect of surface soils and geologic characteristics is known to be highly important to flood peaks. Depending on the permeability of the land, part of the rainfall during its initial period infiltrates into the ground. Rainfall up to 0.5 in. may be absorbed into the soil in humid regions, provided the soil was relatively dry prior to the beginning of the rainfall. The rate of infiltration depends on the soil cover, land use, and cultivation type. Soil infiltration rates, S_i, in terms of inches per hour are often available from state conservation services, the U.S. Department of Agriculture, and similar organizations.

Flood discharges for small and large watersheds have been measured on a regular basis at many locations. The correlation of measured flood magnitudes with the corresponding rainfalls, drainage areas, main channel slopes, soil infiltration coefficients, and other pertinent variables generally results in an equation of the type

$$Q_T = a\, A^b S^c I^d S_i^e \qquad (7.6)$$

Here Q_T in cubic feet per second is the peak discharge of a flood with a given recurrence interval T. The superscripts a through e are constants determined by correlation and are valid for particular geographic regions; A is the drainage area in square miles; S is the main channel slope expressed in feet per mile; I is the rainfall intensity in inches per hour corresponding to the recurrence interval T sought for the flood Q_T; and S_i is the soil infiltration in inches per hour. In equations similar to Equation 7.6 the duration of the critical rainfall is usually taken to be the 24 hr intensity rainfall with a $T = 2.33$ years duration. An example for Equation 7.6 is given in Table 7.1, which gives values of the coefficients computed on the basis of a large-scale multicorrelation of data collected in northeastern Ohio.

When data of the type shown in Table 7.1 are unavailable, flood magnitudes may be estimated by what is known as the *rational formula*.

---

[6]G. A. Hathaway, *ASCE Transactions*, Vol. 110: 697–730, 1945.

TABLE 7.1   Flood Discharge Formulas for Northeastern Ohio

Drainage Basin Type	T years	a	b	c	d	e
Small watershed	2.33	1.9	0.84	0.15	4.03	0.43
(less than 40 m²)	10	3.2	0.81	0.14	4.14	0.41
	20	4.6	0.80	0.15	4.00	0.43
	50	7.3	0.79	0.15	3.70	0.40
Large watershed	2.33	5.2	0.86	0.39	1.62	−0.15
	10	12.0	0.82	0.36	1.58	−0.15
	20	12.2	0.82	0.39	1.65	−0.15
	50	16.5	0.80	0.35	1.72	−0.17

$$Q_T = aA^bS^cI^dS_i^e$$

Note: I is the intensity in inches per day with a constant recurrence interval of 2.33 years.

The rational method assumes a direct relationship between discharge, critical rainfall intensity, and drainage area. It is expressed by the equation

$$Q_T = cI_TA \tag{7.7}$$

where Q_T (acre-inch per hour)[7] is the flood discharge for a recurrence interval T, I_T is the rainfall intensity in inches per hour for a duration corresponding to the time of concentration T_c obtained from Equation 7.5 or by other means, and A is the area of the drainage basin in acres. The coefficient c in Equation 7.7 reflects the percentage of the rainfall that will run off from the watershed. Its magnitude depends on the land surface and is obtained from Table 7.2.

TABLE 7.2   Coefficients for Use with the Rational Formula (Equation 7.7)

Description of Area	Runoff Coefficients
Business	
Downtown	0.70 to 0.95
Neighborhood	0.50 to 0.70
Residential	
Single-family	0.30 to 0.50
Multiunits, detached	0.40 to 0.60
Multiunits, attached	0.60 to 0.75
Residential (suburban)	0.25 to 0.40
Apartments	0.50 to 0.70
Industrial	
Light	0.50 to 0.80
Heavy	0.60 to 0.90
Parks, cemeteries	0.10 to 0.25
Playgrounds	0.20 to 0.35
Railroad yard	0.20 to 0.35
Unimproved	0.10 to 0.30

[7]One acre-inch per hour nearly equals 1 ft³/sec.

The range of values given in Table 7.2 indicates the degree of uncertainty inherent in the rational method. Because of its simplicity, however, its use is rather popular for the design of small hydraulic structures, particularly in urban areas.

An extension of the rational method is incorporated in the U. S. Soil Conservation Service's TR-55 computer program. The refinements incorporated in that procedure allows the use of the rational formula for midsize watersheds.

7.5 THE UNIT HYDROGRAPH METHOD

Of the several other methods used to determine the magnitude of floods caused by rains of specific intensity and duration the *unit hydrograph method* stands out as an illustrative and practical tool.[8] It is used in several versions, from simplified forms to those of various degrees of refinement.

A *hydrograph* is a plot of the discharge at the outlet point of the watershed versus the time elapsed after a specific rain has started. It is generally assumed that the rain is constant in intensity and uniform over the whole drainage area. Usually the *excess rainfall* is considered: what remains after the amounts lost to evaporation and infiltration are subtracted. The fundamental assumption in the use of hydrographs is, therefore, that rainfalls identical in intensity and duration produce identical outflow hydrographs from the drainage basin considered. Because it is difficult to obtain hydrographs from gaging stations for design purposes, and design hydrographs are often needed at ungaged sites, various techniques have been developed over the years that estimate the hydrograph for a given rainfall duration and intensity. The unit hydrograph method is the most fundamental of these techniques.

Barring the existence of natural and artificial storage locations within the watershed, the total volume of the excess rainfall over the drainage basin equals the computed area under the hydrograph plot. For this computation the so-called base flow, the discharge flowing in the channel before and during the rain contributed by exfiltrating groundwater, is excluded. When computing the total volume, care must be taken to observe the units in which the discharge and the time were measured and plotted. The total discharge is obtained by summing the products of the Q and time interval Δt over the whole time period while the runoff from the specific rain is being observed. Figure 7.4 depicts the procedure. In this graph the measured rain for a specific duration and the associated runoff hydrograph are first plotted. The rainfall depth is then reduced to 1 in. The same reduction ratio is applied to the runoff hydrograph throughout the time scale, assuming that the time to peak, t_p, remains the same for the unit hydrograph that results.

The average rainfall over the drainage basin may be computed by dividing the total volume of runoff as determined from the hydrograph by the area of the basin. The result is usually converted to express the average rainfall depth in inches. The conversion table, Table A.4, in the appendix will aid in this process.

---

[8]In fact, it is the only tool a hydraulic designer can apply in underdeveloped countries where rainfall and flood flow information is unavailable.

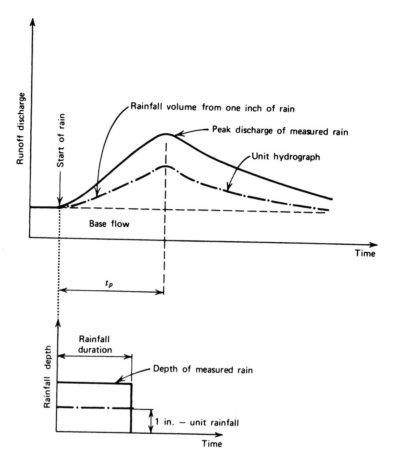

FIGURE 7.4 Hydrograph.

The unit hydrograph is defined as *the hydrograph that results from 1 in. of excess rainfall due to a storm of specified duration.* It may be obtained from any measured hydrograph by first computing the average rainfall depth of the storm and then reducing the measured discharge values proportionately so that the resulting hydrographs will be that of a 1 in. rain of the same duration. For example, if a 6 hr storm produces a certain total runoff volume that represents 3 in. of average rainfall depth over the drainage area, then the unit hydrograph of a rainfall of 6 hr duration is obtained by dividing all measured discharge values by 3. To produce unit hydrographs for various rainfall durations we have to search rainfall records to identify storms of specific durations and then locate the corresponding hydrographs from stream flow records. Base flows have to be deducted from the latter before their total runoff volumes are computed.

The fundamental assumption of the unit hydrograph method is that the runoff of a given storm is independent of the runoffs due to previous storms. This concept allows the generation of composite hydrographs to represent storms of any desired rainfall pattern, simply by adding or subtracting unit hydrographs.

The unit hydrograph method thus provides the hydraulic designer with two important pieces of information concerning the nature of the runoff from a specific storm falling over a given drainage area. One is the magnitude of the peak flow; another is the time when it occurs after the rain has started. Floods of the most critical magnitude generally result from rainfalls whose duration is equal to the time of concentration of the given watershed. This occurrence is explained by a fundamental fact of rainfall statistics: Usually, rainfalls of high intensity are those of short duration. Experience has shown that the time to peak (t_p) can be reasonably approximated by the formula

$$t_p = \frac{D}{2} + 0.6T_c \qquad (7.8)$$

in which D is the duration of the storm.

7.6 THE SCS METHOD

Rainfall–runoff relations for small watersheds (less than 2000 acres) may be represented by the curve number method developed by the U.S. Soil Conservation Service (SCS). This technique is based on extensive analytical work and on a wide range of statistical data concerning storm patterns in the United States, rainfall and runoff characteristics, and other observations. Computer-generated design charts are available,[9] making the application of the method very easy. Use of this technique for the determination of design discharge of most hydraulic installations is often required by state authorities.

If the rainfall and runoff are measured for a series of storms on a certain watershed, their relationship may be plotted in the manner shown in Figure 7.5.

FIGURE 7.5 Rainfall versus direct runoff.

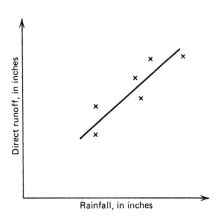

[9]Such as SCS Technical Paper No. 149.

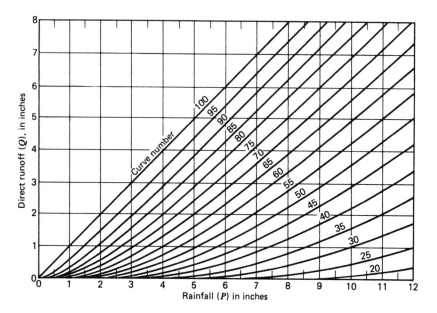

FIGURE 7.6 Curve numbers for rainfall–runoff relationships.

Each point on the graph represents a particular storm. The horizontal coordinate shows the rainfall in inches, and the vertical coordinate is the corresponding direct runoff in inches, computed from the total runoff volume divided by the area of the watershed. Depending on events preceding the rain, the portion of the rainfall that runs off the watershed may vary somewhat, but for average conditions a curve may be drawn through the points that relate direct runoff to rainfall. The slope of this curve will generally depend on the land use and the soil type. A meadow of highly permeable soil may allow no more than a 30 percent runoff, whereas a well-cultivated cornfield made up of clayey soil may have 90 percent runoff. The proportion of runoff also depends on the quantity of the rainfall: In the initial period of the rain more will be lost to infiltration. The curve in Figure 7.5, therefore, will not be a straight line, except for the idealized case when 100 percent of the rainfall runs off directly. That case is represented by curve number 100 on Figure 7.6, which gives curve numbers for various small watersheds. The SCS developed this graphical form of the runoff equation. By measuring a few sets of rainfall–runoff data points and marking them on Figure 7.6, one can determine the curve number that best represents the watershed in question. When no rainfall–runoff data are available, the appropriate curve number may be estimated from Table 7.3. In this table soil group A represents highly permeable soils that drain well even when wet, and D refers to highly impermeable soils. Groups B and C represent intermediate soil types.

Studies of storm patterns throughout the United States have shown that there are significant differences in the 24 hr rainfall distributions in different parts of the country. In Southern California's coastal region, in Hawaii, and in Alaska's interior

TABLE 7.3   Runoff Curve Numbers for Hydrologic Soil-Cover Complexes

Land Use and Treatment or Practice	Hydrologic Condition	Hydrologic Soil Group			
		A	B	C	D
Fallow					
Straight row	—	77	86	91	94
Row crops					
Straight row	Poor	72	81	88	91
Straight row	Good	67	78	85	89
Contoured	Poor	70	79	84	88
Contoured	Good	65	75	82	86
Contoured and terraced	Poor	66	74	80	82
Contoured and terraced	Good	62	71	78	81
Small grain					
Straight row	Poor	65	76	84	88
Straight row	Good	63	75	83	87
Contoured	Poor	63	74	82	85
Contoured	Good	61	73	81	84
Contoured and terraced	Poor	61	72	79	82
Contoured and terraced	Good	59	70	78	81
Close-seeded legumes or rotation meadow					
Straight row	Poor	66	77	85	89
Straight row	Good	58	72	81	85
Contoured	Poor	64	75	83	85
Contoured	Good	55	69	78	83
Contoured and terraced	Poor	63	73	80	83
Contoured and terraced	Good	51	67	76	80
Pasture or range					
No mechanical treatment	Poor	68	79	86	89
No mechanical treatment	Fair	49	69	79	84
No mechanical treatment	Good	39	61	74	80
Contoured	Poor	47	67	81	88
Contoured	Fair	25	59	75	83
Contoured	Good	6	35	70	79
Meadow	Good	30	58	71	78
Woods	Poor	45	66	77	83
	Fair	36	60	73	79
	Good	25	55	70	77
Farmsteads	—	59	74	82	86
Roads[a]					
Dirt	—	72	82	87	89
Hard surface	—	74	84	90	92

Source: Soil Conservation Service Technical Paper No. 149.

[a]Including rights of way.

the 24 hr rainfall distribution (Type I storm) is significantly different from that in the Northwestern coastal regions (Type IA storm), and again from the distribution in the rest of the country (Type II storm). Thus, different standards for runoff discharge determination have been developed for different geographic locations.

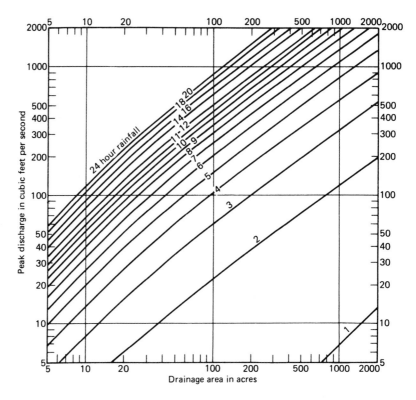

FIGURE 7.7 Peak runoff discharge from small watersheds. A typical Soil Conservation Service design chart. (From *Urban Hydrology for Small Watersheds*, Technical Release No. 55, Soil Conservation Service, 1975)

Using the regional 24 hr storm pattern characteristics allowed SCS hydrologists to eliminate the need to determine the time of concentration for each watershed. In the SCS method the regional 24 hr rainfall depth is used.

The size and shape of the watersheds also were found to influence the runoff peak discharge. The SCS standards were developed for flat (0–3%), moderate (3–8%), and steep (8%+) slopes, interpreted as average slope in the direction of the overland flow over the whole watershed.

The SCS developed a set of 21 charts that provide a direct graphical solution for an otherwise rather tedious computational procedure. Each of these charts is made for a given storm distribution, curve number, and slope. A representative chart is reproduced in Figure 7.7. It gives the peak discharge for a watershed. The solution requires the determination of the watershed area, curve number (e.g., from Table 7.3), the average watershed slope, and the 24 hr rainfall depth. The last factor allows consideration of return periods or maximum probable rainfall in the design.

There are several other ways to obtain the magnitude of flood discharges pertaining to a selected recurrence interval. These include semigraphical techniques,

FIGURE 7.8 Subdivision flooded by the Mississippi River. (Courtesy of U.S. Army Corps of Engineers)

such as the unit hydrograph method, the standard project storm method used by the U.S. Army Corps of Engineers, which is based on storm patterns found to give critical floods in certain geographic areas, the Chicago hydrograph method used for sewerage design, and others. Information on these is available in the technical literature.

7.7 FLOOD ROUTING

Once the flood waters of a rain begin flowing in a channel, the flood wave soon passes the area over which the rain has fallen. The analysis of the flow of a flood wave in a channel is called *flood routing*. Precise routing of floods is very important because it allows timely evacuation of threatened flood plains and enables the preparation of protective measures. Figure 7.8, showing a flooded subdivision near the Mississippi River, proves that such measures are not always successful. Often, floods cause major damage to hydraulic structures that were designed to handle them. Figures 7.9 and 7.10 show a concrete spillway of a small dam in Ohio, before and after a flood. The former is taken from the upstream side; the latter, from downstream.

Routing of a flood means determining the variation in the discharge at the downstream end of a portion of a certain length of channel, called a *reach*, if the upstream discharge variation is known. Dividing a stream into reaches results in a series of "snapshots" showing the downward movement of the flood wave throughout the channel in successive—say, 24 hr—time periods.

FIGURE 7.9 Concrete spillway with energy dissipator at its end in earthen dam. (Courtesy of Ohio Department of Natural Resources)

FIGURE 7.10 Same as Figure 7.9, photographed from the downstream end after a flood. Failure was caused by seepage-induced erosion below concrete lining. (Courtesy of Ohio Department of Natural Resources)

At the edge of a catchment area over which the rain has fallen, the maximum peak discharge can be computed by the methods described in the preceding sections. Knowing the stream's discharge before the rain—the so-called base discharge—and knowing the approximate value of the time of concentration, we may construct a curve showing the approximate relationship between the discharge and time. A nearly S-shaped curve between the base discharge Q_B and the peak discharge Q_p over the time T_c is a reasonable approximation when field gage measurements are lacking. Figure 7.11 shows such a relationship, called an inflow hydrograph of the reach below. The portion of the curve beyond T_c was drawn as the mirror image of the first part of the curve.

The reach may include a reservoir below for which a hydraulic structure is to be analyzed. Alternatively, the reach can include a portion of a river followed by subsequent reaches. The methods to be outlined work for both cases, giving the outflow hydrograph for a reach on the basis of the known inflow hydrograph. The outflow hydrograph becomes the inflow hydrograph of the subsequent reach downstream.

Knowing the length, surface area, and the shape of the reservoir or the general channel section of the reach in question, we can construct a graph showing the depth of the water versus the storage in the channel, as shown in Figure 7.12. By continuity we know that for any given time period Δt the difference between the flow entering and leaving the reach must equal the change of storage in the reach. Denoting the flow at the beginning of Δt by subscript 1 and at the end of Δt by subscript 2 of the inflow and outflow discharges, we can write our continuity relationship as

$$\frac{I_1 + I_2}{2} \Delta t - \frac{O_1 + O_2}{2} \Delta t = \Delta S = S_2 - S_1 \tag{7.9}$$

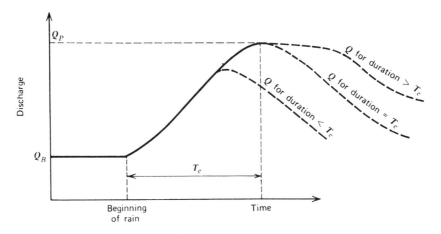

FIGURE 7.11   Inflow hydrograph.

FIGURE 7.12 Storage as a function of depth.

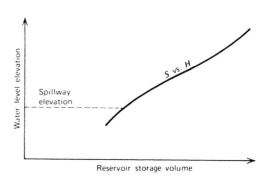

in which ΔS refers to the change in storage within the reach during the period Δt.

The inflow hydrograph is known in Equation 7.9. The initial value of the outflow discharge O_1 is known to be equal to I_1, since the base flow is constant before the rain. The initial value of the storage, S_1, is known, too. The time period Δt is selected to be a convenient value, for example, 6 hr. Rearranging Equation 7.9 to group the unknown values on the right side, we obtain

$$I_1 + I_2 + \frac{2S_1}{\Delta t} - O_1 = \frac{2S_2}{\Delta t} + O_2 \qquad \textbf{(7.10)}$$

This gives us one equation with two unknowns, the outflow discharge and the storage at the end of the time period. To solve the problem, we need an additional equation.

The additional equation needed for the solution of the routing problem may come from two sources: A hydraulic structure along the stream provides one additional relationship. For a spillway the discharge may be determined as a function of the height of the water over the spillway crest, using the weir formula. The relationship is shown by the so-called rating curve of the spillway, an example of which is shown in Figure 7.13. The water elevations shown in Figures 7.12 and 7.13

FIGURE 7.13 Spillway rating curve.

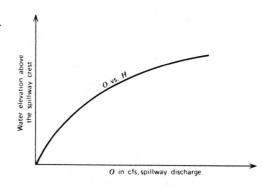

FIGURE 7.14 Combined reservoir storage spillway outflow graph to solve routing equation.

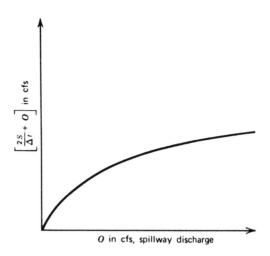

$\left[\frac{2S}{\Delta t} + O\right]$ in cfs

O in cfs, spillway discharge

are identical. Hence, the two graphs give a relationship between the two unknown variables, S and O, in Equation 7.10. For convenience in computations a new graph may be constructed using Figures 7.12 and 7.13, selecting a few water elevations to form corresponding values of $[(2S/\Delta t) + O]$ as shown in Figure 7.14. The graph now simplifies Equation 7.10 by making its right side a function of the outflow only.

The outflow hydrograph, that is, the value of O_2 for all subsequent time periods Δt, can now be computed in a step-by-step manner, using the value of O_2 as O_1 for the next time period.

The flood routing computation for rivers with no control structures (spillways or weirs) is a more involved procedure, because the problem is mathematically complex. For rough computations one may apply the technique developed by the U.S. Army Corps of Engineers known as the *Muskingum[10] method*. It requires that the designer have at least one set of field data on a flood for the reach in question. In particular, the time difference between flood peaking at the inflow and outflow sections must be known. With this time lag K, the relationship between the storage and the inflow and outflow values can be written as

$$S = K[xI + (1 - x)O] \tag{7.11}$$

This equation states that the storage is a function of the inflow as well as the outflow according to a weighting coefficient x. The value of this weighting coefficient is usually taken to be about 0.35 for rivers in the Appalachian region. To determine K and x for a particular reach a set of corresponding inflow and outflow values must be known. For each time period the storage S in the reach can be determined

---

[10]The procedure was first developed for the Muskingum River project in Eastern Ohio, which was built in the 1930s.

from the equation $S = \Sigma (I - O)\Delta t$. Plotting these results against the bracketed term in Equation 7.11 for various assumed values of x will result in graphs resembling hysteresis loops. The graph in which the rising and receding lines are the closest is the one that was plotted with the best weighting factor x. The value of the corresponding time lag, K, is the slope of the best-fit straight line drawn through the selected graph. With the assumed K and x values the outflow O_2 at subsequent time periods Δt can be computed step by step using the formula

$$O_2 = c_0 I_2 + c_1 I_1 + c_2 O_1 \qquad (7.12)$$

where the c coefficients are calculated for the reach before the computations begin, from the equations

$$c_0 = \frac{-Kx + 0.5\,\Delta t}{K - Kx + 0.5\,\Delta t}$$

$$c_1 = \frac{Kx + 0.5\,\Delta t}{K - Kx + 0.5\,\Delta t} \qquad (7.13)$$

$$c_2 = \frac{K - Kx - 0.5\,\Delta t}{K - Kx + 0.5\,\Delta t}$$

(Note that $c_0 + c_1 + c_2$ must equal 1.) To route a flood for a river using this approximate method, we start at the upstream end, where all I values and the O_1 value are known. Using Equation 7.12 we can compute O_2. For subsequent steps the subscripts are reduced by 1, changing O_2 to O_1 and I_2 to I_1, and so on.

7.8   RIVER FLOW

In the large watershed of a major river, rainfall events are often limited to specific areas rather than being spread over an entire area. The rains brought about by different weather patterns generate flood waves that move down the river channel in various ways. At any specific point along the river the stage or discharge hydrograph represents the sum of several components. One is the base flow that results from the infiltration of ground water, itself caused by long-past rains. The remainder of the hydrograph may be the sum of several flood waves: the receding part of an earlier rain runoff, a flood wave just peaking at the stage, and the rising part of the most recent rain's flood wave from another part of the watershed. Forecasting a peak design discharge is done by statistical analysis of a long series of flood peaks. In such an analysis one of the basic assumptions is that each flood peak is independent of the others. The critical flood peak, however, is often the one caused by the coincidence of two independent events, a case that often brings about catastrophic results.

For most hydraulic design problems the yearly distribution of peaks or ebbs is of little interest. One may want to know the expected highest flow, the expected lowest flow, and the percentage distribution of the flows in between. This informa-

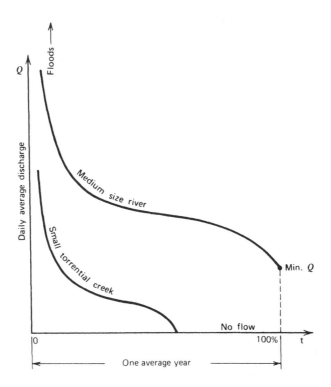

FIGURE 7.15   Discharge-duration graphs for yearly flow of rivers.

tion can best be represented by *discharge-duration* curves, as shown in Figure 7.15. They are created by selecting daily average discharges for 365 days of a year and arranging them according to their magnitude. The source of this data may be the complete series of available flow data reaching back several decades. The result will show the number of days in a year in which a certain discharge was exceeded. For large rivers this curve is relatively flat, peaking toward the left, representing occasional major floods. For small creeks the curvature of the graph is more pronounced, and during part of the year the discharge may be zero.

The usefulness of a discharge-duration curve may best be represented by its use in estimating the hydraulic power that may be generated at a particular site. The site conditions will probably determine the highest possible raising of the water by a dam; therefore, the power that may be generated will depend on the discharge, according to Equation 2.18. If there is no significant storage capacity in the reservoir above the dam, we speak of a "run of the river" type of power generation. In such a case the flow through the dam is equal to or less than the daily discharge. If the power station is designed to provide a constant amount of electricity throughout the year, then the minimum discharge is the design discharge for the turbines. This is the 100 percent flow in the discharge-duration graph. Any discharge greater than this minimum is allowed through the spillway, because no excess generating capacity is thereby wasted from the standpoint of power generation. If the power station is designed for a discharge greater than the minimum, then part of the generating capacity is not utilized during some parts of the year for lack of running water. Because electric power nowadays is fed into a power grid by many power generating stations, a much larger amount of hydraulic energy can be utilized than before. Selecting the optimum capacity of hydraulic power stations is therefore the subject of a complex economic analysis that is largely based on the available hydrologic information.

7.9 COMPUTATIONAL HYDROLOGY

Since the advent of digital computers a great deal of effort has been spent on the development of computer programs to manipulate the huge amount of data required in hydrologic computations. The earliest one, HEC-1, was a computer program developed by the Hydrologic Engineering Center (HEC) of the U.S. Army Corps of Engineers to assist in the analysis of river basin responses to rainfall events. The original program was developed in 1967 by Leo R. Beard and other members of the HEC staff. Since that time various modifications have been made to the program to simplify the input requirements and to increase the capabilities of the program. The HEC-1 program is written in FORTRAN, and versions are available for the PC and a number of mainframe computers. The program is a batch input and output program, requiring the user to prepare a file containing topographic as well as hydrologic and other information.

The general methodology of the program divides the river basin into an interconnected system of stream network components, namely, (1) land surface runoff, (2) river routing, (3) reservoir, (4) diversion, and (5) pump. A number of calculation techniques are available for each of these components.

Precipitation for the hydrologic simulation can be generated by specifying information for a standard project storm, by estimating the probable maximum precipitation, or by using a table of depth duration data. Provisions for snowfall and snowmelt are provided.

The land surface runoff component allows for the following interception/infiltration models: initial and uniform loss, exponential loss rate, SCS curve number method, Horton loss rate, and combined snowmelt and rain losses. These infiltration models can be combined with a number of hydrograph routing techniques. All these techniques are forms of the unit hydrograph method. The unit hydrograph can be entered directly or generated by using a number of synthetic hydrograph generation methods. The available synthetic hydrograph generation methods are the Clark unit hydrograph, the Snyder unit hydrograph, and the SCS dimensionless unit hydrograph. Overland flow can also be calculated by using kinematic wave equations and the Muskingum-Cunge routing techniques. These techniques are described in detail in the HEC-1 documentation.

Infiltration losses can be calculated in the river and reservoir components by a number of routing techniques available in the program, namely, the Muskingum,

Muskingum-Cunge, Modified Puls, level-pool routing, the working R and D method, and the standard orifice and weir equation.

The HEC-1 model is designed to be used by a trained hydrologist or civil engineer. The input may require various iterations to calibrate the hydrologic model. Engineering judgment will need to be applied in most cases.

Another program developed by the U.S. Army Corps of Engineers' Hydrologic Engineering Center is HEC-4, which analyzes monthly stream flows at interrelated stations. It statistically analyzes the stream flow data and generates a sequence of hypothetical flows for watersheds with similar characteristics, for any desired time period. The program will also reconstitute missing stream flow data on the basis of concurrent flows observed at other locations.

Flood flow frequencies may be analyzed by the HECWRC program. It performs a statistical probability analysis using the log-Pearson type III distribution[11] to compute the frequency curve.

Another popular computer program used in hydrologic analysis is the Technical Release 20 (TR-20). This is a computer program written to implement the hydrologic routing methods outlined in the National Engineering Handbook-4 (NEH-4), which is published by the SCS. The program is written in FORTRAN and utilizes batch input and output.

The SCS method is used throughout the TR-20 program. Watersheds are delineated into homogenous subwatersheds from which hydrographs are generated. Actual and synthetic rainfall distributions can be used in the generation of the watershed hydrographs. Watershed hydrographs can be combined and routed to a reservoir or a reach. Reservoir routing uses the storage-indication method, and reach routing uses the modified attenuation-kinematic (Att-Kin) method to route the hydrograph flow to an outlet.

Another popular program, Technical Release 55 (TR-55), is considerably simpler than those mentioned earlier. It is designed to calculate storm runoff volumes, peak discharge rates, hydrographs, and required storage volumes for small urban watersheds. TR-55 was written by the SCS and originally released in 1975. Rainfall information provided at the level of U.S. counties is generated using a 24 hr rainfall distribution; four distributions are included with the program. Depending on the requirements of the user, the program can provide peak discharge or an output hydrograph. For peak discharge calculations the graphical peak discharge method is used. In situations where a hydrograph is required a tabular hydrograph method is used. Infiltration and soil conditions are modeled using the SCS curve number method.

A commercial computer program, POND-2, encapsulates the methodology of TR-55 with a menu-driven interface. POND-2 adds the capability to route hydrographs through outlet structures such as weirs and culverts. Also included is the ability to route the hydrographs through ponds using inventory routing. Ponds are modeled using a stage-storage relationship.

---

[11]The reader is referred to the literature on statistics concerning distribution functions.

Of the many other hydrologic computer programs, Stormwater Management Model (SWMM) is perhaps the most extensive hydrologic and hydraulic routing model produced. It has a diverse array of computational routines for computing runoff, analyzing sanitary and storm sewer networks, and modeling treatment of sewage and pollutants. The program can generate hydrographs and estimate runoff pollutant, sediment, and sewage loads. Introduction of these computer solutions is beyond the scope of this text.

EXAMPLE PROBLEMS

Example 7.1

Determine the return period and intensity of a rainfall in the Columbus, Ohio, area for a rain lasting 2 hr if the total rainfall depth was measured to be 3 in.

Solution. By definition the rainfall intensity is the rainfall depth divided by its duration in hours; hence,

$$I = \frac{3}{2} = 1.5 \text{ in./hr}$$

From Figure 7.1 a 2 hr duration rain at 3 in. rainfall depth corresponds to a 50-year return period. □

Example 7.2

Estimate the maximum possible 48 hr rainfall over a 20 mi^2 area in the northwest corner of Ohio.

Solution. The maximum possible rainfall at this location for 24 hr of rainfall over a 200 mi^2 area is indicated, in the map of Figure 7.2, to be 22 in. By using Figure 7.3 we find that the 48 hr rainfall for a 20 mi^2 area is 130 percent of the 22 in., or 28.6 in. □

Example 7.3

If a 25 mi^2 drainage area north of Columbus, Ohio, has a mean channel slope of 2 percent, and the soil infiltration is 0.6 in./hr determine the expected maximum flood that has a recurrence interval of 10 years.

Solution. From Equation 7.6

$$Q_T = a A^b S^c I^d S_i^e$$

in which

$$A = 25 \text{ mi}^2$$
$$S = 0.02 \text{ ft/ft} = 0.02 \times 5280 = 105.6 \text{ ft/mi}$$

From Figure 7.1

$$I = 2.5 \text{ in./24 hr} \qquad \text{at 2.33 years frequency}$$
$$S_i = 0.6 \text{ in./hr}$$
$$T = 10 \text{ years}$$

From Table 7.1 $a = 3.2$, $b = 0.81$, $c = 0.14$, $d = 4.14$, $e = 0.41$. Then,

$$Q_{10} = 3.2(25)^{0.81}(105.6)^{0.14}(2.5)^{4.14}(0.6)^{0.41}$$
$$= 3.2(13.56)(1.92)(44.4)(0.81)$$
$$Q_{10} = 3000 \text{ cfs} \qquad \square$$

Example 7.4

A 30-acre single family residential development is built on gently sloping land. The general slope of the land is 6 ft/mi. The maximum length of travel for the runoff water is 1500 ft. Estimate the 5-year design discharge for a drainage channel, assuming that the local rainfall statistics correspond to the data in Figure 7.1.

Solution. The time of concentration T_c is computed by Equation 7.4 in customary American units as follows:

$$S = \frac{6 \text{ ft}}{1 \text{ mi}} = \frac{6}{5280} = 0.0011 \text{ ft/ft}$$

$$T_c = 0.0078 \left(\frac{1500}{\sqrt{0.0011}} \right)^{0.77} = 30 \text{ min}$$

From Figure 7.1 the 30 min duration rain at 5-year frequency corresponds to a rainfall depth of 1.29 in. of rain. The intensity of this rain is $I_5 = 2.58$ in./hr.
By using Equation 7.7, we get

$$Q_T = cI_T A$$

With $c = 0.35$ from Table 7.2 we get

$$Q_5 = 0.35 \times 2.58 \times 30$$
$$= 27.1 \text{ acre in./hr}$$

Using the conversion table in the appendix, Table A.3, we find that this discharge equals

$$Q = \frac{27.1}{12} \times 24 = 54.2 \text{ acre ft/day}$$
$$= 54.2 \times 0.504$$
$$= 27.3 \text{ cfs} \qquad \square$$

Example 7.5

A rainfall lasting 1 hr produced a runoff hydrograph from a 10 mi^2 drainage basin as shown in the line labeled "Measured Q" in the following table of computations. The discharge in cubic feet per second was measured at the beginning of each of twelve 4 hr time periods. At the beginning of the storm the base flow for the basin was 20 cfs. Determine the unit hydrograph of the basin for 1 hr storms.

Solution. First, we deduct the base flow from all measured discharges, as shown in the line labeled "Reduced Q." Next, we compute average discharges for each time period by averaging the discharges recorded at the beginning and end of each period; these data are on the line labeled "Average Q." We multiply the average discharge for each period by its duration (4 hr = 14,400 sec); these data are on the line labeled "Period Total" and represent millions of cubic feet. We determine the total volume of surface runoff by adding all the volumes for each time period. The result is 40.1 million ft³. We divide the total volume by the basin area to determine the equivalent depth of water over the basin, which yields

$$\frac{40,100,000}{10(5280)^2} = 0.14 \text{ ft, or } 1.7 \text{ in.}$$

This is the average amount of rain that fell on the watershed during the 1 hr storm. To obtain the unit hydrograph for storms of 1 hr duration, we divide the "Reduced Q" values by 1.7 in.; these data are on the last line of the table.

Table of Computations for Example 7.5

4 hr Time periods	1	2	3	4	5	6	7	8	9	10	11	12	
Measured Q (cfs)	20	90	430	960	650	410	210	110	60	40	30	20	
Reduced Q (cfs)	0	70	410	940	630	390	190	90	40	20	10	0	
Average Q (cfs)	35	240	675	785	510	290	140	65	30	15	5	0	
Period total (10^6 cfs)	0.5	3.5	9.7	11.3	7.3	4.2	2.0	0.9	0.4	0.2	0.1	0.0	Sum = 40.1
Unit hydrograph (cfs)	0	41	241	553	371	229	112	53	24	12	6	0	

Example 7.6   Over the same drainage area considered in Example 7.5 a rainstorm falls in the following manner: 0.5 in./hr for 2 hr, followed by 2 in./hr for 1 hr. Using the unit hydrograph developed in Example 7.5, construct a runoff hydrograph for the storm.

Solution. First, we multiply the unit hydrograph values for Example 7.5 by 0.5 because 0.5 in. of rain fell the first hour. The second hour is a repeat of the first except it is delayed 1 hr. During the third hour 2 in. of rain fell, so we multiply the unit hydrograph by 2; this hydrograph lags the first by 2 hr. The runoff hydrograph is the net result obtained by combining the flows of the three hydrographs just described. The result is shown in Figure E7.6. The peak discharge is 1590 cfs; it occurs 15 hr after the beginning of the storm.

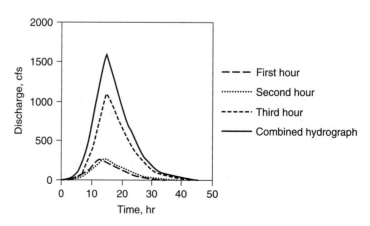

FIGURE E7.6

Example 7.7

A 58-acre meadow at La Jolla, California, is on steep clayey land. Hydrologic analysis of several recent storms indicates that 81 percent of the rainfall runs off from the watershed. According to the U.S. Weather Bureau, the 24 hr maximum probable precipitation in the region is 14 in. What is the expected maximum peak runoff from the land?

Solution. First, we check the result of our hydrologic analysis using the data in Table 7.3. For clay (hydrologic soil group D) the appropriate curve number for meadows is 78; therefore, our result of 81 is close enough. La Jolla is in the California coastal region, subject to storm patterns defined as "Type I." We assume that the SCS design chart shown in Figure 7.7 applies. For 58 acres and 14 in. of rainfall over 24 hr the resulting peak runoff is 400 cfs.

□

Example 7.8

Floods from a 5.87 mi² drainage area enter a reservoir that has a 50-acre surface area bounded by steep shores such that the surface area is essentially constant for small variations of the water level. The water leaves the reservoir through a weir with a width of 39.6 ft and a coefficient of 0.707. From hydrologic analysis it was determined that the 50-year flood results in a peak discharge of 835 cfs. The time of concentration was computed to be 100 min. Route this flood through the reservoir under the assumption that the reservoir is full to the crest of the weir at the beginning of the rain and that the duration of the rain is 100 min.

Table of Computations for Example 7.8

1	2	3	4	5	6	7
T	Q_{in}	$V_{(in)}$	$V_{(out)}$	S	H_1	Q_{out}
min	cfs	ft³	ft³	ft³	ft	cfs
0	0	0	0	0	0.0000	0.000
10	10	3	0	3	0.0014	0.011
20	70	24	0.003	27	0.0124	0.31
30	200	81	0.096	108	0.050	2.48
40	370	171	0.84	278	0.128	10.25
50	510	264	4	538	0.247	27.60
60	620	339	11	866	0.398	56.3
70	710	399	25	1240	0.569	96.5
80	775	446	46	1639	0.753	146.7
90	820	479	73	2045	0.94	204
100	835*	497	105	2436	1.12	266
110	810	494	141	2788	1.28	326
120	780	477	177	3088	1.42	379
130	720	450	211	3327	1.53	424
140	660	414	241	3500	1.61	458
150	580	372	265	3607	1.66	479
160	470	315	281	3641	1.67	486*
170	350	246	289	3598	1.65	477
180	230	174	289	3483	1.60	454
190	220	135	279	3338	1.53	426
200	210	129	264	3202	1.47	401

Columns 3, 4, and 5 are in terms of 1000 ft³.

*Starred discharges represent peak flows.

Solution. Column 1 is the time in increments of 10 min. Columnn 2 is the corresponding inflow hydrograph with a peak flow, Q_{max}, of 835 cfs at 100 min from the beginning of the rainfall. Column 3 is the volume that flowed into the reservoir during the current time increment. It is calculated by averaging Q_{in} at the beginning and end of the current time period and multiplying by the duration of the period (10 min = 600 sec). Column 4 is the volume that flowed out of the reservoir during the previous time period. It cannot be calculated for the current time step because the discharge over the weir has not yet been computed. V_{out} is calculated similarly to V_{in} except it uses the average of Q_{out} at the beginning and end of the previous time period. Column 5 is the reservoir storage. It is calculated by adding the storage from the previous time step plus the volume that flowed into the reservoir during the current time period minus the volume that flowed from the reservoir. Column 6 is the head over the weir. It is calculated by dividing the storage by the area of the reservoir [50 acres (43560 ft/acre) = 2,178,000 ft³]. Column 7 is the outflow hydrograph calculated by the weir formula (Equation 3.23)

$$Q_{out} \text{ (cfs)} = 0.707 \,(39.6)(64.4)^{1/2} \, H_1^{3/2} = 224.7 \, H_1^{3/2}$$

where H_1 is the height of water in the reservoir over the crest of the weir.

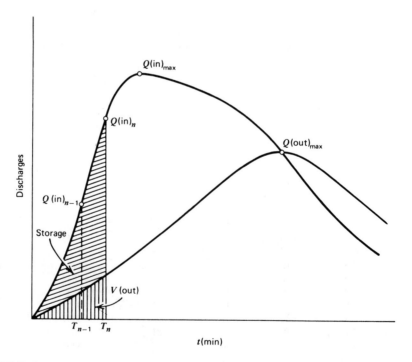

FIGURE E7.8

Figure E7.8 shows the inflow and outflow hydrographs as a function of time. It shows that the reservoir serves to moderate the flood peak. Thus, the outflow hydrograph from the reservoir has a smaller peak and a longer duration than the inflow hydrograph. The maximum inflow of 835 cfs occurs 100 min after the initiation of the rainfall. The rain also stops after 100 min. The maximum outflow of 486 cfs occurs 60 min after the peak inflow (or 60 min after the storm stops). This represents a 42% reduction in the flood peak discharge. For the case when the duration of the rain exceeds the time of concentration of a basin, the inflow hydrograph should be adjusted by extending horizontally the peak flow of the inflow hydrograph for the rain duration.

In some instances the outflow hydrograph may be computed with the aid of a graph showing the relationship of water elevation in the reservoir and spillway dishcarge (see Figure 7.13). Also a graph of the water level elevation as a function of reservoir storage volume is usually necessary (see Figure 7.12). Such graphs are constructed from topographic data. In this problem, however, the surface area of the reservoir is considered to be constant for small variations of H_1, making such a graph unnecessary.

To design an outflow structure such as the spillway width, we must repeat the routing procedure several times using various spillway sizes. Usually the controlling factor of such designs is the maximum allowable water elevation in the reservoir. Computers are of inestimable value in performing such analyses. □

PROBLEMS

7.1   Determine the rainfall depth of a 30 min rain with a 10-year frequency at Columbus, Ohio (Figure 7.1). What is the intensity of this rainfall?

7.2   Determine the maximum possible precipitation at New York City that may be encountered on a 100 mi^2 drainage area over a 96 hr period.

7.3   Determine the time of concentration for a tract of agricultural land where the main channel slope is 0.0041 and the longest drainage distance is 0.8 mi.

7.4   Compute the 50-year flood discharge on a large creek in northeastern Ohio if the total discharge area is 75 mi^2. The average soil infiltration is 0.5 in./hr; the main channel slope is 0.008 (use Figure 7.1 for the rainfall intensity).

7.5   A 20-acre parking lot for which the runoff coefficient may be taken as 0.9 is located near Columbus, Ohio. Determine the 5-year flood discharge using the rational formula with a time of concentration of 15 min.

7.6   A 2.5 hr long uniform rainfall resulted in a runoff hydrograph that has the shape of an equilateral triangle. The base of the triangle is 64 hr; the peak flow is 535 cfs; the drainage area is 5.7 mi^2. Determine the corresponding unit hydrograph if the base flow for the channel is 10 cfs.

7.7   A 300-acre forested watershed in fair hydrologic condition located on the clayey flat coastal region of Southern California receives 4.0 in. of rain in a 24 hr period. Determine the corresponding flood peak of the runoff using the SCS curve number method.

7.8   Based on the 50-year flood discharge computed in Problem 7.4 and a time of concentration of 4 hr, route the flood through a reservoir with a surface area of 57 acres that remains essentially constant for all water levels. The outflow from the reservoir is controlled by a weir. The discharge through this weir is dependent on the height of water over it, H, according to the formula $Q_{outflow} = 100(2g)^{1/2} H^{3/2}$. What is the peak flow over the weir, and when does it occur?

7.9   Use the Muskingum method to route a flood through a reach of a river. The base flow is 100 cfs, and the incoming flood flow increases 50 cfs every 2 hr over a 6 hr period and then subsides at the same rate. From past observations the time lag between peaks at the inlet and outlet of the reach is 2.5 hr, and the x coefficient is 0.35. What is the peak discharge at the reach outlet, and when does it occur?

7.10 By continuous discharge measurements the discharge in a stream was determined over a 3-year period. After the data were assembled, it was found that the discharge exceeded 120 cfs 3 percent of the time; it exceeded 60 cfs 35 percent of the time; it exceeded 20 cfs 80 percent of the time; and the minimum discharge was 5 cfs. Construct a discharge-duration graph for the average year. What discharge is available half of the time?

MULTIPLE CHOICE QUESTIONS

7.11 Depth of rainfall divided by the duration of the rain is called
  A. Duration.
  B. Return period.
  C. Frequency.
  D. Intensity.
  E. Rainfall depth.

7.12 Rainfall frequency is
  A. Expressed in years.
  B. A statistical quantity representing the return period of a specified rainfall.
  C. Dependent on the duration and intensity of rains specified.
  D. All of the above.
  E. None of the above.

7.13 A "24 hr rainfall" is
  A. A rain with 24 hr duration.
  B. All rain measured over a 24 hr period.
  C. A rainfall with low intensity.
  D. All of the above.
  E. None of the above.

7.14 Studies show that rains of high intensity generally are concentrated on a smaller area; conversely, rainfalls of long duration and low intensity are usually spread over a wide area.
  A. True.
  B. False.

7.15 The watershed is defined as the
  A. Boundary separating two adjacent rainfall collection basins.
  B. Surface preventing infiltration into the soil.
  C. Contiguous land surface with a single outflow point.
  D. Continental divide.
  E. None of the above.

7.16 A watershed's "time of concentration" means the
  A. Time it takes for the first raindrop to flow down the steepest descent.
  B. Time it takes to drain the area after a rain stops.
  C. Time for the most distant point to drain its first raindrop.
  D. Longest time needed to drain off the area after a rain stops.
  E. Time that is equal to the duration of the rain.

7.17 According to the rational formula, flood discharges depend on the
   A. Infiltration coefficient, watershed area, and time of concentration.
   B. Runoff coefficient, area, and rainfall duration.
   C. Area, rainfall depth, and duration of rain.
   D. Runoff coefficient, area, and rainfall intensity.
   E. Intensity, duration, time of concentration, area, and runoff coefficient.

7.18 The unit hydrograph is a hydrograph that results from 1 in. excess rainfall of a storm of specified duration.
   A. True.
   B. False.

7.19 Two important data obtained from unit hydrograph analysis are
   A. Base flow and duration.
   B. Frequency and duration.
   C. Time of concentration and runoff coefficient.
   D. Time to peak and peak flow.
   E. Total runoff and base flow.

7.20 A reach is
   A. A distance associated with time of concentration.
   B. The distance between adjacent gaging stations.
   C. The length of a reservoir.
   D. A portion of a river analyzed as a single unit.
   E. The distance between inflow and outflow hydrographs.

7.21 Flood routing techniques are used in
   A. Forecasting of floods downstream.
   B. Providing adequate flow cross sections for floods.
   C. Determining outflow hydrographs for reaches.
   D. Constructing spillway rating curves.
   E. All of the above.

7.22 The Muskingum method is
   A. Used to estimate unit hydrographs.
   B. Used to determine the time of concentration.
   C. For routing floods over a spillway.
   D. Based on weighting coefficients.
   E. Developed for small watersheds.

7.23 The area under the discharge-duration curve represents the
   A. Total potential energy of the site.
   B. Average volume runoff during a year.
   C. Average flood peak in a year.
   D. Variation of discharge in the period.
   E. None of the above.

7.24 The dimension of the x coefficient in the Muskingum method is
   A. Length.
   B. Velocity.
   C. Volume.
   D. Discharge.
   E. Nothing.

7.25 The most likely value for a suburban business district's runoff coefficient is
  A. 0.1–0.3.
  B. 0.3–0.5.
  C. 0.5–0.7.
  D. 0.7–0.9.
  E. 0.8–1.0.

8

Open Channel Flow

The concepts relating to flow in channels with a free surface are certainly the most complex ones in the science of hydraulics. Yet their practical relevance is boundless to understanding the behavior of water in creeks, rivers, and human-made channels. This chapter surveys the fundamental open channel concepts that are essential for the practicing hydraulician. Solutions for gradually varied flow are explained in detail. Concepts concerning erosion control and sediment transportation are introduced.

8.1 INTRODUCTION

When flow takes place in a channel or pipe such that the water has a free surface exposed to the atmosphere, we speak of open channel flow. Rivers, creeks, sewers, irrigation ditches and drainage channels, culverts, spillways, and similar human-made structures are designed and analyzed by the methods of open channel hydraulics.

Historically speaking, the use—hence, the design—of open channels is perhaps the oldest hydraulic activity. Irrigation was central to the existence of ancient civilizations in Mesopotamia, Egypt, and China. Water distribution, although practiced by the Greeks and the Egyptians, rose to new heights under the Romans.[1] On the North American continent Hohokam Indians (ancestors of the Pima Indians) living in the Phoenix, Arizona, area between 300 B.C. and A.D. 1200 built a system of irrigation canals totaling about 250 mi, some of which are still visible. In the

---

[1]Sextus Julius Frontinus (Rome, A.D. 40–103), Rome's commissioner of water described it in great pride and detail in his two-volume book *De Aquaeductibus Urbis Romae Commentarius*, published in A.D. 97.

large Mayan archaeological site of Copan, Honduras, an underground aqueduct, built between A.D. 400 and 580, has been unearthed.

The primary difference between confined flow in pipes and open channel flow is that in open channels the cross-sectional area of the flow is a variable that depends on many other parameters of the flow. For this reason hydraulic computations related to open channel flow are more complicated.[2]

Because the pressure on the open surface is always atmospheric, the hydraulic grade line in open channels coincides with the water surface. In pipe flow the hydraulic grade line is independent of the position of the pipe; the pressure causes the movement of the water. In an open channel there is no pressure energy to drive the water. Because the only energy causing flow in an open channel is elevation energy, gravitational force drives the water from higher to lower elevations.

In channels other than those of rectangular shape the depth of the water is variable in each cross section along the flow line. For this reason it is convenient for computational reasons to introduce the concept of mean depth. *Mean depth*, or as it is sometime called, *hydraulic depth*, is computed by dividing the cross-sectional area of the flow by the width of the water surface. Both values are measured in a plane that is perpendicular to the direction of the velocity. Hence, the mean depth is defined as

$$y_{\text{mean}} = \frac{A}{B} = \frac{\text{flow area}}{\text{surface width}} \tag{8.1}$$

By such conversion irregular channels may be considered to be rectangular in shape. The velocity in any channel is variable across the flow area because of the boundary layer effects along the channel boundaries. In practice the magnitude and direction of the velocity measured at various points in the cross-sectional area are rarely found to be uniform and parallel to the direction of the channel. For the purposes of hydraulic computations an *average flow velocity* is used; this is defined as the discharge of the flow divided by its cross-sectional area.

Most of this chapter deals with flow conditions in open channels with a known and constant discharge. Solutions are generally directed toward finding the depth of the flow under certain conditions. For a given discharge there exist some characteristic depth values that serve as guideposts describing the flow. For instance, when only the effect of gravitation is considered, the law of minimum energy manifests itself in the form of the *critical depth,* which will be introduced in the next section. Another set of guideposts are the *conjugate depths,* which refer to the two possible depths at which a given discharge may flow given the same amount of hydraulic energy. The third important characteristic depth is the *normal depth.* This one refers to the depth at which the given discharge would flow down the channel such that the energy gained by gravity would balance the energy lost by friction. In hydraulic design these values are often needed as parts of the computational procedure.

---

[2]There is a strong similarity between the mechanics of open channel flow and that of the flow of compressible fluids.

However, they should not be construed to be actual flow depths associated with the discharge given. The real depth of the flow at any point of the channel is controlled by upstream and downstream flow conditions in addition to the local conditions that are characterized by these significant variables.

8.2   ENERGY EQUATION AND CRITICAL FLOW

The *total available energy*, *E*, of the still water of a lake or reservoir equals the elevation of the water surface with respect to any arbitrary reference level or datum plane below. This base elevation may be the mean sea level or any conveniently determinable elevation below the deepest point of the channel to be analyzed. In a flowing river or other kind of open channel the total available energy with respect to the reference base elevation may be separated into three distinct components: the relative elevation Y of the bottom above the base line, the depth y of the water, and the kinetic energy head $v^2/2g$ plotted above the water surface.

To gain an insight into the complex relationships among energy, discharge, and depth in an open channel, let us consider the total available energy to be measured with reference to the bottom of the channel at a certain cross section of rectangular shape with B width. This may be written as

$$E = \frac{v^2}{2g} + y = \frac{Q^2}{2gB^2y^2} + y \qquad (8.2)$$

This equation, often called the *specific energy equation,* may be plotted in two different ways, either as in Figure 8.1, with the energy held constant, or as in Figure 8.2, with the discharge held constant.[3]

Looking at Figure 8.1 we may note that for any fixed amount of total available energy a given amount of water may flow through the channel cross section with two different depths, except when the discharge reaches a certain maximum value. At this maximum discharge value the depth is called *critical depth*. In accordance with the law of optimum energy, mentioned at the end of Chapter 2, at critical depth the least amount of energy is utilized per mass of fluid flowing; therefore, this is an important natural condition of open channel flow. For instance, when water enters a spillway from a lake, the depth of flow over the spillway crest will reduce to the critical depth with respect to the constant energy level provided by the lake's surface. The discharge will be the maximum possible under the given circumstances, the *critical discharge*. The spillway of a dam provides another, somewhat more complex, example: In this case the water level above the dam will rise until the total available energy at the upstream side (elevation of the water surface plus the kinetic energy) is sufficient to pass the river's discharge over the spillway. Barring the presence of a high downstream water level that would force the water to back up, the discharge will be critical.

---

[3]This type of representation was first introduced by Alexander Koch (Germany, 1852–1923) who was an early proponent of scientific hydraulics.

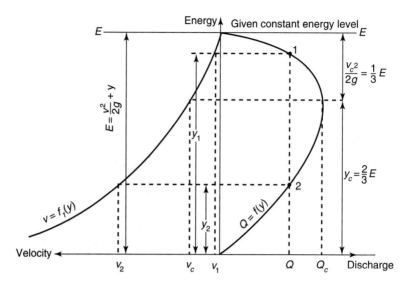

FIGURE 8.1 Specific energy equation plotted for constant energy.

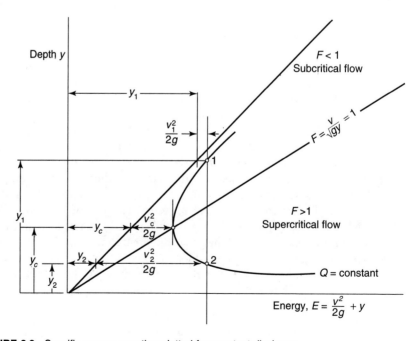

FIGURE 8.2 Specific energy equation plotted for constant discharge.

In Figure 8.2, as in Figure 8.1, we see that a given constant discharge may flow in a channel at two distinct depths, unless the energy for this given discharge is at its minimum value. In the latter case, again, the flow is critical.

The critical flow condition creates two distinct ranges in which flow may take place in a channel. With the same discharge maintained, if the velocity is less than the critical velocity, then one speaks of *subcritical* flow. Continuity requires that the depth of subcritical flow must exceed the critical depth. If the velocity of the flow exceeds the critical velocity, the flow takes place in the *supercritical* range. In supercritical flows the flow depth is less than the critical depth. Most river flows as well as flows in human-made channels are subcritical or *tranquil* flows. Supercritical flows are also called *rapid* flows. Flows over spillways, in chutes, in steep mountain creeks, or rainwater rushing toward storm inlets on pavements are supercritical.

The point representing critical conditions in Figure 8.2 can be derived by differentiating Equation 8.2 with respect to y and setting the result to equal zero. This process locates the vertical tangent at y_c, leading to

$$\frac{Q_c^2}{g} = \frac{A^3}{B} \qquad \text{when } y = y_c \tag{8.3}$$

where A is the cross-sectional area of the flow channel of arbitrary shape, and B is the width of the surface. Table 8.1 gives critical discharges and velocities as well as geometric parameters for common channel geometries.

In using Figure 8.2 one must recall that it was created for a rectangular flow channel. For channels of nonrectangular shape the function plots differently. For a rectangular channel, dividing Equation 8.3 by A^2 and rearranging, we get

$$v_c = \sqrt{gy_c} \tag{8.4}$$

which is the velocity of small surface waves in shallow waters and is called wave *celerity*.[4] If a small stone is dropped into the water, it will cause small ringlike waves around the point of impact. As long as the velocity of the water is less than critical, these rings will travel upstream. At critical velocity the ring waves will not be able to travel upstream, since their velocity equals that of the stream itself. At velocities higher than critical the ringlets will not be able to form; they will be washed away, forming a V-shaped shock wave on the water surface.

The dimensionless parameter that indicates whether a flow is subcritical, critical, or supercritical is the Froude number, introduced in Chapter 3. It may be found by dividing the average velocity of the flow by the celerity,

$$\mathbf{F} = \frac{v}{\sqrt{gy}} \tag{8.5}$$

---

[4]The term was introduced by Jean-Claude Barre de Saint-Venant (France, 1797–1886), who is well-known for his contributions in the field of elasticity.

TABLE 8.1 Channel Section Geometry

	Area A	Wetted Perimeter P	Hydraulic Radius $R = \dfrac{A}{P}$	Top Width B	Depth $\dfrac{A}{B}$	Critical Discharge Q_c	Critical Velocity v_c
(rectangular section)	By	$B + 2y$	$\dfrac{By}{B + 2y}$	B	y	$\sqrt{g}\, By_c^{1.5}$	$\sqrt{gy_c}$
(trapezoidal section)	$(b + zy)y$	$b + 2y\sqrt{1 + z^2}$	$\dfrac{(b + zy)y}{b + 2y\sqrt{1 + z^2}}$	$b + 2zy$	$\dfrac{(b + zy)y}{b + 2zy}$	$\dfrac{\sqrt{g[(b + zy_c)y_c]^{1.5}}}{(b + 2zy_c)^{0.5}}$	$\sqrt{\dfrac{b + zy_c}{b + 2zy_c}\, gy_c}$
(triangular section)	zy^2	$2y\sqrt{1 + z^2}$	$\dfrac{zy}{2\sqrt{1 + z^2}}$	$2zy$	$0.5y$	$0.71\sqrt{g}\, zy_c^{2.5}$	$\sqrt{\dfrac{gy_c}{2}}$
(circular section)	$0.125(\theta - \sin\theta)\, D^2$	$0.50\,\theta D$	$0.25\left(\dfrac{\theta - \sin\theta}{\theta}\right) D$	$2\sqrt{y(D - y)}$	$0.125\left(\dfrac{\theta - \sin\theta}{\sin\frac{1}{2}\theta}\right) D$	$\dfrac{0.044\sqrt{g}(\theta - \sin\theta)^{1.5}}{(\sin\frac{1}{2}\theta)^{0.5}}\, D^{2.5}$	Use Figure 8.3.
Parabola	$\dfrac{2}{3} By$	$B + \dfrac{8y^{2*}}{3B}$	$\dfrac{2B^2 y}{3B^2 + 8y^2}{}^*$	$\dfrac{3A}{2y}$	$\dfrac{2}{3} y$	$0.54\sqrt{g}\, By_c^{1.5}$	$\sqrt{\dfrac{2}{3} gy_c}$

*For $y < B/4$.

288

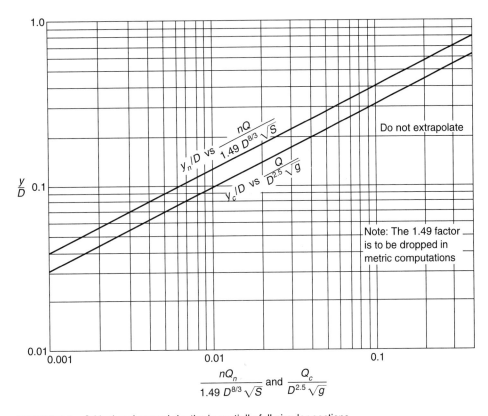

FIGURE 8.3   Critical and normal depths in partially full circular sections.

Equation 8.5 is for rectangular channels. Using this definition, we represent critical flows by $\mathbf{F} = 1.0$, subcritical flows by $\mathbf{F}$ values less than 1.0, and supercritical flows by $\mathbf{F}$ values exceeding 1.0.

By combining Equations 8.1 and 8.3 and taking the square root of the result, we obtain

$$\frac{Q}{\sqrt{g}} = A \sqrt{y_{mean}} = \sqrt{\frac{A^3}{B}} = Z \tag{8.6}$$

in which Z is the so-called shape or section factor of the channel. For natural channels of irregular shape the section factor plotted against depth can be used to simplify the computation of the critical depth for any given discharge. Once Z values are computed for a few depths, a graph may be constructed. Using this graph with a given discharge, y_c can be obtained. The solution of Equation 8.6 for given channel cross sections is often difficult because of the relative complexity of the geometric function defining A. To ease the difficulties, numerous graphs are available in the hydraulic literature. One typical graph showing the function of critical depth for partially full circular sections is reproduced in Figure 8.3. For pipe diameters D, the critical depth y_c is represented by the lower curve in the graph.

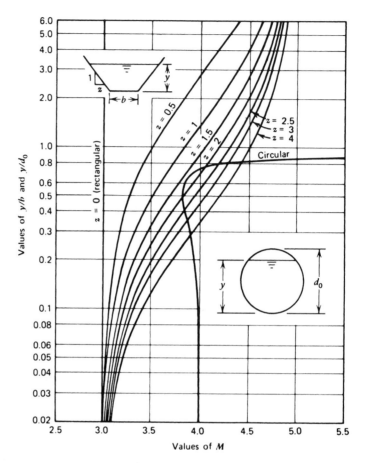

FIGURE 8.4 Critical flow exponents for rectangular, trapezoidal, and circular channels. (From Ven Te Chow, *Open Channel Hydraulics.* Copyright © 1958, McGraw-Hill Book Company. Used with permission of McGraw-Hill Book Company)

Comparing the formulas given for critical discharges in Table 8.1, we see that they can be put in the form

$$Z^2 = \frac{Q^2}{g} = f(y^M) \qquad (8.7)$$

meaning that the critical discharge is expressible as a function of the depth y raised to some exponent M. The exponent M is called the *critical flow exponent* and may be expressed analytically for any cross-sectional shape, as shown in Figure 8.4.

Critical flow is an important parameter in open channel flows. Locations in a channel where a critical flow condition is known to exist are used as *control sections*. At these locations the discharge can be determined by a single measurement of depth. In the determination of water surface profiles over long sections, control sections are fixed control points. Depending on the relative location of these control points, one may speak of *downstream* or *upstream control.*

8.3 HYDRAULIC JUMP

Figure 8.1 shows that a given discharge may flow in a channel at two alternative depths, except in the case of critical flow. One of these depths is always greater, the other less, than the critical depth. The two corresponding depths are called *conjugate depths.* If one is known, the other can be calculated by writing the sum of the depth and kinetic energy terms for both cases and equating these. For rectangular channels this yields

$$y_2 = \frac{y_1}{2}\left(\sqrt{1 + \frac{8q^2}{gy_1^3}} - 1\right) \tag{8.8}$$

where q equals Q/B.

Under certain conditions supercritical flow may suddenly change into subcritical flow. This sudden change is called a *hydraulic jump.*[5] It occurs after a stream of water rushes down a spillway, particularly if the downstream conditions are such that they impede further supercritical flow, as by the presence of energy-absorbing devices or by a high water level pool. The hydraulic jump means that the water level suddenly changes from the supercritical conjugate depth y_1 to the subcritical conjugate depth y_2. Equation 8.8 shows the correspondence between these two depths, but for practical use it is convenient to convert this equation by introducing the Froude number for the supercritical flow, $\mathbf{F}_1$, as a variable. This mathematical conversion results in a convenient dimensionless formula,

$$\frac{y_2}{y_1} = \frac{1}{2}(\sqrt{1 + 8\mathbf{F}_1^2} - 1) \tag{8.9}$$

Depending on the magnitude of the Froude number at the beginning of the jump, the strength of the jump increases with $\mathbf{F}_1$. For $\mathbf{F}_1$ less than 1.7, experiments indicate that the jump appears as a series of undulating waves along the downstream surface. For $\mathbf{F}_1$ between 1.7 and 2.5 a series of rollers appear on the surface, but the presence of the jump does not significantly disturb the downstream water level, and it is considered weak and unstable in its position. For Froude numbers of 2.5 to 4.5 the jump grows in strength, causing forceful random waves to travel downstream. For this range the elevation difference between the two sides of the jump is the greatest. For $\mathbf{F}_1$ of 4.5 to 9.0 the position of the jump stabilizes in the channel and is less influenced by the tailwater depth. For $\mathbf{F}_1$ larger than 9.0 the jump again causes downstream waves, but their effect is less destructive, as the energy dissipation in

---

[5]The first scientist to study this phenomenon was Giorgio Bidone (Italy, 1781–1839), a professor of hydraulics at the University of Turin. The formal solution of the problem using momentum equilibrium is due to Jean Baptiste Belanger (France, 1789–1874), who also presented an elementary formulation for gradually varied flow.

such jumps is high, because a hydraulic jump is associated with high turbulence, vortices, and swirls. As a result there is considerable loss of total available energy in hydraulic jumps. The energy loss through a hydraulic jump may be computed by the formula

$$\frac{E_{loss}}{E_1} = \frac{2 - 2y_2/y_1 + \mathbf{F}_1^2 \left(1 - A_1^2/A_2^2\right)}{2 + \mathbf{F}_1^2} \qquad (8.10)$$

where A_1 and A_2 are cross-sectional areas associated with y_1 and y_2, respectively. The energy-absorbing character of the hydraulic jump is further discussed in the following chapter. Equation 8.10 is valid for any channel cross sections.

Because the discharge and the two conjugate depths are inherently related by Equation 8.8, the relationship can be utilized to determine the discharge by measuring the depths before and after the hydraulic jump in a rectangular channel of B width and substituting them into the formula

$$Q = B \left[\frac{y_1 + y_2}{2} \left(g y_1 y_2 \right) \right]^{1/2} \qquad (8.11)$$

There is no corresponding sudden change of water level when the flow changes from subcritical to supercritical. Hence, there is no great energy loss associated with such changes. Hydraulic jumps are possible only when the velocity is reduced from supercritical to subcritical.

8.4 THE CONCEPT OF NORMAL FLOW

In regard to the computational difficulties associated with the analysis of flow in open channels, engineers seeking a simple method for discharge calculations have developed formulas for the case when the energy grade line is assumed to be parallel with the bottom slope of the channel. In this *simplified* case the energy gained by the water at any point equals exactly the energy lost through friction. Since there is no acceleration or deceleration along the channel, the depth and the kinetic energy term $(v^2/2g)$ remain constant. In turn this means that the slope of the energy grade line S_e must equal the slope of the channel bottom, S; that is, we assume for such flow that

$$S_e = \frac{H_2 - H_1}{L} \qquad (8.12)$$

in which H_1 and H_2 are the upstream and downstream channel bottom elevations, respectively, and L is the length of the channel.

In nature the channel bottom slope is rarely uniform; the roughness and the cross-sectional area of the channel vary from section to section. Thus, it is clear that this zero-acceleration condition rarely occurs in practice. However, the savings in computation as well as the uncertainty associated with the determination of the true expected discharge undoubtedly makes the approach worthwhile. Flow with no acceleration or deceleration is called *normal flow*.

Normal flow in open channels is computed by the Chézy[6] formula,

$$v = C\sqrt{RS} \qquad (8.13)$$

where C is Chézy's resistance coefficient, v is the average velocity, and S is the slope of the channel. Because we are dealing with steady flow, S is the same as the energy grade line S_e.

The depth corresponding to such flow is called normal depth y_n. In Equation 8.13 R is called the *hydraulic radius*, which is not a radius in the geometric sense but rather

$$R = \frac{A}{P} \qquad (8.14)$$

where P is the wetted perimeter of the channel, that is, the length of the line along which the water is in contact with the channel bottom in the cross-sectional area A of the flow. This geometric parameter indicates the hydraulic efficiency of the cross section; A is a positive and P is a negative contributor to the movement of the water: The more bottom surface there is to create frictional resistance, the more the flow is retarded. Conversely, the larger the flow area is compared with P, the more easily the water will move. In channel design an attempt is made to select the best possible hydraulic radius, that is, the one that will carry the design discharge with the least amount of digging. Selection of the optimum hydraulic radius for trapezoidal or other channels is a rather cumbersome procedure.

In general practice the determination of the *design depth* of the channel is sometimes simplified by neglecting optimum conditions for the hydraulic radius and selecting the depth as a function of the cross-sectional area alone. Such formulas may be written generally as

$$y = \sqrt{\frac{A}{e}} \qquad (8.15)$$

where the constant e is taken to be 4 by the U.S. Bureau of Reclamation and 3 in irrigation channel design practice in India, for instance.

In the design of drainage or irrigation channels the bottom width b is usually determined first, and then y is calculated from the equations. In some cases, depth considerations are important, such as when the water level is to be kept near the groundwater level, as in the case of drainage and swamp reclamation design, the depth is predetermined and the base width b of the channel is computed. In wide rivers where the surface width B is 10 or more times larger than the depth d, the hydraulic radius is assumed to equal the depth of the flow. This simplifies the computations.

C in Equation 8.13 is Chézy's channel resistance coefficient, an experimentally determined factor with the dimension given by Equation 3.48. Based on a great

---

[6]Antoine Chézy (France, 1718–1798) developed his resistance formula in connection with the design of a channel constructed to improve the water supply of the city of Paris. Independently, the formula was also published in 1801 by Johann Albert Eytelwein (Germany, 1764–1848) with 50.9 for C in metric units.

deal of measurements in the field and in laboratory channels since the beginning of the nineteenth century, the value of C has been determined in metric units to be

$$C = \frac{1}{n} R^{1/6} \qquad (8.16)$$

in which n is referred to as Manning's[7] coefficient in American engineering practice, a resistance factor referring to the condition of the channel. Table 8.2 gives values of n for various channel conditions. It is based on several hundred field experiments performed during the nineteenth century.

Converting Equation 8.16 to conventional American units we have

$$C = \frac{1.49}{n} R^{1/6} \qquad (8.17)$$

with C now in feet$^{1/2}$/sec and R in feet. The number 1.49 is the conversion factor, enabling the designer to use the metric n values without unit conversion. Combining Equations 8.13 and 8.16, we obtain the Chézy-Manning formula in metric units:

$$Q = \frac{A}{n} R^{2/3} S^{1/2} \qquad (8.18)$$

where Q is in cubic meters per second, and R and A are in meters and square meters, respectively. Combining Equation 8.13 with Equation 8.17 gives the Chézy-Manning formula in conventional American units:

$$Q = \frac{1.49}{n} A R^{2/3} S^{1/2} \qquad (8.19)$$

where Q is in cubic feet per second, and A and R are in square feet and feet, respectively. Figure 8.5 shows a popular nomograph by which Equation 8.19 may be solved. In the case of normal flow in partially full circular channels, the computations are facilitated by the graph given in Figure 8.3.

For an arbitrary cross section the Chézy-Manning equation may be written as

$$\frac{Q}{\sqrt{S}} = K \qquad (8.20)$$

where K is called the *conveyance*. By Equation 8.18 K in metric units is given by

$$K = \frac{1}{n} A R^{2/3} \qquad (8.21)$$

---

[7]Robert Manning (Ireland, 1816–1897) discarded formulas that later were associated with his name, proposing instead a dimensionally homogeneous flow formula. Instead, two Swiss engineers, Emile Oscar Ganguillet (1818–1894) and Wilhelm Rudolf Kutter (1818–1888), developed a popular open channel formula that introduced n as a roughness factor. An improved formula for C was introduced later by Nicolai N. Pavlovsky (Russia, 1884–1937), who replaced the exponent of R with a variable depending on n and R.

TABLE 8.2   Roughness Coefficients for Open Channels

Description of Channel	n
Exceptionally smooth, straight surfaces: enameled or glazed coating; glass; lucite; brass	0.009
Very well planed and fitted lumber boards; smooth metal; pure cement plaster; smooth tar or paint coating	0.010
Planed lumber; smoothed mortar ($\frac{1}{3}$ sand) without projections, in straight alignment	0.011
Carefully fitted but unplaned boards; steel troweled concrete, in straight alignment	0.012
Reasonably straight, clean, smooth surfaces without projections; good boards; carefully built brick wall; wood troweled concrete; smooth, dressed ashlar	0.013
Good wood, metal, or concrete surfaces with some curvature, very small projections, slight moss or algae growth or gravel deposition; shot concrete surfaced with troweled mortar	0.014
Rough brick; medium quality cut stone surface; wood with algae or moss growth; rough concrete; riveted steel	0.015
Very smooth and straight earth channels, free from growth; stone rubble set in cement; shot, untroweled concrete; deteriorated brick wall; exceptionally well excavated and surfaced channel cut in natural rock	0.017
Well-built earth channels covered with thick, uniform silt deposits; metal flumes with excessive curvature, large projections, accumulated debris	0.018
Smooth, well-packed earth; rough stone walls; channels excavated in solid, soft rock; little curving channels in solid loess, gravel, or clay with silt deposits, free from growth and in average condition; deteriorating uneven metal flume with curvatures and debris; very large canals in good condition	0.020
Small, human-made earth channels in well-kept condition; straight natural streams with rather clean, uniform bottoms without pools and flow barriers, cavings, and scours of the banks	0.025
Ditches; below-average human-made channels with scattered cobbles in bed	0.028
Well-maintained large floodway; unkept artificial channels with scours, slides, considerable aquatic growth; natural stream with good alignment and fairly constant cross section	0.030
Permanent alluvial rivers with moderate changes in cross section, average stage; slightly curving intermittent streams in very good condition	0.033
Small, deteriorated artificial channels, half choked with aquatic growth; winding river with clean bed, but with pools and shallows	0.035
Irregularly curving permanent alluvial stream with smooth bed; straight natural channels with uneven bottom, sand bars, dunes, few rocks and underwater ditches; lower section of mountainous streams with well-developed channel with sediment deposits; intermittent streams in good condition; rather deteriorated artificial channels, with moss and reeds, rocks, and slides	0.040
Artificial earth channels partially obstructed with debris, roots, and weeds; irregularly meandering rivers with partly grown-in or rocky bed; developed flood plains with high grass and bushes	0.067
Mountain ravines; fully ingrown small artificial channel; flat flood plains crossed by deep ditches (slow flow)	0.080
Mountain creeks with waterfalls and steep ravines; very irregular flood plains; weedy and sluggish natural channels obstructed with trees	0.10
Very rough mountain creeks; swampy, heavily vegetated rivers with logs and driftwood on the bottom; flood plain forest with pools	0.133
Mudflows; very dense flood plain forests; watershed slopes	0.22

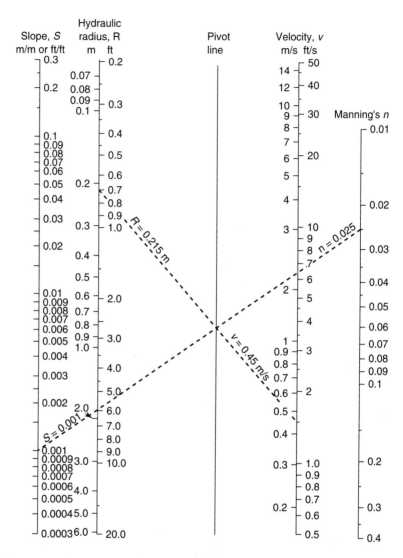

FIGURE 8.5 Nomograph for the solution of the Manning formula. (From USDT Federal Highway Administration, 1973)

in which both A and R depend on the geometry.[8] With the cross section of the channel known, one may plot the value of K with respect to the depth y. With this conveyance graph at hand, the normal depth may be read off directly for any discharge with the use of Equation 8.21.

---

[8]The conveyance method for pipe flow was developed from this idea.

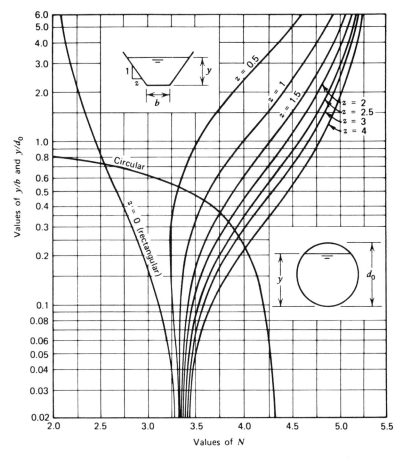

FIGURE 8.6   Normal flow exponents for rectangular, trapezoidal, and circular channels. (From Ven Te Chow, *Open Channel Hydraulics.* Copyright © 1958, McGraw-Hill Book Company. Used with permission of McGraw-Hill Book Company.)

Just as in the case of critical flow, it may be proved that the formula for normal discharge in any channel of arbitrary shape is a function of the depth y raised to a power. Indeed, for any given channel geometry, Equation 8.20 can be expressed by the channel depth y such that

$$K^2 = my^N = f(y^N) \tag{8.22}$$

where m is a coefficient depending on roughness and geometry, and N is called the *normal flow exponent.* Equation 8.22 is comparable to Equation 8.7, which expresses a similar concept for critical flow. Both of these hydraulic exponents are important parameters in design for conditions when normal flow cannot be assumed to occur. Values of N are shown in Figure 8.6 for common trapezoidal and circular channels. When the discharge, the bottom slope, and the roughness are known, both the normal depth and the critical depth can be calculated from formulas already presented in this chapter. If the critical depth is shown to be smaller than the normal depth, the normal flow is *subcritical.* Conversely, if the critical depth is larger than the normal depth, the normal flow is *supercritical.* The slope at which the critical depth equals the normal depth is the *critical slope.* Most natural watercourses as well as human-made channels conduct the flow under subcritical conditions. These concepts will be studied in detail in the following section.

Just as in the case of critical flow, the normal flow condition can also serve as a control section. The difference is that the water surface on a long channel of constant slope approaches normal depth asymptotically.

8.5   NATURAL FLOW CONDITIONS

Whether we consider a natural stream or a human-made channel, we find that our theoretical approach rarely resembles the natural flow conditions. In engineering practice we design our watercourses for maximum flows: those 5-, 10-, 50-, or even 100-year floods for which we would wish the channel to be adequate. The actual flows are made up of a random succession of flood waves due to rains of various durations and intensities. Between these flood waves the channel carries only a relatively small *base flow*, generated by the contributions of exiting or reentering groundwater along the stream. Unprotected channels tend to erode during high flows, particularly if there is no natural protection by firmly established grass on the banks. During low flows the sediment carried by the water settles out on the bottom. Hence, the periodic movement of the sediment followed by silting continually change the channel's shape and alignment. To maintain a channel so that it is always ready to carry the maximum design flood is a continuous process, requiring supervision and periodic maintenance.

The *slope* of the channel is rarely constant over even relatively short distances. It is dependent on the geology of the region. Less erodible soils, rock outcrops, ledges, and faults alternate with soft, erodible alluvial layers. The assumption of constant slope is therefore a crude but necessary approximation. The flowing water must overcome boundary-induced resistances because water is a viscous fluid. Hence, the flowing water gradually loses its energy along its path. If the energy loss along the channel is less than the energy gained by the decreasing elevation, the energy content of the discharge increases; therefore, the flow accelerates. Conversely, if the frictional losses along the channel exceed the energy gained by the downward flow, the flow decelerates.

In the case of *steady flow*, acceleration along the channel results in decreasing depth, and deceleration results in increasing depth. In either case the water surface will not remain straight and parallel with the bottom of the channel but will be slightly curved.

To make our hydraulic computations tractable, we are required to make an assumption that our channel shape does not change along the path of the flow. Channels of constant shape are called *prismatic channels.* A prismatic channel of

FIGURE 8.7 California's San Luis Canal. (Courtesy of U.S. Army Corps of Engineers, Sacramento, CA)

constant slope can carry a given discharge either in a decelerating, accelerating, or—rarely—in a "normal" manner. In channels specifically built to deliver water for irrigation and other purposes the flow is normal as long as the utilization is constant. Figure 8.7 shows a section of the San Luis Canal in California built for such a case. Figure 8.8 depicts a water diversion structure on the California Aqueduct. Its operation causes the flow in the channels to vary. The cross-sectional area of the flow in the channels increases, decreases, or remains constant. Concepts introduced in the previous sections enable us to compute the critical depth and the normal depth of the flow of a given discharge, if all other variables are known. These two characteristic depths, however, will not tell us at what depth the *actual flow* will occur. That is controlled by the two end conditions of the reach considered and by the relationship between the energy available and the energy required to overcome the losses that are encountered along the channel.

Let us assume that for a given channel and given discharge we can vary the slope of the channel. This condition can be easily fulfilled in a hydraulics laboratory equipped with a *tilting flume*. Calculating critical and normal depths for various slopes, we find that there is one particular slope at which the flow throughout the channel is critical. This occurs when the normal depth equals the critical depth. This slope is called the *critical slope.*

To determine the critical slope, S_c, for a given discharge and for known channel conditions, we equate Equation 8.4 with Equation 8.13, resulting in

FIGURE 8.8 Diversion structure on the California Aqueduct. (Courtesy of U.S. Army Corps of Engineers, Sacramento, CA)

$$\sqrt{gy} = C\sqrt{RS} \tag{8.23}$$

For wide channels where R equals the depth Equation 8.23 may be rearranged to express the critical slope as

$$S_c = \frac{g}{C^2} \tag{8.24}$$

For channels where this simplification cannot be made we make use of the formula for the C coefficient. For SI units the Chézy coefficient may be replaced by Equation 8.16. Squaring C and substituting, we obtain

$$S_c = \frac{gyn^2}{R^{4/3}} \tag{8.25}$$

where R is the hydraulic radius. In a wide rectangular channel R may be replaced by the depth; therefore,

$$S_c = \frac{gn^2}{y^{1/3}} \tag{8.26}$$

We may recall that in critical flow the Froude number, Equation 8.5, equals unity. Denoting the discharge of a unit width of channel by q enables us to put Equation 8.5 in the form

$$q = y\sqrt{gy} = \sqrt{g}\,y^{3/2} \tag{8.27}$$

Because we assumed critical flow, y is the critical depth. From this it follows that

$$y_c = \left(\frac{q}{\sqrt{g}}\right)^{2/3} \tag{8.28}$$

Substituting Equation 8.28 into Equation 8.26 gives us the critical slope, in SI units, as

$$S_c = g^{1.11} \frac{n^2}{q^{0.22}} \tag{8.29}$$

Parametric analysis of this formula shows that for larger discharges the critical slope tends to be somewhat less steep and that for rougher channels it tends to be steeper. The issue is further complicated if one considers that for larger discharges the roughness coefficient may decrease. It has also been proved that there exists a minimum critical slope for any given channel at a certain critical discharge. These considerations, however, are of little practical interest.

For a given discharge, slopes that are less than critical are called *mild* slopes; those steeper than critical are *steep* slopes. Of course, flow can occur on horizontal slopes as well as in adversely sloping channels. In the latter the water will not, of course, flow upward: The upstream water level will rise until the necessary energy is available to overcome the obstacle of the high bottom.

Calculating the normal and critical depths for mild and steep channels, we find that for steep channels the critical depth is higher than the normal depth; hence, the normal flow is supercritical (see Figure 8.1 or 8.2). On mild slopes the normal flow depth exceeds the critical depth; therefore, the normal flow is subcritical.

Depending on the conditions at the ends of the reach the actual depth of the flow may be at any position, regardless of the normal and critical depths. For mild and steep slopes three zones may be differentiated.[9] Zone 1 is where the actual depth exceeds both normal and critical depths; zone 2 is where the actual depth is between the normal and the critical depths; and zone 3 is where the actual depth is less than either. In these zones the actual depth of the water is either accelerating or decelerating: This factor determines the shape of the water surface within the reach. For example, a dam at the end of the reach forces the actual water level to rise higher than both the normal and the critical depth, and the slope is mild. The water surface will be of convex shape, and the flow will decelerate along the channel as the cross-sectional area gradually increases. The primary purpose of many dams, such as the one built at Jonesville, Louisiana, on the Mississippi River, shown in Figure 8.9, is to raise the water level for many miles upstream to facilitate navigation. The picture shows the navigation lock, the spillway, and the gates that are opened to allow floods to pass through.

Figure 8.10 illustrates the shape of the water surface in reaches of different slopes. In this figure the notations are self-explanatory: The different possible water surface curvatures are labeled according to the slope and zone in which the water

---

[9]This classification was first introduced by Emmanuel Joseph Boudin (Belgium, 1820–1893).

FIGURE 8.9   Typical lock and dam facility on the Mississippi River, Jonesville, Louisiana. Note the operating spillway, the five closed gates, and the navigation lock. (Courtesy of U.S. Army Corps of Engineers)

surface is located. For complete understanding remember that whether a slope is steep, mild, or critical depends on the discharge encountered.

The most common use of the water surface profiles shown in Figure 8.10 is for sketching the water surface around and through a hydraulic structure. The position of the water surface with respect to the critical and normal depths, as well as its general curvature, is an important first step in the analysis. Figure 8.11 shows an example of this type of sketch. Later we will learn how to compute the water surface elevation at points along the channel. This example shows how to connect the points computed by a qualitative approach.

Computation of the shape of a water surface is made by the so-called *backwater flow* formula. The term *backwater* refers to an $M1$ curve, but the mathematical formula applies to any position of the actual water surface, so the method is quite general. It applies in every case unless the flow crosses critical conditions, such as in the case of a hydraulic jump.

In the usual applications the discharge is given and a series of depths are computed to define the shape of the water surface. But, with a little more computational difficulty, the problem can be reversed: We can determine the discharge from the elevations of the water level at the two ends of a reach. Let us consider the cases represented in Figure 8.12.

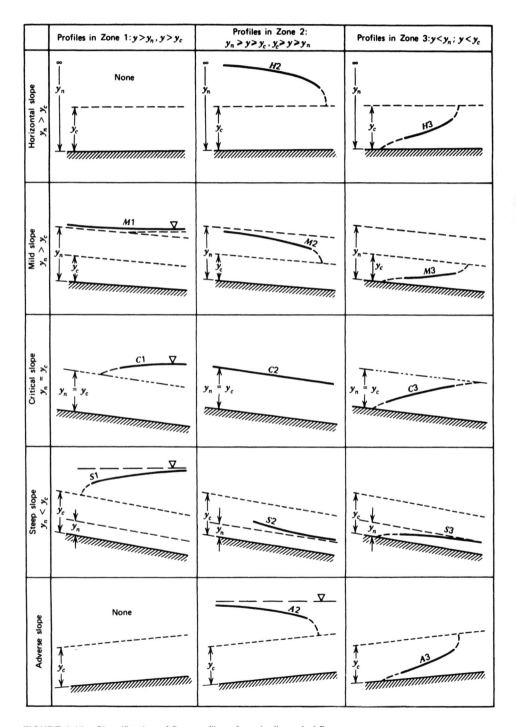

FIGURE 8.10 Classification of flow profiles of gradually varied flow.

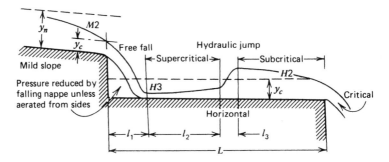

FIGURE 8.11   Water surface profiles for flow through a drop structure.

In the situations shown in these two illustrations the water flows from an upper to a lower reservoir through a channel that has an arbitrary but constant configuration along the flow path. In both cases the channel bottom is of constant slope, and the water surface elevations are also the same regardless of the flow in the channel. In Figure 8.12a the depth of the water in the upper reservoir is always at the same elevation, while the depth in the lower reservoir is varied for the sake of illustrating the principle in question. As long as the water level in the lower reservoir is the same as that in the upper reservoir no flow will take place in the channel. If the water level is reduced in the lower reservoir, the water will begin to flow. The energy gain causing the flow depends strictly on the difference between the water surface elevations in the reservoirs.

Looking at the drawing we can recognize that as long as S_e is not parallel to S, the bottom slope of the channel, the available energy in the flow at any particular point along the channel will be different. Because the discharge along the channel is assumed to be constant, the kinetic energy and the depth of the flow will vary from point to point, resulting in a curved water surface connecting the upper reservoir level to the lower reservoir level.

As the water level in the lower reservoir is lowered, more and more energy is made available to cause flow. This increases the flowing discharge. With the upper water level remaining constant, the discharge in the channel reaches a maximum value when the depth of the flow at the lower end of the channel reaches critical depth. For this case, when the flow is controlled by a critical condition at the exit, the flow in the channel will depend on the position of the water level at the entrance of the channel. There are two possible cases. In the first case the depth in the upstream reservoir results in an entrance depth in the channel that is less than the critical depth based on the exit condition. In this case the flow throughout the channel will be supercritical, with the depth of the water gradually rising because of deceleration caused by viscous shear losses. The other possible case arises when the upper reservoir level is such that the entrance depth is larger than the critical depth at the downstream end. In this case the flow throughout the channel will be subcritical, with the depth of the water slowly subsiding along the flow, characterizing an accelerating condition.

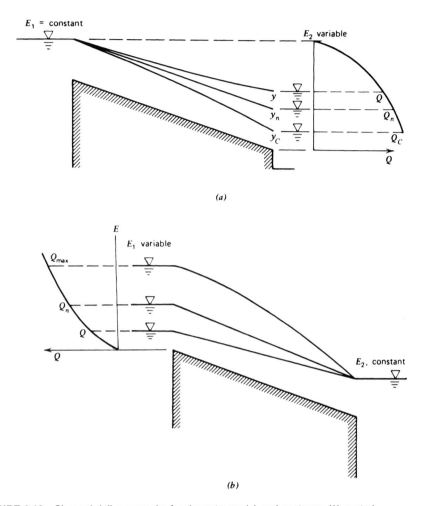

FIGURE 8.12 Channel delivery graphs for downstream (*a*) and upstream (*b*) control.

Figure 8.12*b* shows physical conditions identical with those depicted in Figure 8.12*a*, the only difference being that now the water level in the lower reservoir is kept constant and the level in the upper reservoir is varied. In this case no flow will occur in the channel regardless of the lower reservoir level until the level in the upper reservoir exceeds the channel bottom's highest point. From there the flow will increase rapidly with increasing upper reservoir depth until the channel reaches its maximum capacity, at which point the flow into the lower reservoir becomes critical.

From here the problem gets more complex and as such is beyond the scope of this book. One complication is that the flow can reach a critical state at any other point in the channel depending on whether the bottom slope is mild, critical, or

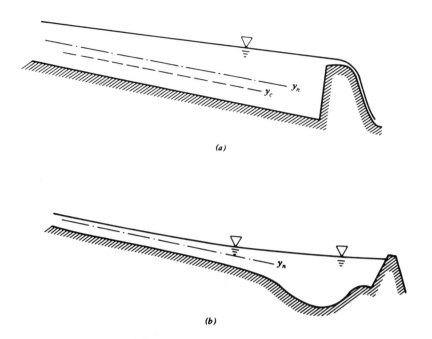

(a)

(b)

FIGURE 8.13 Examples of backwater flows. (a) Backwater above dam. (b) Backwater in tributary due to flood in main channel.

steep. Another complicating aspect is that in nature neither end of the channel is endowed with a constantly fixed water level. This results in an increase in the number of variables controlling the flow. Therefore, the *channel delivery graph* is composed of a family of curves, each depicting a particular set of circumstances controlling the flow. A third difficulty is that in nature the bottom slope, or even the channel cross section, varies along the path of the flow. This last complicating aspect arises because in practice the discharge in the channel rarely remains steady. Computations related to unsteady discharges are in the realm of flood routing, a topic of considerable complexity.

8.6 BACKWATER COMPUTATIONS

The limitations on the use of the normal flow concept were made clear in the discussion of the delivery of channels. Whenever there is a barrier in the downstream portion of the flow that causes the water to rise before it continues on its path, the water surface cannot be assumed to be parallel to the bottom of the channel. In accordance with the gradual increase of the water depth toward the barrier, the available cross-sectional area enlarges, causing the velocity, hence the kinetic energy, to drop. Figure 8.13 shows two such cases frequently found in practice. One is a dam that causes the flooding of the reservoir behind it, giving rise to the classic case of backwater flow. The other problem shown is conceptually identical with

FIGURE 8.14 Notations for the derivation of the backwater formula.

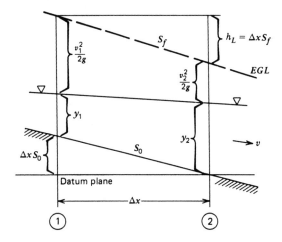

the first, except that the water rise is caused by a flood in the recipient water body, such as a major river to which the channel being analyzed is a tributary. Such problems commonly occur, resulting in flooding of the lower portions of tributary rivers and creeks. To properly plan for the clearing and deepening of such tributary channels and the raising of bridges, roads, and sewer manholes or to determine the height requirements for projected levees around such tributaries, computations of backwater heights are necessary.

The derivation of the formula for gradually varied flow is rather straightforward. First, using the notations shown in Figure 8.14, we write the Bernoulli equation for two adjacent points in the channel at a Δx distance apart. Because the flow may accelerate or decelerate between these points, the slope of the channel bottom, S_0, and the energy grade line, S_f, may be different. Bernoulli's equation takes the form

$$\Delta x \, S_0 + y_1 + \frac{v_1^2}{2g} = \Delta x \, S_f + y_2 + \frac{v_2^2}{2g} \tag{8.30}$$

To simplify, we may introduce $\Delta y = y_2 - y_1$ and $\Delta(v^2/2g) = v_2^2/2g - v_1^2/2g$. Multiplying by S_0 and rearranging Equation 8.30 results in

$$\frac{\Delta x}{\Delta y} = \frac{1 + \Delta(v^2/2g)/\Delta y}{S_0(1 - S_f/S_0)} \tag{8.31}$$

From Equation 8.20 we may write $S_f = Q^2/K^2$, but if $S_f = S_0$, then $S_0 = Q^2/K_n^2$, where K_n is at normal depth. Hence,

$$\frac{S_f}{S_0} = \frac{K_n^2}{K^2} \tag{8.32}$$

Furthermore, by using Equation 8.22, this may be put in the form of

$$\frac{S_f}{S_0} = \left(\frac{y_n}{y}\right)^N \tag{8.33}$$

where N is the normal flow exponent.

The kinetic energy term in the numerator of the right-hand side of Equation 8.31 represents the change in kinetic energy that comes about by the acceleration or deceleration of the flow. In most methods proposed for simplified computation of gradually varied flow this term is either neglected or approximated in one way or another. To properly account for this effect, one should remember that the change in velocity is caused by the change in the cross-sectional area of the flow. The change in depth, Δy, equals the change in area divided by the width of the water surface, B. By introducing these terms, differentiating, and substituting Equations 8.6 and 8.7 (for details, see Chow[10]) we may express the change in kinetic energy as

$$\frac{\Delta(v^2/2g)}{\Delta y} = -\frac{Z_c^2}{Z^2} = -\left(\frac{y_c}{y}\right)^M \tag{8.34}$$

where M is the critical flow exponent. By introducing a new variable in the form of

$$u = \frac{y}{y_n} \tag{8.35}$$

into Equations 8.31, 8.33, and 8.34 and rearranging, we obtain

$$\Delta x \frac{y_n}{S_0}\left[1 - \frac{1}{1-u^N} + \frac{u^{N-M}}{1-u^N}\left(\frac{y_c}{y_n}\right)^M\right] \tag{8.36}$$

which is the *differential equation of the water surface for gradually varied flow*. It gives the slope of the surface at specified flow depths for a given discharge and known channel conditions.

The development of this gradually varied flow equation was initiated by Dupuit,[11] who used a series integration. The first to present an analytic solution for the problem was Bresse,[12] whose tabulated results are still in use today.

The solution of Equation 8.36 is difficult as long as the hydraulic exponents, N and M, are strongly dependent on the variable depth of the flow. This is the case with flow in partially filled pipes, such as culverts and sewers. But in rectangular channels, like wide streams, N and M can be assumed constant. Their value may be obtained from Figures 8.4 and 8.6 for the average depth along the whole length of the backwater curve, obtained as

$$y_{ave} = \frac{y_0 + 1.01 y_n}{2} \tag{8.37}$$

The depth y_0 here is the initial depth at the beginning of the curved surface line, such as the water elevation at a dam. There are many solution methods available

[10]V. T. Chow, *Open Channel Hydraulics*, New York: McGraw-Hill, 1959, p. 220.
[11]Known already from his work on the well discharge formulas.
[12]Jacques Antoine Charles Bresse (France, 1822–1883).

in the literature for solving Equation 8.36. Most involve some simplifying assumptions. We review three different methods here. The first is a computer solution using numerical integration. The second is the classic longhand solution introduced by Boris A. Bakhmeteff [13] (Russia, 1880–1951), and the third is a simple approximation method. The first step in each of these solution methods is to break up the whole length of the backwater curve into *reaches between predefined depths.* The length of each reach between these arbitrarily selected depths is then computed by Equation 8.36 or some other formula.

Without a computer at hand, we must resort to the Bakhmeteff backwater formula. It requires the use of a new constant, formed from the hydraulic exponents,

$$J = \frac{N}{N - M + 1} \tag{8.38}$$

and another new variable,

$$v = u^{N/J} \tag{8.39}$$

Bakhmeteff's solution is a tabular one—a natural for a spreadsheet program. First, intermediate flow depths are selected in an arbitrary manner, between the control depth, the depth at the downstream end, and an upstream end value. The latter is usually selected as 5 or 10 percent above the normal depth of the flow. The reason for this is that the upstream end of a backwater curve approaches normal flow asymptotically and reaches it only at infinity. Once the intermediate flow depths, y_1, y_2, and so on, are determined, the corresponding u and v values are computed. Bakhmeteff's equation gives the distance x of a given y flow depth from the theoretical origin of the function in the form

$$x = \frac{y_n}{S} [u - F(u, N) + BF(v, J)] + \text{constant} \tag{8.40a}$$

where the constant B is given by

$$B = \frac{J}{N} \left(\frac{y_c}{y_n} \right)^M \tag{8.41}$$

In Equation 8.40a the term F is called Bakhmeteff's varied flow function. A table of this variable, extended and corrected by V. T. Chow, is included in Appendix A, Table A.6. Appropriate F values are selected corresponding to u and N, and v and J values, as determined for the various flow depths and hydraulic exponents.

In practical applications the distance between the two ends of a reach is desired, which is the distance between two predefined y depth values. Therefore, Equation

---

[13]He was Russia's ambassador to the United States before the Russian Revolution and remained in the country. Besides working as a well-known hydraulics consultant, he made a fortune manufacturing matches.

8.40a may be rewritten in the form

$$L = x_2 - x_1 = \frac{y_n}{S}[(u_2 - u_1) - (F(u_2, N) - F(u_1, N)) + B(F(v_2, J) - F(v_1, J))] \quad \textbf{(8.40b)}$$

in which the subscripts refer to the adjacent depth values. L is the length of the reach sought. In the tabular solution the successive reach lengths are computed individually and then added together. From these results, the depth of flow versus the distance from the downstream end of the backwater curve may be plotted. Recall that a backwater curve pertains to a particular discharge. For other discharges new backwater computations are required.

Water surface profiles may be generated by computer using the HEC-2 program. It computes over 80 different output variables such as the discharge, flow depth, flow area, velocity, energy gradient, computed and critical water surface elevations, energy losses, width of submerged area, and volume of excavation for channel improvements. It handles subcritical as well as supercritical flows and can model obstructions such as bridges, culverts, weirs, and flood-plain structures. The SCS uses the SWP-2 program for determining water surface profiles and flow characteristics based on the standard step method. The presentation of computer methods is beyond the scope of this text.

A *rough approximation* for a backwater shape in a small channel may be made by using the *conveyance method* expressed by Equations 8.20 and 8.21. If small distances are considered between stations, it may be assumed that the change in kinetic energy is negligible. As a result, the water surface is assumed to be parallel to the slope of the channel bottom for distances not exceeding 200 to 300 m. In this case the conveyance equation gives

$$\Delta L = \frac{K^2}{Q^2}\, \Delta y \quad \textbf{(8.42)}$$

For a channel of known shape, y may be plotted against K^2/Q^2. For round numbers of y values this function can be read from the graph, and the corresponding distance between the stations can be determined by Equation 8.42.

Most often the gradually varied flow formula is used to compute $M1$-type surface profiles. The methods presented are equally applicable to the other types of surface profiles shown in Figure 8.10. In doing so the reader must first consult the figure to determine two things: One is the direction in which the computation must proceed. To begin the curve, one must have a *control point* where the elevation of the surface is known. In the case of Figure 8.11, for instance, through length l_3 the calculation starts from the critical section at the right and proceeds toward the left. S_0 is at the $M2$ curve on the left. The free fall section, l_1, may be determined from a technique shown in the next chapter. The supercritical segment, l_2 can be computed only after l_3 is known at the right of the jump, where the depth is the conjugate of the depth at the right end of the $H3$ profile. Therefore, the $H3$ segment will be computed from right to left. The second thing to check before the actual computations is the *curvature of the surface*, shown also in Figure 8.10. Whether the curve is convex or concave will determine if the surface is raised or reduced at each step.

The backwater formula provides a solution for open channel flow problems under the most general conditions. Therefore, its use is not limited to the determination of backwater-caused flooding only. As an example of the broad application of Equation 8.40, one may consider its use in the solution for the *delivery of channels*. This question may be an important one in the design of power canals feeding water to turbines. Certainly, the largest amount of delivered discharge is not the one computed by the equation for normal flow. If we refer to Figure 8.12 showing the delivery problem for a channel of L length under the conditions when the water level is kept constant in the upper reservoir and is varied in the lower reservoir, we see that the discharge variation due to the downstream water elevation may be determined. Three points are easy to obtain from this discharge-depth curve. These are the maximum at the point where the depth at the lower reservoir is critical, the normal discharge when the depths at both ends of the channel are equal, and the zero discharge when the lower water elevation is the same as the upper elevation. The discharge-depth curve can easily and reliably be approximated by determining at least one more point on the curve.

Unfortunately, there is no direct solution for a problem of this type. The backwater formula is written such that it results in a length L for a known discharge and its associated normal depth. In the delivery problem, on the other hand, the length L is known, and either y_2 or Q is the unknown variable. Because of the complexity of Equation 8.40, there is no way to rearrange it to give either unknown in an explicit manner. Therefore, the only way to solve the problem is to use a trial-and-error approach.

In the *trial-and-error computation*, first a discharge value is selected that is somewhat less than the value of the normal discharge. A discharge of about 75 percent of the normal discharge is a reasonable start. Next, a trial value is selected for the downstream depth, y_2. Approximating this to be about halfway between the normal depth and the elevation of the upper reservoir is reasonable. Starting with these values, we may carry out the procedure shown for backwater computation, keeping in mind that y_2 is arbitrarily selected. The next step is to compute a new normal depth using Chézy's equation for the selected discharge. This will result in a new value for the normal flow exponent, N. The critical flow exponent, M, may then be calculated for the algebraic average of y_2 and y_1. With these values at hand, u, v, and J can be determined.

In this computation the upstream depth y_1 is the depth in the channel at the upper reservoir. After proper substitution into Equation 8.40b, a trial value of L is obtained. Chances are that this computed L will not be equal to the actual length of our channel, unless by chance we selected y_2 correctly. The computation has to be repeated for a newly selected y_2, keeping the discharge Q the same as before. For the second trial, y_2 should be selected such that if L_1 computed was larger than L actual, the channel length y_2 of the new trial should be smaller than its first assumed value.

TABLE 8.3   Channel Stability Data for Various Soils

Soil Type	Erosion Coefficient, G	Tractive Force, T N/m^2	Allowable Side Slope, z
Sand	2.5	1.7–4.0	2
Loose, sandy clay	2.5	1.9–8.0	3
Firm clay	3.5	4.0–11.0	1.5
Muck and peat	—	2.0–12.0	0.25
Rock protection	—	150–250	1.0
Soft sandstone	3.8	—	0

The backwater computation procedure must be repeated with the new y_2 value to give a new L_{II} value. Unless the two L values are significantly different from the actual length of the channel, further trial computations are not warranted. Instead, the proper value of y_2 associated with the selected Q may be determined by a linear approximation using the formula

$$y_2 = \frac{y_{II} - y_I}{L_{II} - L_I}(L_{actual} - L_I) + y_1 \qquad (8.43)$$

in which the subscripts I and II refer to the trial solutions. The resulting y_2 value will be a reasonably correct approximation for the downstream depth for which Q was selected to be 75 percent of the original Q_n. With four points known on the discharge delivery curve, a reasonably correct hydrograph may be drawn.

8.6   EROSION AND SEDIMENTATION

Channels built in erodible earth are susceptible to scouring[14] as well as to deposition of sediment carried by the water. Field studies by the U.S. Bureau of Public Roads resulted in useful practical information for the design of stable earth channels. According to these studies, scouring in an earth channel will not occur as long as the Froude number of the flow does not exceed 0.35.

Permissible maximum design velocities with regard to erosion of unlined earth channels in various soils were found to be

$$v_{maximum} = Gy_n^{0.2} \qquad (8.44)$$

where y_n is the normal depth in feet, resulting in a velocity dimension of feet per second. For these units, values of *the erosion coefficient G* for various soils are given in Table 8.3. If the velocity is calculated to be too high to maintain a stable

---

[14]The earliest studies on erosion were written by Domenico Guglielmini (Italy, 1655–1710), a professor of hydraulics who later switched to medicine to support himself. Although Dupuit wrote insightful comments on sediment transportation, real scientific progress did not occur until Hubert Engels (Germany, 1854–1945) built his tilting flume at the University of Dresden specifically to study movable bed models in 1898.

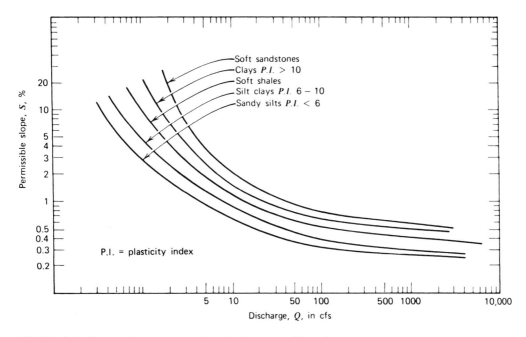

FIGURE 8.15   Permissible maximum channel slope in erodible soils.

channel in an erodible soil, roadside ditches are built either with a concrete lining or with a flatter slope. The latter requires the construction of *drop structures*, which are usually small concrete or rock-fill weirs. The distance L between these drop structures is computed from

$$L = \frac{h}{S_1 - S_2} \tag{8.45}$$

in which h is the height of the drop at the weir, S_1 is the slope of the original channel at which scour would occur, and S_2 is the new flattened slope.

The *permissible maximum channel slope* for different soils is shown in Figure 8.15 for various design discharges. This graph indicates that the selection of the proper design discharge is of great significance in designing a stable channel. Experience obtained by observing the performance of roadside ditches along interstate highways has shown that scouring of channels begins at the deepest points below the minimum permanent flow, the trickling discharge due to ground-water contributions between rains. For this reason the recent tendency in design is to provide a partial concrete lining limited to the bottom of the channel. This will prevent bottom scouring in places where grass cannot develop because of the permanent presence of water. On higher portions of the channel if strong grass growth is allowed, it usually will prevent scouring during occasional high discharges.

Another way of determining the limiting channel slope beyond which the bed will erode is the *tractive force method.* The tractive force T of the moving water is computed by the equation

$$T = \gamma R S \tag{8.46}$$

where γ is the unit weight of water, R is the hydraulic radius, and S is the channel slope. Typical values of T for various soils are given in Table 8.3.

Deposition of silt does not occur as long as the Froude number is 0.12 or larger unless the discharge is very large. In sandy silt and silt-clay soils, the most susceptible to erosion and deposition, the design velocity must be kept above a minimum value,

$$v_{\text{minimum}} = 0.63 y_n^{0.64} \tag{8.47}$$

where y_n, the normal depth, is in feet. The resulting minimum velocity is obtained by Equation 8.47 in feet per second.

The allowable side slope of human-made ditches in various soils and linings is given in Table 8.3. Because most such channels carry discharges intermittently, side slopes are usually covered with grass, if not lined with rock or concrete. Grassy slopes are protected from erosion by the grass, although newly made channels may be eroded if high waters are encountered before a strong grass cover is allowed to develop. Various temporary protective measures can be used to prevent such erosion, such as netting, asphaltic sprays, and the like. The stability of channel side slopes is in the domain of soil mechanics. In addition to the usual slope stability considerations, it should be noted that the soil in the vicinity of the slope is fully saturated. Sudden reduction of the water level in the channel creates a considerable hydraulic gradient in the groundwater, which increases the seepage force in the soil. The combined effects of the increased weight of the saturated soil and the reduction of the shear strength due to soil moisture and buoyancy effects on the soil grains often lead to slope failures. The danger of erosion can often be reduced by decreasing the slope of the channel. In cases of design, S may be fixed by a predetermined alignment of the channel, such as for roadside ditches. Even in these cases the slope can occasionally be reduced by using drop structures along the channel at certain intervals. In other cases when only the two ends of the channel are fixed in space, the slope can be decreased at will by following land contours with a predetermined slope, in which case the length of the channel is determined from

$$L = \frac{H_1 - H_2}{S} \tag{8.48}$$

where H_1 is the upper available energy line elevation, H_2 is the elevation of the water level at the recipient water body at the lower end of the channel, and S is the predetermined channel slope. The channel length, of course, enters into the design considerations as an economic factor, giving rise to the total required amount of land acquisition and excavation along with the required cross-sectional area A and the so-called free board, the depth of the water level below the top of the channel. The magnitude of the free board, u, may be determined from

$$u = \sqrt{cy} \qquad (8.49)$$

in which the free board is obtained in feet for a given depth of water in feet. The U.S. Bureau of Reclamation recommends that the factor c selected be 1.5 for a small canal of 20 cfs capacity and 2.5 for canal capacities of 3000 cfs or more.

In nature the slope and the discharge vary, which causes a continuous change in erosion and sedimentation both in time and in location along the channel. The quantity of sediment carried in the water is the greatest during ascending floods and the least during base flow. Another way to look at this phenomenon is that discharge and slope may be combined in a function that defines the *sediment carrying capacity* of a stream. As either variable changes, the sediment content of the water either increases or decreases. Water tends to erode the channel banks to increase its sediment load or deposit some of it to maintain an optimum sediment content under the circumstances of time and place.

Nature's law that a stream always "fills" its sediment carrying capacity was demonstrated soon after the completion of the Aswan dam on the Nile River. The fertilizing effects of the Nile's yearly floods were lost as the new reservoir caused the sediment to settle out. Below the dam, erosion caused major damages that were totally unanticipated by the designers.

Sediment is carried by the water in two forms: suspended and as bed load. *Suspended sediment* is carried along when the turbulence of the water is high enough that the vertical, upward turbulent velocity components exceed the settling velocity of the particles. *Bed load* is carried along by the shear forces generated along the wetted perimeter of the channel.[15] Suspended sediment load is measured by taking water samples and boiling off the water to leave the remaining solids. Waterborne sediment is usually expressed in terms of weight, such as tons per year. The average density of sediment in reservoirs may be taken to be 1500 kg/m³. Bed load is measured by a variety of sediment traps lowered to the bottom of the channel. The combined sediment load can then be expressed as a percentage of the discharge. By frequent measurement of the sediment load of a stream over an extended period, the annual sediment load of a stream can be estimated. This figure is closely related to the annual discharge; therefore, if statistical information is available on the flow of water, it may be applicable to the sediment load also. Another way to determine the annual sediment load is to study nearby reservoirs in similar watersheds and to measure the amount of sediment collected since their completion.

---

[15]One of the earliest formulas for sediment transportation was published by Paul Francois Dominique du Boys (France, 1847–1924). Although his basic assumptions were quite wrong, his bed load formula has since been used as a basis for many empirical studies.

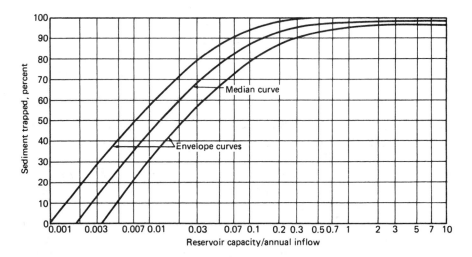

FIGURE 8.16 Sedimentation rate of reservoirs. (After G. M. Brune, Trap Efficiency of Reservoirs, *Transactions of the American Geophysical Union*, Vol. 34, pp. 407–418, June 1953)

The phenomenon described makes it inevitable that reservoirs behind dams will fill with sediment at least to some extent over a period of time. Depending on the climate, the geology of the region, and on the relative size of the reservoir with respect to the yearly discharge of the stream some reservoirs may lose very little of their storage capacity; others may have only a short useful life.

All other variables being equal, the rate of sedimentation of a reservoir depends on its *capacity/annual inflow* ratio. Let us assume, for example, that the average daily discharge of a stream is 15 m³/s. This will result in 473 million m³ (384,000 acre-ft) of yearly inflow. If a reservoir is built such that its initial storage capacity is 10 million m³ (8110 acre-ft), then its capacity/inflow ratio is 0.021. Based on a study of a number of reservoirs, the SCS offers a graph shown in Figure 8.16, to estimate the *sediment trapping efficiency* depending on the capacity/inflow ratio.[16] For the given case 60 percent of the sediment carried in the water will be deposited in the reservoir during the first year. Let us further assume that the water, on average, contains 0.2 percent sediment by volume. On a yearly basis, this is 0.946 million m³ of sediment. If 60 percent of this is entrapped, 568,000 m³ will be lost from the initial storage capacity of the reservoir by the end of the first year. This is 5.68 percent of the initial storage capacity of the reservoir, leaving only 9.43 million m³ for the second year. However, with this reduced storage capacity the new capacity/inflow ratio and the percentage of the entrapped sediment will be somewhat less in the second year. In practice, one should also consider the degree of consolidation of the sediment in the reservoir, which tends to reduce its volume.

[16]*National Engineering Handbook*, Section 3, 2d ed., 1983, after G. M. Brune.

For trap efficiency values between 20 to 85 percent the median curve in Figure 8.16 is approximated by

$$E(\%) = 100 \left(1 - \frac{1}{1 + 100\,C/I}\right)^{1.2} \tag{8.50}$$

where E is the reservoir's trap efficiency in percent, C is its volume, and I is the annual inflow of water, in identical units. Using this formula reservoir sedimentation problems can be computed directly.

Several computer programs have been developed for the analysis of sediment movement. One example is HEC-6, which evaluates the sediment transport in rivers and reservoirs on a long-term basis. HEC-6 simulates the transportation of sediment in a stream and can determine both volume and location of sediment deposits. The program uses any one of 12 different theoretical sediment transport functions in its calculations. It can analyze dredging operations, shallow reservoirs, and scour and deposition effects in rivers and can also model the rise and fall of movable bed material during several flood cycles. The presentation of such methods is beyond the scope of this text.

EXAMPLE PROBLEMS

Example 8.1   A channel of nonuniform cross section, as shown in Figure E8.1, delivers 100 cfs. The channel width is 30 ft at the water level. For an estimated average channel depth of 5 ft, estimate the average velocity.

FIGURE E8.1

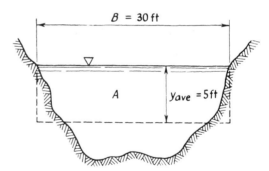

Solution. From Equation 8.1

$$A = B y_{\text{ave}}$$
$$= 30 \times 5 = 150 \text{ ft}^2$$

Therefore, the average velocity can be calculated from

$$v = \frac{Q}{A} = \frac{100}{150} = 0.67 \text{ fps} \qquad \square$$

Example 8.2   A trapezoidal channel with a bottom width of 4 m and side slopes of 1:4 (one vertical to four horizontal) carries water at a depth of 2 m (Figure E8.2). Compute the wetted perimeter and the average depth.

FIGURE E8.2

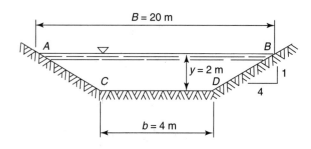

Solution. The cross-sectional area of the channel is

$$A = by + zy^2 = 4 \times 2 + 4 \times 2^2 = 24 \text{ m}^2$$

The width of water surface $B = b + 2zy = 20$ m.
The wetted perimeter is

$$P = 2\overline{AC} + \overline{CD}$$
$$= 2\sqrt{2^2 + 8^2} + 4$$
$$= 20.49 \text{ m}$$

Then, from Equation 8.1 the average depth is

$$y_{ave} = \frac{A}{B} = \frac{24}{20} = 1.2 \text{ m} \qquad \square$$

Example 8.3

If the discharge in the channel of Example 8.2 is 50 m³/s, calculate the average velocity and the discharge per unit width.

Solution. The average velocity is

$$v = \frac{Q}{A} = \frac{50}{24} = 2.08 \text{ m/s}$$

The width is

$$B = 20 \text{ m}$$

Therefore,

$$q = \frac{Q}{B} = \frac{50}{20} = 2.5 \text{ m}^3\text{/s/m} \qquad \square$$

Example 8.4

A 7.9 m wide rectangular channel carries 22.0 m³/s discharge. Determine the critical depth and the critical velocity.

Solution. By using Equation 8.6 we compute the critical shape factor to be

$$\frac{Q}{\sqrt{g}} = \frac{22}{\sqrt{9.81}} = 7.02 = Z$$

Next, we select a series of arbitrary depths and compute the corresponding Z values from the right-hand side of this equation to be

$$Z = 7.9y^{1.5} = A\sqrt{y_{mean}}$$

The results are

y(m)	0.1	0.5	1.0	2.0
Z =	0.25	2.79	7.9	22.3

The tabulated values are plotted in Figure E8.4.

FIGURE E8.4 Shape factor versus depth.

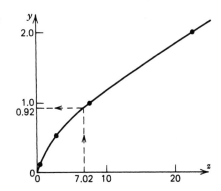

Entering the graph at the critical value calculated above, we find the critical depth to be

$$y_c = 0.92 \text{ m}$$

The critical velocity is

$$v_c = \sqrt{gy_c} = 3 \text{ m/s}$$

This example demonstrates the solution procedure for a channel with an arbitrary cross section. For a rectangular channel Equation 8.6 can be solved directly:

$$\frac{Q}{B\sqrt{g}} = y_c^{1.5} \quad \text{or} \quad y_c = \left(\frac{22}{7.9\sqrt{9.8}}\right)^{2/3} = 0.92 \text{ m} \qquad \square$$

Example 8.5

A 12 m wide rectangular channel carries 10 m³/s discharge. Calculate the critical depth of the flow.

Solution. From Table 8.1 the critical discharge in a rectangular section is

$$Q_c = \sqrt{g}By_c^{1.5}$$

By Equation 8.6 the section factor is

$$\frac{Q}{\sqrt{g}} = Z$$

Comparing the two formulas, we find that

$$Z = By_c^{1.5} = \frac{Q_c}{\sqrt{g}}$$

Substituting the given values for B and Q_c, we obtain

$$12y_c^{1.5} = \frac{10}{\sqrt{9.81}}$$

Solving for the critical depth, we get

$$y_c = 0.41 \text{ m} \qquad \qquad \square$$

Example 8.6

Determine the critical flow exponent for a trapezoidal channel of 1 : 4 side slopes and 16 ft bottom width when the depth of the flow is 3 ft (Figure E8.6), and compute the differences between corresponding critical flow exponents if the depth increases to 6 and 9 ft.

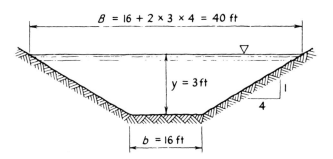

FIGURE E8.6

Solution. From Equation 8.7

$$\frac{Q^2}{g} = f(y^M)$$

$$M = \text{critical flow exponent}$$

From Figure 8.4 the relation of z and y/b determines M. We have $y/b = 3/16 = 0.1875$, $z = 4$. Entering these values in Figure 8.4, we obtain $M = 3.65$. If the depth changes from 3 ft to 6 ft and 9 ft, we obtain the following values:

y	3	6	9
y/b	0.1875	0.375	0.563
z	4	4	4
M	3.65	4.05	4.25

M values were determined from Figure 8.4 corresponding to y/b and z. Note how small the change in *M* is as compared with the change in depth.   ☐

Example 8.7

In a 12 m wide rectangular chute, water flows in supercritical condition at a rate of 150 m³/s. At the end of the chute, on a horizontal concrete apron, the pool level of the downstream water is 3 m above the apron as a result of a hydraulic jump. Analyze the jump.

Solution. The discharge per unit width is

$$q = \frac{Q}{B} = \frac{150}{12} = 12.5 \text{ m}^3/\text{s/m}$$

The Froude number at the front of the jump is

$$\mathbf{F}_1 = \frac{v}{\sqrt{gy_1}} = \frac{q/y_1}{\sqrt{gy_1}} = \frac{q}{\sqrt{gy_1^{3/2}}}$$

Using Equation 8.9 to express the conjugate depth y_1 for the known $y_2 = 3$, we have

$$\frac{y_2}{y_1} = \frac{1}{2}(\sqrt{1 + 8\mathbf{F}_1^2} - 1)$$

$$= \frac{1}{2}\left(\sqrt{1 + 8\frac{q^2}{gy_1^3}} - 1\right)$$

From this, we get

$$y_2 y_1^2 + y_1 y_2^2 - \frac{2q^2}{g} = 0$$

and if we substitute all known values,

$$3y_1^2 + 9y_1 - \frac{2(12.5)^2}{9.81} = 0$$

or

$$y_1^2 + 3y_1 - 10.62 = 0$$

Solving this quadratic equation, we find that

$$y_1 = 2.09 \text{ m}$$

Substituting into the expression for the Froude number, we get

$$\mathbf{F} = \frac{q}{\sqrt{gy_1^{3/2}}} = \frac{12.5}{\sqrt{9.81}\,(2.09)^{3/2}} = 1.32$$

which is less than 1.7. As described, the hydraulic jump will take the form of a series of undulating waves along the downstream water surface. The location of the jump is rather ill defined, and shore protection is required due to the wave action.

The total available energy downstream is

$$E_2 = y + \frac{q^2}{2gy_2^2}$$

$$E_2 = 3 + \frac{12.5^2}{2(9.81)3^2} = 3.88 \text{ m}$$ ☐

Example 8.8

A rectangular channel 10 ft wide delivers 320 cfs of water at 1.8 ft depth before entering a hydraulic jump (Figure E8.8). Compute the downstream water level and the critical depth.

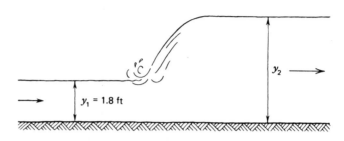

FIGURE E8.8

Solution.

$$q = \frac{\text{discharge}}{\text{width}} = \frac{320}{10}\,\text{cfs/ft}$$

$$q = 32\,\text{cfs/ft}$$

$$y_1 = 1.8\,\text{ft}$$

From Equation 8.8,

$$y_2 = \frac{y_1}{2}\left(\sqrt{1 + \frac{8q^2}{gy_1^3}} - 1\right)$$

$$= \frac{1.8}{2}\left(\sqrt{1 + \frac{8(32)^2}{32.2 \times (1.8)^3}} - 1\right)$$

$$y_2 = 5.11\,\text{ft} = \text{downstream water level}$$

To find the critical depth, from $q_c = y_c\sqrt{gy_c} = \sqrt{gy_c^3}$

$$y_c = \left(\frac{q^2}{g}\right)^{1/3}$$

$$= \left[\frac{(32)^2}{32.2}\right]^{1/3}$$

$$y_c = 3.17\,\text{ft}$$

□

Example 8.9

The downstream depth after a hydraulic jump in a 12 ft wide rectangular channel is 5 ft (Figure E8.9). Compute a rating curve showing the discharges and the corresponding conjugate upstream depths.

Solution. From Equation 8.11

$$Q = B\left[\frac{y_1 + y_2}{2}(gy_1y_2)\right]^{1/2}$$

If we substitute

$$Q = 12\left[\frac{y_1 + 5}{2}(32.2 \times 5y_1)\right]^{1/2}$$

$$= 107.666[y_1(y_1 + 5)]^{1/2}$$

y_1 (ft)	1.0	1.5	2.0	2.5	3.0	3.5	4.0	4.5	5.0
Q (cfs)	264	336	403	446	527	587	646	704	761

Plotting these data will give us the rating curve sought.

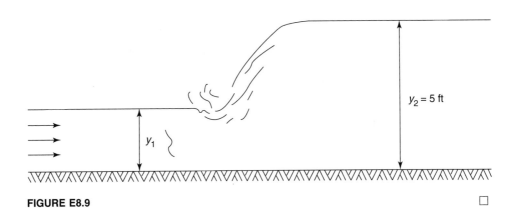

FIGURE E8.9

Example 8.10   The slope of the channel described in Examples 8.2 and 8.3 is $S = 0.002$ (Figure E8.10). Determine the roughness coefficient n of the channel assuming that normal flow is encountered.

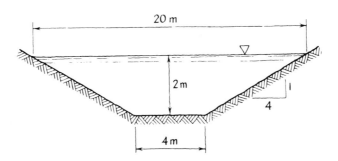

FIGURE E8.10

Solution. From Equation 8.13

$$v = \frac{Q}{A} = C\sqrt{RS}$$

where

$$R = \frac{A}{P} = \frac{24}{20.49} = 1.17$$

Therefore, Chézy's coefficient is

$$C = \frac{2.08}{\sqrt{1.17(0.002)}} = 43$$

From Equation 8.16

$$C = \frac{1}{n} R^{1/6}$$

$$n = \frac{R^{1/6}}{C}$$

$$= \frac{(1.17)^{1/6}}{43} = 0.024$$

A roughness coefficient of 0.024 in Table 8.2 corresponds to small humanmade earth channels in well-kept condition. □

Example 8.11

For a triangular channel of $z = 2$ side slopes that carries a discharge of 100 cfs at a depth of 5 ft, determine the channel slope S under the assumption that normal flow exists and that n equals 0.022.

Solution. The surface width B is

$$B = 2zy = 2(2)5 = 20 \text{ ft}$$

Hence, the cross-sectional area A is

$$A = \frac{1}{2} By = \frac{1}{2} (20)5 = 50 \text{ ft}^2$$

The velocity is

$$v = \frac{Q}{A} = 2 \text{ fps}$$

The hydraulic radius from Equation 8.14 is

$$R = \frac{A}{P}$$

$$P = 2\sqrt{y^2 + \frac{B^2}{4}} = 2\sqrt{25 + 100} = 22.4 \text{ ft}$$

$$R = \frac{50}{22.4} = 2.2 \text{ ft}$$

From Equation 8.17

$$C = \frac{1.49}{n} R^{1/6} = \frac{1.49}{0.022} 2.2^{0.17} = 77.4 \sqrt{\text{ft}}/\text{sec}$$

From Equation 8.13, we get

$$v = C\sqrt{RS}$$
$$2 = 77.4\sqrt{2.2S}$$

From which

$$S = \left(\frac{2}{77.4}\right)^2 \frac{1}{2.2} = 3 \times 10^{-4} \text{ ft/ft}$$

or 3 ft of drop over 10,000 ft. □

Example 8.12

A well-kept earth channel is laid out on a slope of $S = 10^{-4}$. The shape of the trapezoidal channel is defined by $z = 2$, $y = 1$ m, and $b = 2$ m. How much is the discharge in the channel if normal flow is assumed?

Solution. From Table 8.1

$$A = (b + zy)y = [2 + 2(1)]1 = 4 \text{ m}^2$$
$$R = \frac{(b + zy)y}{b + 2y\sqrt{1 + z^2}} = \frac{4}{2 + 2(1)\sqrt{1 + 2^2}} = 0.618 \text{ m}$$

From the Chézy-Manning formula (Equation 8.19)

$$Q = \frac{1.49}{n} AR^{2/3}S^{1/2} = \frac{1.49}{0.025}(4)(0.618)^{2/3}(10^{-4})^{1/2} = 1.7 \text{ m}^3/\text{s}$$

where $n = 0.025$ from Table 8.2. □

Example 8.13

A trapezoidal channel with a bottom width of 20 ft and side slope $z = 3$ carries a flow of 400 cfs at normal conditions with a depth of 3.3 ft from a point with elevation of 786 ft above sea level to a recipient water body of 680 ft water level (Figure E8.13). Determine the required channel length if $n = 0.025$.

FIGURE E8.13

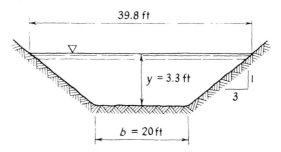

Solution.

$$Q = 400 \text{ cfs}$$
$$z = 3$$
$$y = 3.3 \text{ ft}$$
$$b = 20 \text{ ft}$$
$$n = 0.025$$
$$B = 20 + 2 \times 3 \times 3.3 = 39.8 \text{ ft}$$

Then,

$$A = \frac{1}{2}(39.8 + 20)3.3 = 98.7 \text{ ft}^2$$

$$R = \frac{A}{\text{wetted perimeter}} = \frac{98.7}{20 + 2 \times 3.3\sqrt{1 + 3^2}} = 2.41 \text{ ft}$$

By using the Chézy-Manning formula, Equation 8.19, we get

$$Q = \frac{1.49}{n} AR^{2/3} S^{1/2}$$

$$S = \left[\frac{400 \times 0.025}{1.49 \times 98.7 \, (2.41)^{2/3}}\right]^2 = 0.00143$$

From Equation 8.12,

$$S = S_e = \frac{H_1 - H_2}{L}$$

Hence, the required channel length is

$$L = \frac{786 - 680}{0.00143} \cong 74{,}100 \text{ ft} \cong 14 \text{ mi} \qquad \square$$

Example 8.14

A rectangular channel is 7.9 m wide and is cut in rock for which n is assumed to be 0.017. The discharge is 22 m³/s. The channel is divided into three consecutive sections of varying slopes. In the first section $S_1 = 0.012$, in the second $S_2 = 0.05$, and in the last $S_3 = 0.006$. Determine the three normal depths, y_1, y_2, and y_3, using Equation 8.21.

Solution. Equations 8.20 and 8.21 will take the form

$$\frac{Q}{S^{1/2}} = \frac{yB}{n}\left(\frac{yB}{2y + B}\right)^{2/3} = K$$

To prepare the plot of K versus y, we select a series of depth values and compute

$$K = \frac{7.9y}{0.017}\left(\frac{7.9y}{2y + 7.9}\right)^{2/3}$$

the results are

y(m)	0.3	0.5	0.7	0.9
K =	59.50	135.19	230.02	340.01

The results are plotted in Figure E8.14.

FIGURE E8.14 Channel conveyance graph.

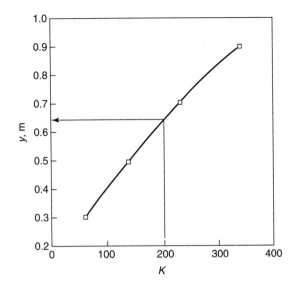

The K values for the three different slopes are as follows:

$$K_1 = \frac{Q}{\sqrt{S_1}} = \frac{(22)}{0.012^{1/2}} = 200.83$$

$$K_2 = 98.39$$

$$K_3 = 284.02$$

From the plot of K versus y, we obtain the three normal depth values sought:

$$y_1 = 0.64 \text{ m}$$

$$y_2 = 0.41 \text{ m}$$

$$y_3 = 0.80 \text{ m} \qquad \square$$

Example 8.15   For the open channel flow problem treated in Example 8.10, determine the normal flow exponent N.

Solution. We use Figure 8.6 with

$$b = 4 \text{ m}$$

$$y = 2 \text{ m}$$

$$\frac{y}{b} = \frac{2}{4} = 0.5$$

$$z = 4$$

Entering the graph at the vertical coordinate 0.5 and turning down at the $z = 4$ curve, we find that

$$N = 4.45 \qquad \square$$

Example 8.16   A rectangular channel 30 ft wide carries a discharge of 450 cfs. The slope of the channel is 0.0017, and its roughness factor is $n = 0.022$. The channel is tributary to a river in which the existing flood stage is 12 ft above the normal depth in the channel. Determine the resulting backwater curve in the tributary channel at four points along the channel (Figure E8.16). Use Bakhmeteff's method.

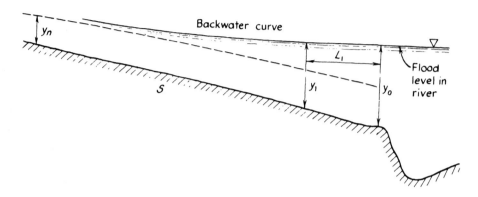

FIGURE E8.16

Solution. Given:

$$Q = 450 \, \text{cfs}$$
$$S = 0.0017$$
$$n = 0.022$$
$$b = 30 \, \text{ft}$$

For a wide rectangular channel it may be assumed that the hydraulic radius equals the depth of the flow; hence, the normal depth can be computed by Equation 8.19 in which $A = by_n$:

$$Q = \frac{1.49}{n} \, Ay_n^{2/3}S^{1/2}$$

$$450 = \frac{1.49}{0.022} \, 30y_n^{5/3}0.0017^{1/2}$$

$$y_n = 2.74 \, \text{ft}$$

The discharge per unit width is

$$q = \frac{Q}{b} = \frac{450}{30} = 15 \, \text{cfs/ft}$$

The critical depth may be computed by Equation 8.3, in the form of

$$q_c = \sqrt{gy_c^3}$$

from which

$$y_c = \left(\frac{q^2}{g}\right)^{1/3} = \left(\frac{15^2}{32.2}\right)^{1/3} = 1.91 \, \text{ft}$$

We select the upstream limit of the backwater curve to be

$$y_4 = 1.05y_n = 2.88 \, \text{ft}$$

The downstream limit, at the entrance of the river, is

$$y_0 = y_n + 12 = 14.74 \text{ ft}$$

The average depth is then

$$y_{ave} = \frac{14.74 + 2.88}{2} = 8.81 \text{ ft}$$

From

$$\frac{y_{ave}}{b} = \frac{8.81}{30} = 0.29$$

and by Figures 8.4 and 8.6 the hydraulic exponents are

$$M = 3.0 \quad \text{and} \quad N = 2.8$$

By Equation 8.38

$$J = \frac{2.8}{2.8 - 3 + 1} = 3.5$$

and

$$\frac{N}{J} = \frac{2.8}{3.5} = 0.8$$

From Equation 8.41 we obtain

$$B = \left(\frac{y_c}{y_n}\right)^M \frac{J}{N} = \left(\frac{1.91}{2.74}\right)^3 \frac{3.5}{2.8} = 0.42$$

We arbitrarily select the intermediate depths sought to be round numbers between the limits of 2.88 and 14.74 ft, as follows:

$$y_0 = 14.74 \text{ ft}$$
$$y_1 = 12.0 \text{ ft}$$
$$y_2 = 8.0 \text{ ft}$$
$$y_3 = 4.0 \text{ ft}$$
$$y_4 = 2.88 \text{ ft}$$

The distances between the locations of these stage readings will be denoted by L_1, L_2, L_3, and L_4 and will be determined by Equation 8.40b. The computations are carried out in a tabular manner as shown in the following table, using Equations 8.35 and 8.39. After determin-

Table of Computations for Example 8.16

y ft	u	v	F(u, N)	F(v, J)	BF(v, J)	x ft	L ft
14.74	5.37	3.84	0.028	0.013	0.0054	8620	0
12.0	4.38	3.26	0.040	0.022	0.0092	7010	1610
8.0	2.92	2.35	0.081	0.048	0.020	4610	4010
4.0	1.46	1.35	0.330	0.225	0.095	1970	6650
2.88	1.06	1.04	0.948	0.750	0.315	690	7930

ing the u and v values, we obtain the F functions for the sake of illustration as the nearest tabular values from Appendix A, Table A.6.   □

Example
8.17

A 2000 ft long, 18 ft wide rectangular channel connects two reservoirs (Figure 8.12*a*). The slope of the channel is 0.0015; its roughness factor is 0.02. The water level in the upper reservoir is constant at 5 ft above the channel bottom at the entrance. The water level in the lower reservoir varies. Develop a discharge delivery curve for the channel as a function of the lower reservoir level.

Solution. Taking the length and the slope of the channel, we can calculate the water elevations with respect to a datum plane at the bottom of the exit of the channel denoted as point 2. At point 1, the entrance, the bottom elevation of the channel is

$$h_1 = LS = 2000 \times 0.0015 = 3 \text{ ft}$$

The upper reservoir elevation is $h_1 + 5 = 3 + 5 = 8$ ft above datum. This corresponds to the zero discharge condition at the lower reservoir level. The maximum discharge in the channel occurs when the depth at the lower reservoir corresponds to critical conditions.

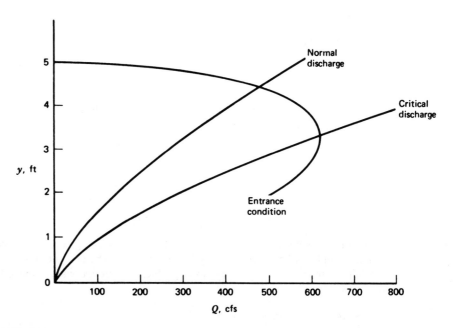

FIGURE E8.17

Normal flow in the channel delivers somewhat less. Normal flow occurs when $y_2 = y_1$. For lower reservoir elevations above the normal depth the discharge gradually decreases until it reaches zero when the lower reservoir elevation equals that of the upper reservoir. Using Equation 8.2, we may compute the entrance conditions as follows:

At the upper reservoir with $E_1 = 5$ ft,

$$E_1 = \frac{Q^2}{2gB^2y_1^2} + y_1$$

$$5 = \frac{Q^2}{(64.4)18^2y_1^2} + y_1$$

$$5 = \frac{Q^2}{20,866y_1^2} + y_1$$

This expression showing the entrance conditions may be computed for several depth values and plotted for future reference, as shown in Figure E8.17.

y_1 (ft)	3	4	4.5	5
Q (cfs)	612.8	577.8	459.6	0

For normal flow in the channel, we should solve Equation 8.19 such that the resulting Q_n and y_n correspond to the entrance requirements just stated. For this solution, again, we must compute Q_n for several y_n values, plot the results, and find the intercept of the two graphs.

$$Q_n = \frac{1.49}{n} AR^{2/3}S^{1/2}$$

$$= \frac{1.49}{0.02}(18y)\left(\frac{18y}{18+2y}\right)^{2/3}0.0015^{1/2}$$

$$Q_n = 356.72\frac{y^{5/3}}{(18+2y)^{2/3}}$$

y (ft)	2	3	3.5	4	4.5
Q_n (cfs)	144	268	337	410	486

Even without plotting the two curves we may note that the normal depth that corresponds to the entrance condition is about

$$y_n = 4.45 \text{ ft}$$

Therefore, the normal discharge that satisfies the entrance condition is

$$Q_n = 475 \text{ cfs}$$

For critical conditions at the exit point, Equation 8.6 must be satisfied. This result must also satisfy the entrance conditions. Writing

$$Q_c = b\sqrt{gy^3}$$
$$Q_c = 18(32.2^{1/2})y^{1.5} = 102.14y^{1.5}$$

and solving it for a series of y values, we have

y (ft)	2	3	3.25	3.5	4
Q_c (cfs)	289	531	598	669	817

Comparing these results with the entrance conditions, we see that for the critical discharge

$$Q_c = 620 \text{ cfs}$$

is a reasonable approximation, if we neglect the acceleration over the channel and the resulting deviation between entrance and exit elevations. In this case the downstream reservoir elevation for maximum discharge is

$$y_c = 3.33 \text{ ft}$$

Now we have three discharge versus downstream elevation points for our delivery curve. For the sake of demonstration we now may compute a fourth one with the aid of the backwater equation.

Let us first assume a discharge that is about 50 percent of the previously computed maximum discharge Q_c, say,

$$Q = 300 \text{ cfs}$$

Now we take a trial value for the depth at the lower reservoir as

$$y_2 = 6 \text{ ft}$$

By inspecting the conditions at the entrance, we find the depth there to be about

$$y_1 = 4.75 \text{ ft}$$

Before carrying out the backwater computations, we need the hydraulic exponents, which we will determine for the average depth,

$$y_{ave} = \frac{4.75 + 6}{2} = 5.37 \text{ ft}$$

from which $y/b = 5.37/18 = 0.298$, and from Figures 8.4 and 8.6,

$$M = 3 \quad \text{and} \quad N = 2.8$$

From these values, by Equation 8.38,

$$J = \frac{N}{N - M + 1} = 3.5$$

The normal and critical depths associated with the discharge of 300 cfs may be taken off the graph in Figure E8.17.

The results are

$$y_c = 2.05 \text{ ft}$$
$$y_n = 3.24 \text{ ft}$$

We are now ready to solve Equation 8.41 for B.

$$B = \left(\frac{y_c}{y_n}\right)^M \frac{J}{N} = \left(\frac{2.05}{3.24}\right)^3 \frac{3.5}{2.8} = 0.32$$

Next, we calculate the u and v values as follows:

$$u_2 = \frac{6}{3.24} = 1.85$$

$$u_1 = \frac{4.75}{3.24} = 1.47$$

For the v values we have $N/J = \dfrac{2.8}{3.5} = 0.8$ and

$$v_2 = (1.85)^{0.8} = 1.64$$
$$v_1 = (1.47)^{0.8} = 1.36$$

From Table A.6 in Appendix A we may find the corresponding F values:

$$F(u_2, 2.8) = 0.198$$
$$F(u_1, 2.8) = 0.325$$
$$F(v_2, 2.8) = 0.252$$
$$F(v_1, 2.8) = 0.393$$

With these values we enter Equation 8.40b for L_1.

$$L_1 = \frac{3.24}{0.0015}[(1.85 - 1.47) - (0.198 - 0.325) + 0.32(0.252 - 0.393)]$$
$$= 2160(0.38 + 0.127 - 0.045]$$
$$L_1 = 998 \text{ ft}$$

is less than the actual channel length of 2000 ft.

For our second trial we attempt to increase the resulting L_{II} beyond 2000 ft by maintaining $Q = 300$ but increasing y at the downstream end to 7 ft.

Repeating the computations we get a new average depth of

$$y_{\text{ave}} = \frac{4.75 + 7}{2} = 5.87 \text{ ft}$$

$$\frac{y}{b} = \frac{5.87}{18} = 0.326$$

$$M = 3 \quad \text{and} \quad N = 2.8$$

Hence, J and B remain essentially the same.

The new u_2 and v_2 values are

$$u_2 = \frac{7}{3.24} = 2.16 \qquad v_2 = 1.85$$

and

$$F(u_2, 2.8) = 0.146 \qquad F(v_2, 2.8) = 0.198$$

Hence,

$$L_{II} = 2160[(2.16 - 1.47) - (0.146 - 0.325) + 0.32(0.198 - 0.393)]$$
$$= 1742 \text{ ft}$$

which approaches $L = 2000$ and is acceptable. To solve the problem, we now use Equation 8.43

$$y_2 = \left[\frac{7-6}{1742-998}(2000-998) \right] + 6$$

$$= 7.35 \text{ ft}$$

which is the lower reservoir elevation for $Q = 300$ cfs discharge.

Summarizing our results for the downstream reservoir elevations, we have, for $y_1 = 5$ ft at the upstream end of the 2000 ft long channel

Q (cfs)	y_2 (ft)
0	8.0
300	7.35
475	4.45
620	3.33

These results may now be plotted in the form of a rating curve for the channel on the assumption that the upper reservoir level remains constant. If the level varies, the computations may be carried out for other upstream reservoir elevations. □

Example 8.18   For the conditions described in Example 8.13 and assuming that the channel is dug in clay, check if it is susceptible to erosion or deposition.

Solution.

$$v = \frac{Q}{A} = \frac{400}{98.7} = 4.05 \text{ fps}$$

1. Check velocity; $v_{min} = 0.63 y_n^{0.64}$, Equation 8.47:

$$v_{min} = 0.63(3.3)^{0.64} = 1.35 \text{ fps} < 4.05 \qquad \text{OK}$$

2. Check velocity; $v_{max} = G y_n^{0.2}$, Equation 8.44. From Table 8.3 for clay, $G = 3.5$. Therefore, the allowable velocity is

$$v_{max} = 3.5(3.3)^{0.2} = 4.44 \text{ fps} > 4.05 \qquad \text{OK}$$

3. Check the permissible maximum channel slope. For clay from Figure 8.15 the maximum permissible slope of the channel is 0.0053, which is greater than 0.0015. OK

4. Using the tractive force method (Eq. 8.46); with $R = 2.41$ ft $= 0.73$ m we have

$$T = 9810(0.73)0.0015 = 10.7 \text{ N/m}^2$$

From Table 8.3 we find that the allowable T value for clay ranges between 4 and 11; hence, our result is still within the safe limit. OK

Therefore, the channel is stable under the given conditions. □

Example 8.19   The initial capacity of a reservoir is 4 million m^3. How long will it take to silt up such that only 1 million m^3 capacity remains if the average annual inflow of water is 60 million m^3, and the average annual sediment inflow is 62,000 metric tons with 1400 kg/m^3 settled density.

Solution. The volumetric rate of the annual sediment flow is

$$S = \frac{62,000 \text{ tonne/year}}{1.4 \text{ tonne/m}^3} = 44,000 \text{ m}^3/\text{year}$$

The volumetric percentage of sediment averages

$$100 \frac{0.044 \times 10^6}{60 \times 10^6} = 0.074 \text{ percent}$$

The rate of sedimentation will be determined in increments of 1 million m³ of lost reservoir capacity. The sedimentation for average reservoir capacities of 3.5 million, 2.5 million, and 1.5 million m³ are shown in the following table. Capacity/inflow ratios are determined for these values, and corresponding trap efficiencies are calculated by using Equation 8.50:

Average Reservoir Volume million m³	C/I	E	Annual Sediment Trapped = $S \times E$	$\dfrac{1 \times 10^6 \text{ m}^3}{S \times E}$ = Years to Fill
3.5	0.058	0.826	36,300	28
2.5	0.042	0.774	34,100	29
1.5	0.025	0.668	29,400	34
				Total = 91 years

$\square$

PROBLEMS

8.1 Water is flowing 3 ft deep in a 20 ft wide rectangular open channel. The average velocity of the flow is 3 fps. Compute the total energy of the water with respect to the bottom of the channel.

8.2 Calculate the critical depth for Problem 8.1.

8.3 A circular culvert with a diameter of 1 m is flowing partially full under critical conditions with $y/D = 0.60$. Compute the critical discharge.

8.4 In a rectangular channel the depth of the water is 2.5 m, and the width is 50 m. Calculate the section factor.

8.5 Assume that the critical discharge in Problem 8.4 is 125 m³/s. What is the corresponding critical depth?

8.6 Determine the critical discharge in a 90° triangular channel if the critical depth is 2.5 ft.

8.7 In front of a hydraulic jump in a rectangular channel the velocity of the water is 7 m/s, and the depth is 30 cm. What is the conjugate depth downstream of the jump?

8.8 Express the Froude numbers before and after the hydraulic jump for Problem 8.7. How much energy was lost by the jump?

8.9 A trapezoidal channel cut into rock has a bottom width of 4 ft, side slopes of 1:1, and a channel slope of 0.0005. Compute the discharge, assuming normal flow, if the water depth is 3 ft.

8.10 Determine the conveyance of the channel in Problem 8.9.

8.11 What is the critical slope of the channel in Problem 8.9?

8.12 The channel described in Example 8.12 is built in fine-grained sand. Using the tractive force method determine if the channel is susceptible to erosion. Is the side slope OK?

8.13 Repeat Example 8.16 if the existing flood stage is 6 ft above the normal depth in the channel.

8.14 For the conditions described in Example 8.11 and assuming that the channel is dug in sand, check if it is susceptible to erosion or deposition.

MULTIPLE CHOICE QUESTIONS

8.15 Critical discharge is the one at which
   A. Erosion of the channel is most likely.
   B. The energy consumed is the largest.
   C. The velocity equals the square root of g.
   D. Energy consumption is optimal.
   E. Hydraulic jump will occur.

8.16 The Froude number is a dimensionless quantity representing
   A. The wave propagation velocity.
   B. Critical flow conditions.
   C. Velocity divided by celerity of waves.
   D. Supercritical flow.
   E. Specific energy conditions.

8.17 The specific energy equation is
   A. Bernoulli's equation written at the channel bottom.
   B. Valid for a certain location of the channel only.
   C. Composed of kinetic and pressure energy terms.
   D. All of the above.
   E. None of the above.

8.18 In subcritical flow in an open channel
   A. The celerity is less than the average flow velocity.
   B. Froude's number is greater than 1.
   C. Reynolds's number is less than 1.
   D. All of the above.
   E. None of the above.

8.19 Supercritical flow in an open channel
   A. Occurs when the slope is steep.
   B. Causes less energy loss than subcritical flow.
   C. Leads to steady flow.
   D. Cannot be normal flow.
   E. None of the above.

8.20 The "section factor" in a channel equals
   A. $Q/\sqrt{g}$.
   B. Z.
   C. $A\sqrt{y}$
   D. All of the above.
   E. None of the above.

8.21 A ''control section'' in an open channel is the site
   A. Where the flow quantity can be controlled.
   B. At which the flow is known to be critical.
   C. Where the discharge can be measured.
   D. Of a hydraulic jump.
   E. Where the specific energy is determined.

8.22 Conjugate depths
   A. Are the two roots of a second-order equation.
   B. Are defined by the equation of specific energy.
   C. Always correspond to the same discharge.
   D. Are dependent on the total available energy.
   E. All of the above.

8.23 The critical flow exponent is
   A. Related to the critical discharge in a given channel.
   B. Independent of the shape of an open channel.
   C. A constant for circular open channels.
   D. Independent of the depth of the flow.
   E. None of the above.

8.24 Normal flow in open channels occurs when
   A. The channel slope and the discharge are both constants.
   B. The channel is prismatic and the flow is unsteady.
   C. The energy gained equals the elevation lost.
   D. The channel shape and roughness are uniform.
   E. All of the above.

8.25 When the critical depth equals the normal depth, the
   A. Flow is subcritical.
   B. Energy grade line is horizontal.
   C. Acceleration is greater than unity.
   D. Flow encounters a hydraulic jump.
   E. Slope is critical.

8.26 Gradually varied flow means that the
   A. Discharge is variable.
   B. Channel is nonprismatic.
   C. Depth is neither critical nor normal.
   D. Hydraulic jump is present.
   E. None of the above.

8.27 The Chézy-Manning equation is valid if
   A. The acceleration of the flow is zero.
   B. The discharge, slope, and depth are constant.
   C. The flow in the channel is normal.
   D. All of the above.
   E. None of the above.

8.28 The approximate value of the roughness coefficient for a clean earthen channel is
   A. 0.2.
   B. 0.1.
   C. 0.005.
   D. 0.02.
   E. 0.07.

8.29 Conveyance in an open channel is a term expressing the
   A. Normal flow exponent.
   B. Roughness and geometry of a channel.
   C. Hydraulic radius.
   D. Critical depth.
   E. 2/3 power of R.

8.30 Erosion in an open channel is not likely to occur if
   A. The Reynolds number is below 0.35.
   B. The Froude number is below 0.35.
   C. The slope is critical.
   D. The side slope is 1 to 1.
   E. G is below 2.5.

8.31 The permissible slope in eroding channels is greater if the discharge is less.
   A. True.
   B. False.

8.32 With respect to the bottom of a channel a $C2$ flow profile in gradually varied flow is
   A. Convex.
   B. Concave.
   C. Parallel.
   D. Nonexistent.
   E. Horizontal.

8.33 Gradually varied flow on a steep slope with an $S2$ profile is
   A. Critical.
   B. Subcritical.
   C. Supercritical.
   D. Nonexistent.
   E. Zero.

8.34 The normal flow exponent refers to the power of
   A. R in Manning's formula.
   B. y in the conveyance formula.
   C. S in Chézy's equation.
   D. R in the Chézy-Manning formula.
   E. None of the above.

8.35 Backwater computations are for solving
   A. Water profiles behind dams.
   B. Flood flow surfaces.
   C. All gradually varied flow surfaces.
   D. Nonnormal flow surfaces.
   E. A, B, and C.

8.36 Bakhmeteff's equation provides
   A. Direct solutions for channel delivery problems.
   B. The distance between water levels.
   C. Solutions for unsteady flow problems.
   D. Solutions for flows in natural channels.
   E. The discharge under known flow surface profiles.

8.37 Backwater formulas are valid only for
   A. Constant channel slopes.
   B. Natural channels.

C. Unsteady flow conditions.

D. All of the above.

E. None of the above.

8.38 Channel delivery curves show the

A. Discharge under known end conditions.

B. Water surface elevations along a known channel.

C. Surface profile for a given roughness coefficient.

D. Upstream levels when downstream data are given.

E. Downstream conditions for given upstream data.

9

Flow through Hydraulic Structures

A large share of the hydraulician's work is spent on designing and analyzing hydraulic structures that control the flow of water in rivers and human-made channels or conduct water through embankments or other barriers. This chapter surveys the hydraulic design principles and methods related to the most common structures. Flows under gates and over weirs and spillways are discussed. Energy dissipators and the design of culverts are reviewed. The general design principles of sewerage concludes this chapter.

9.1 INTRODUCTION

The material discussed in the previous eight chapters provides the general conceptual foundations for the design of the various components of hydraulic structures. In contrast to the design process in other fields, there is very little standardization in hydraulics; almost every new hydraulic structure is unique. The hydraulic designer cannot refer to design codes and standard specifications. The basic structure of hydraulic design consists of the fundamental formulas of fluid mechanics, refined by field- or laboratory-tested correction factors and seasoned by the collective practical experience of generations of designers. The successful combination of these three components, in addition to the uncertainty inherent in the determination of the design discharge, brings an element of art into the science of hydraulics.

Hydraulics is not a field in itself but a common component in several separate professions related to water resources. Being protected from or making use of water has always been an essential part of life. Beyond the most obvious issues, like protection from floods or provision of water during droughts, water is essential for our daily living in many indirect manners. For instance, before a ton of steel is shipped to the user, 60,000 gal of water are used by industry; to produce 1 bushel of wheat requires 15 gal of water; it takes 150 gal of water to produce a pound of cotton. The major uses of water in America are in agriculture (47 percent), industry (44 percent), and cities or residences (9 percent). Seventy-five percent of this water is taken from lakes or rivers, and 25 percent from wells.

The influence of hydraulic activity on the world's civilization cannot be underestimated. In the past little attention was paid to the environmental effects of the spread of civilization. For instance, from the time the state of Ohio was settled, 90 percent of its wetlands were destroyed. Draining the swamps used to be considered a laudable contribution of engineers. Having been recognized as natural filters of groundwater, wetlands today are federally protected. The increase of the world's population places heavy demands on water resources. Two-thirds of this demand is for agricultural uses. Since 1950 the number of high dams (over 50 ft in height) has increased from just over 5000 in 1950 to some 38,000 in 1995. Fifteen percent of the world's renewable water supply is held by dams. The increased use of these waters has altered river patterns. Many rivers do not reach the sea anymore because their waters are entirely used up. For much of the past 25 years the Colorado River, for instance, has not reached the sea. The Salt and Gila Rivers in Arizona used to join west of Phoenix. Today they dry up east of the city because farms divert their waters. The Aral Sea of Central Asia was once the world's fourth largest freshwater lake. Now it is drying out, because the two rivers that used to feed it, the Amu Dar'ya and Syr Dar'ya, were diverted for the irrigation of the "virgin lands" of the Soviet utopia. The quality of the remaining water in the Aral has deteriorated to such an extent that it is toxic. Similar examples may be found worldwide. Public attention to the extent of water pollution reached a peak on June 22, 1969, when the Cuyahoga River in Cleveland, Ohio, caught fire. That event spawned a series of environmental regulations. Examples like these emphasize the importance of careful planning, conservation, and environmental regulation.

There are excesses on the side of the regulators, too. Construction of the $110 million Tellico Dam in Tennessee was held up in 1978 for years by environmentalists' concern that the dam would destroy the alleged sole habitat of the snail darter, a tiny fish, causing its extinction. The U.S. Supreme Court sided with the environmentalists, but Congress voted for the dam. A year after the completion of the dam, snail darters were discovered living 60 miles below the dam.

The various fields where hydraulics is used include the following:

Agriculture

Irrigation (storage, distribution application)
Surface drainage
Groundwater control
Reclamation and wetlands maintenance
Food production (dairies, fisheries, etc.)

Flood Control

Levees (inland drainage, pumping stations)
River training
Flood control reservoirs
Reforestation (erosion control measures)

Water Power Generation

Mountain reservoirs and power stations
Run of river power stations
Pumped storage systems
Tidal power generation
Geothermal and ocean-thermal schemes

Navigation

River navigation (locks and dams, dredging)
Navigation canals
Ocean transport (harbors, breakwaters, sea walls)
Offshore structures and construction
Naval architecture (ship design)

Water Supply and Waste Disposal

Water treatment processes
Distribution systems
Sewerage systems
Waste treatment processes
Storm sewers
River pollution modeling and control
Hazardous waste (groundwater pollution control)

Coastal Engineering

Ecological management (coastal wetlands)
Erosion and sedimentation control
Harbors and navigation structures
Saltwater intrusion control
Storm damage mitigation

Most of these separate technical fields have their own professional societies, organizations, publications, and standards. The material covered in this chapter includes the most common elements of hydraulic structures. For specific information on design procedures related to the various fields of water resources the reader is urged to consult the literature of each field.

Hydraulic structures are built to control or transmit discharges and to maintain water levels in streams and channels. In contrast with previous chapters that deal with the quantitative and qualitative analysis of the flow of water along an essentially uniform channel cross section, this chapter treats localized effects concentrated on

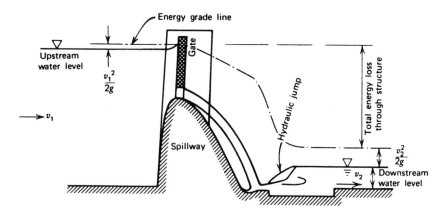

FIGURE 9.1 Energy parameters in flow through a hydraulic structure.

or around human-made structures. The types of these structures are many, yet they can be broadly classified into two groups. In one group the flow takes place under pressure through a definitely fixed cross section, in a manner somewhat analogous to pipe flow. Flow through orifices, nozzles, short pipes, sluiceways, or under gates is of this kind. In the other group flow occurs through an initially undetermined cross section, as in open channels. Examples include flow over weirs, spillways, chutes, and drop structures, and through culverts and sewers.

In both cases the flow is primarily controlled by the *upstream* water level. However, if the energy grade line at the downstream side is relatively high, it also exerts influence on the flow through the structure. The difference in total available energy level before and after the structure represents an energy loss. The lost hydraulic energy is dissipated on or immediately downstream of the structure, as shown in Figure 9.1, in the form of viscous shear between water particles or within the water and the surrounding structural elements. This effect should be seriously considered in the design because it exerts forces on the structure and may cause erosion. Erosion and underestimation of maximum encountered discharge are the most common causes of failure of hydraulic structures.

The analysis of flow through hydraulic structures is based on the energy equation. Localized effects of inertia and viscous shear are commonly included in the form of experimentally determined flow coefficients. For simple structures many of these coefficients are readily available in the hydraulic literature. A representative collection of these data are included later in this chapter. For more complicated structures usually no reliable data are available. For structures of particularly large size and cost the designers often rely on laboratory studies of reduced-scale models.

9.2 ORIFICES AND SLUICEWAYS

In the case of a barrier placed in a stream in which the flow takes place through a geometrically fixed opening located under the upstream water level, the flow is

FIGURE 9.2 Northern California's Shasta Dam. Note operating sluiceways at the center. The power station with the penstocks is seen at the left. (Courtesy of U.S. Army Corps of Engineers, Sacramento, CA)

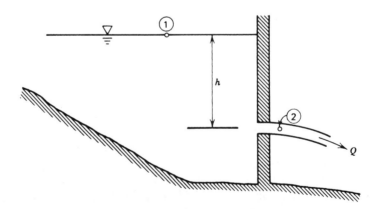

FIGURE 9.3 Notations for the orifice problem.

analyzed by the orifice formula. Such is the case of the water exiting from the three outlets shown in Figure 9.2, which depicts Shasta Dam in northern California. Consider a small opening called an *orifice* in the side or the bottom of a container as shown in Figure 9.3. The total available energy at the center of this opening equals the depth of the water h. Under the influence of this energy a jet of water will issue out of the orifice with a velocity v. Writing Bernoulli's equation for two points, one at the water surface of the container at point 1, another at the center of the jet at point 2, we obtain

$$\frac{v_1^2}{2g} + \frac{p_1}{\gamma} + z_1 = \frac{v_2^2}{2g} + \frac{p_2}{\gamma} + z_2 \tag{9.1}$$

Here the following assumptions may be made: Because of the large size of the container the approach velocity at point 1 is negligible compared with the velocity of the jet at point 2; hence, v_1 is zero. We further assume that there are no energy losses. The pressure at the surface as well as in the thin jet is atmospheric; hence, $p_1 = p_2$. Substituting $z_1 - z_2 = h$, the depth of the orifice below the surface, we get

$$h = \frac{v_2^2}{2g} \tag{9.2}$$

or, by expressing for the velocity of the jet,

$$v_{\text{jet}} = \sqrt{2gh} \tag{9.3}$$

which is the Torricelli equation. The discharge through an orifice can be calculated if the area of the orifice A is sufficiently small with respect to h, such that the variation of h across the opening is negligible. For this case the velocity of the flow throughout A can be considered constant, and the discharge is

$$Q = cA\sqrt{2gh} \tag{9.4}$$

The c in this equation is called the *discharge coefficient*. It accounts for the effects of the inertia and the viscosity of the water. Portions of the fluid approaching the hole sideways cause a curvature in the stream, reducing its size, as shown in Figure 9.4. The narrowest portion of the jet stream is called the *vena contracta*. Furthermore, the viscosity of the fluid causes it to cling to solid surfaces, and this creates a boundary layer effect along the edges of the orifice, shown in Figure 9.5. These two effects combine to reduce the discharge. Indeed, experimental results have shown that the actual discharge through an orifice is considerably less than the discharge given by Equation 9.4. This reduction of discharge is expressed by the discharge coefficient. Based on many experiments with small, round, and square orifices discharging into air, the discharge coefficient ranges from 0.60 to 0.68. Experimental results further indicate that the discharge coefficient is larger with smaller diameters and larger heads.

FIGURE 9.4 The development of the *vena contracta.*

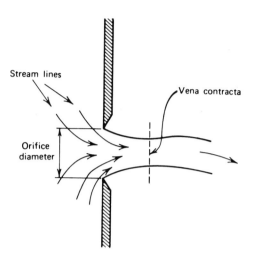

In the case where the jet issuing through the opening exits into the downstream water, the value of *h* in Equation 9.4 is the difference between the two water levels. For actual structures where the orifice cannot be considered infinitely thin but rather is a *short tube*, the value of the discharge coefficient increases. Figure 9.6 shows discharge coefficients based on experimental data on short tubes of various materials and configurations.

A practical application of this concept is sluiceways under dams. A *sluiceway* is a pipe or tunnel that passes through a dam to allow for the removal of water from the reservoir as needed. Drainage of the upstream water is often provided by sluiceways during construction. Generally, part of the work of a hydraulic designer is

FIGURE 9.5 Velocity distribution due to viscous shear loss at the orifice edge.

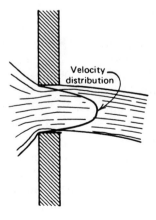

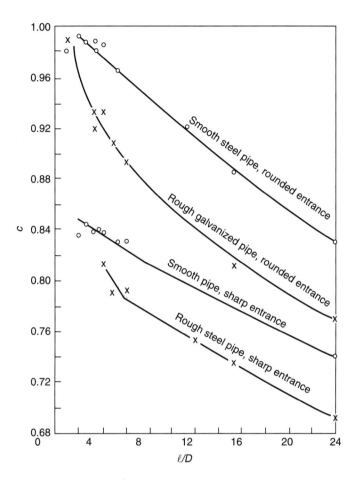

FIGURE 9.6 Discharge coefficients for short tubes.

to provide for the passing of floods during the construction period to prevent potential but often unexpected disasters.[1] The location of the sluiceway is dependent on the size and type of the dam. Concrete or masonry dams may be equipped with one or more sluiceways at a location determined by reservoir control considerations. In earthen dams or rock-fill dams the sluiceway usually is located outside the limits of the embankment for stability considerations. Because the earthen dam causes the soil below to settle, sluiceways located there could bend and break. Seepage along the outside of the pipe walls must be minimized by placing projecting collars on the pipe at certain intervals along its length. The size and number of these collars should increase the seepage path by at least 25 percent.

---

[1]Several half-built dams have been partially destroyed by such unexpected floods during construction that could have been forecast with considerable certainty. Inadequate planning for passing winter floods through construction sites of dams is a major cause of disasters.

9.3 FLOW UNDER GATES

Flow under a *vertical gate* can be defined as a square orifice problem as long as the opening height a under the gate is small compared with the upstream energy level H_0. Furthermore, we assume here that the downstream water level H_2 does not influence the flow. Using the concepts in Equation 9.4 we may write

$$Q = b\, a\, c\, \sqrt{2\, g\, (H_0 - H_1)} \qquad (9.5)$$

where b is the width of the gate, and the other terms are as shown in Figure 9.7. Direct use of this equation is difficult in practice because of the uncertainty in the determination of H_1, the depth of water at the vena contracta. Because H_1 depends on the opening height a, we may write

$$H_1 = \psi\, a \qquad (9.6)$$

Experimental values for ψ were found to depend on H_0/a and are shown in Figure 9.8. Equation 9.6 can be put in the form

$$Q = b\, a\, c\, \sqrt{2\, g}\, \frac{H_0}{\sqrt{H_0 + \psi a}} \qquad (9.7)$$

Experimentally determined values for the discharge coefficient c for various H_0/a values are also shown in Figure 9.8. For H_0/a values that exceed the range of the graph, the c value asymptotically approaches 0.70, and ψ approaches 0.62.

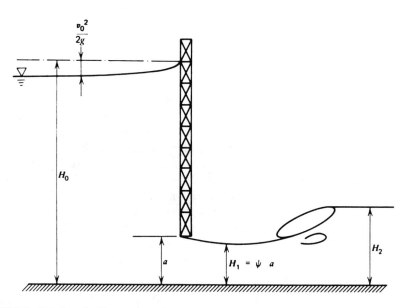

FIGURE 9.7 Notations for flow under gates.

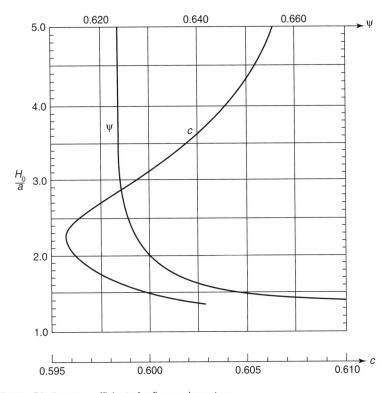

FIGURE 9.8   Discharge coefficients for flow under gates.

Equation 9.7 is valid only if the downstream water level is not influencing the flow, such as when a hydraulic jump exists below the gate. This condition requires that the conjugate depth of H_2, the downstream water level, be equal to or larger than H_1. The concept of conjugate depth was introduced in Chapter 8. It may be calculated by using Equation 8.8 with $y_2 = H_2$, and y_1 as the depth immediately in front of the jump. Instead of carrying out the computations, one may utilize Figure 9.9, which gives the relationship between H_0/a and H_2/a at the condition when the free outflow becomes retarded by the downstream water level. The discharge under a gate in the case of such *partial downstream control* is computed from

$$Q_{\text{retarded}} = k\,Q \tag{9.8}$$

where Q is to be obtained from Equation 9.7, and k is a coefficient that may be found in Figure 9.10 for various H_0/a and H_2/a values.

When $H_0 - H_1$ is equal to or less than H_2 one speaks of *absolute downstream control*. In this case H_1 is completely submerged, and the discharge computation is based on $h = H_0 - H_2$. In this case the discharge formula is

$$Q = b\,a\,c\,\sqrt{2\,g\,h} \tag{9.9}$$

For metric values this formula may be solved by the graph shown in Figure 9.11.

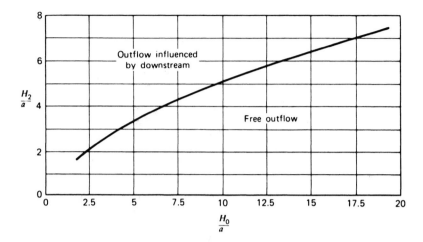

FIGURE 9.9 The range of downstream influence on flow under gates.

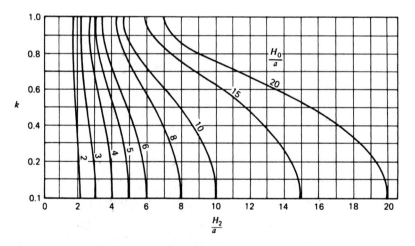

FIGURE 9.10 Retardation coefficient for partial downstream control of flow under gates.

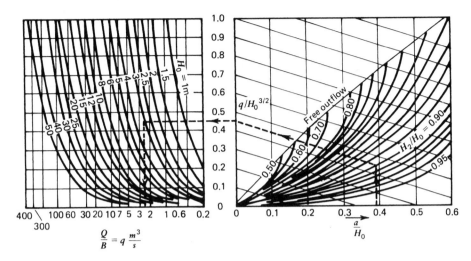

FIGURE 9.11 General solution for the discharge formula for flow under gates. (After Chertousov, courtesy of Ö. Starosolszky)

As the water rushes under a partially lifted gate the high velocity causes low pressures at the gate's bottom edge. In designing lifting mechanisms, this excess load should be carefully considered. Additional design considerations must include the possibility of floating debris (even runaway barges, in some cases) lodging against the gate, ice packs, and so on, that may prevent the raising of the gate just when it is most needed. Such considerations brought about the development of various types of gates such as segment gates and drum gates.

9.4   WEIRS

Flow taking place over a hydraulic structure under free surface conditions is analyzed with the *weir formula*. Generally speaking, all barriers on the bottom of the channel that cause the flow to accelerate in order to pass through can be considered weirs. More specifically, weirs are constructed with openings of simple geometric shapes. Rectangular, triangular, or trapezoidal shapes are the most common. In each case the bottom edge of the opening over which the water flows is called the *crest*, and its height over the bottom of the reservoir or channel is known as the *crest height*. Triangular outflow weirs at a wastewater treatment plant are shown in Figure 9.12. Figure 9.13 depicts a standard SCS-type drop inlet. The spillway is hidden by the grating that encloses the weir. The French term *nappe* ("sheet") is often applied to the overfalling stream of water. Weirs where the downstream water level is below the crest allow the water to pass in a free fall. Under this condition weirs are good flow measuring devices, particularly if their wall at the crest and sides is thin. Weir shapes other than rectangular are almost exclusively used in flow

FIGURE 9.12 Flow through triangular outflow weirs at a wastewater treatment plant. (Courtesy of District of Columbia, Department of Public Works, Washington, DC)

FIGURE 9.13 Typical drop inlet spillway designed by the U.S. Soil Conservation Service. (Courtesy of Ohio Department of Natural Resources)

measurement. The use of *sharp-crested weirs* in flow measurement was discussed in Chapter 3. For practical purposes as long as their crest thickness is more than 60 percent of the nappe thickness weirs are considered broad-crested. Flow over *broad-crested weirs* is significantly influenced by viscous drag, which causes a boundary layer to form in the velocity profile of the overflow. This effect is quantified in the form of a discharge coefficient that depends on the shape of the crest and on the upstream energy level. If the width of the upstream channel is greater than the width of the weir opening, we speak of *contracted weirs.* The side contraction in weirs results in an additional contribution to the discharge coefficient by reducing the flow further. In weirs whose width equals the width of the upstream channel surface the side contraction is suppressed. These weirs are called *suppressed weirs.*

The general weir formula was derived already in Chapter 3 as

$$Q = c\, b\, \sqrt{2g}\, H_1^{3/2} \qquad\qquad \textbf{(9.10)}$$

where Q is the total discharge over a rectangular weir of b width under a head of H_1, and c is the discharge coefficient. We may combine the discharge coefficient with $2g$ to obtain, in metric terms, the *metric weir coefficient M*, so that

$$Q = M\, b\, H^{3/2} \qquad\qquad \textbf{(9.11)}$$

Figure 9.14 provides a graphical solution for Equation 9.11.

An extensive collection of various broad-crested weir shapes with their metric weir coefficients determined by laboratory experiments are listed in Table 9.1. To obtain the c coefficient for computations with conventional American units, the M values found in the table are divided by 4.43.

When the downstream water level H_2 exceeds the crest height, it may influence the discharge over the weir. In this case the water is prevented from passing by free fall. As long as the downstream water level is below the midpoint of the nappe over the crest, a hydraulic jump will form over a weir, as shown in Figure 9.15a. The nappe in this case will consist of a *supercritical portion* leading the hydraulic jump. With further increase of the downstream water level the jump will initially only be *submerged.* Additional increase of the downstream water level will cause a rise in the upstream water level; the water will be backed up, and the jump will be *washed out.* For this case, shown in Figure 9.15b, the discharge can still be computed by Equation 9.10 with the stipulation that H_1 in the equation be replaced by the difference between the upstream and downstream water levels. In all cases when the free fall over the weir is prevented by the height of the downstream water level, the discharge coefficient will be reduced. The actual ranges of various flow patterns over weirs depend on the configuration of the weir. For one typical weir shape the ranges of free overfall, free or submerged hydraulic jump, and subcritical overflow are shown in Figure 9.16.

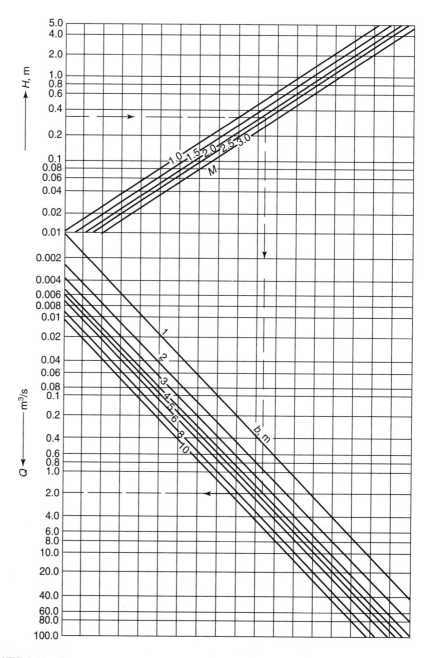

FIGURE 9.14 Nomograph for the solution of the weir discharge formula, Equation 9.11.

TABLE 9.1 Discharge Coefficients for Broad-Crested Weirs (Slopes are vertical: horizontal)

Cross Section[a]	Upstream Head $h(m)$							
	0.15	0.30	0.45	0.60	0.75	0.90	1.20	1.50
(1) 0.8, 2:1, 0.2, 1:2	1.61	1.86	1.98					
(2) 0.8, 2:1, 0.2, 1:3	1.60	1.80	1.90					
(3) 0.8, 2:1, 0.2, 1:5	1.58	1.75	1.79					
(4) 0.8, 2:1, 0.4, 1:2	1.53	1.64	1.77					
(5) 0.8, 2:1, 0.4, 1:6	1.54	1.62	1.69					
(6) 0.8, 1:2, 0.2, 1:2	1.72	1.88	1.98					
(7) 0.8, 1:1, 0.2, 1:2	1.65	1.88	2.00					
(8) 0.8, 0.2, 1:2	1.53	1.80	1.93					
(9) 1.6, 1:2, 0.1, 1:2				1.96	1.96	1.97	1.99	2.02
(10) 1.6, 1:2, 0.1, 1:5				1.94	1.92	1.89	1.92	1.97
(11) 1:2, 0.1, 1.6		2.12	2.10	2.08	2.08	2.06	2.04	2.00
(12) 1:2, 0.2, 1.6		1.88	1.96	2.01	2.04	2.05	2.05	2.05
(13) 1:3, 0.2, 1.6				1.96	1.96	1.96	1.96	1.96
(14) 1:5, 0.2, 1.6				1.86	1.86	1.86	1.86	1.86

[a] All dimensions are in meters. Tabulated values represent metric weir coefficients.

TABLE 9.1 continued

Cross Section[a]		Upstream Head $h(m)$							
		0.15	0.30	0.45	0.60	0.75	0.90	1.20	1.50
(15)	0.8, 1:1, 0.15, 1:1	1.91	2.00						
(16)	R = 0.1, 1.5, 0.83		1.56	1.60	1.65	1.70	1.74	1.84	1.92
(17)	1.5, 0.1, 2.15		1.56	1.56	1.55	1.55	1.55	1.55	1.54
(18)	0.8, 1:1	2.13	2.13	2.13					
(19)	0.8, 1:2	1.93	1.94	1.94					
(20)	0.54, 1:2	1.94	1.98	1.97					
(21)	0.8, 1:5	1.69	1.73	1.73					
(22)	0.54, 1:1, 1:1	2.28	2.25	2.06					
(23)	0.54, 1:1, 1:2	2.08	2.12	2.12					
(24)	0.54, 1:1, 1:3	1.92	1.93	1.92					
(25)	0.54, 1:2, 1:2	2.10	2.13	2.13					
(26)	0.54, 2:1, 1:2	2.03	2.03	2.01					
(27)	0.54, 3:1, 1:2	2.03	2.03	2.01					
(28)	0.8, 2:1, 0.2, 1:1	1.65	1.94	2.10					

[a] All dimensions are in meters. Tabulated values represent metric weir coefficients.

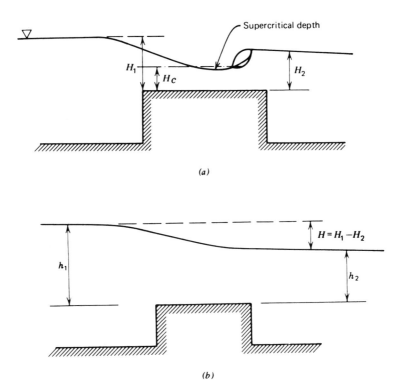

FIGURE 9.15 Noncritical flows over broad-crested weirs. (a) Weir with hydraulic jump and (b) weir with subcritical flow.

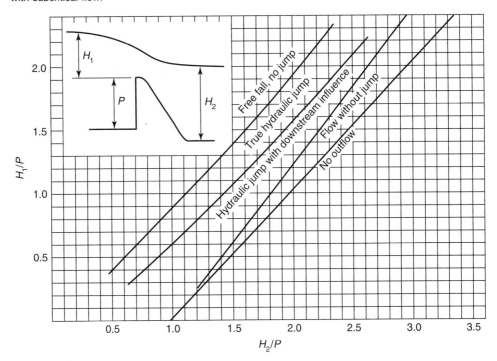

FIGURE 9.16 Ranges of flow types over weirs.

9.5 SPILLWAYS

Dams are equipped with spillways to allow the excess water to pass over the dam in a safe manner. A *spillway* is a rectangular concrete channel connecting upstream to downstream over which the water flows at a supercritical velocity. The ideal shape of a spillway is that of the underside of the overfalling nappe of water. Theoretically it is of parabolic configuration with an inversely curved lower portion—the bucket—that serves to deflect the falling water smoothly toward the downstream side. It is important that for all discharges the descending water be in contact with the spillway surface. Otherwise, if the water is allowed to shoot away from the spillway, it may exhibit hydraulic instability. Unless the nappe is aerated from below, the rushing water causes negative air pressure to develop, which tends to pull the nappe down until the outside air breaks through. The resultant cyclic pounding on the structure has significant destructive effects.

The discharge of spillways may be calculated by the broad-crested weir formula, with proper selection of the discharge coefficient. For flows less than the *design discharge,*[2] the discharge coefficient of the spillway will be smaller.

As the nappe falls over the spillway it is subjected to gravitational acceleration. As the velocity increases downward the thickness of the nappe decreases. The shape of the nappe in free fall may be determined, on a theoretical basis, assuming no friction on the weir. The computation is facilitated by the coordinates developed by Bazin[3] and Craeger on the basis of the dynamics of free fall. The coordinates are listed in Table 9.2. The notations of this table correspond to those shown in Figure 9.17 with $X = xH_d$, and $Y = yH_d$. The design head, H_d, is used here as a scaling factor, y_1 and y_3 refer to the lower and upper coordinates of the nappe, and y_2 refers to the spillway shape recommended by Bazin and Craeger. Knowing the shape of the nappe is important in the design of the spillway side walls as well.

Numerous experimental and theoretical studies on spillway nappe profiles resulted in several standard spillway shapes. A typical one, developed by the U.S. Army Corps of Engineers' Waterways Experiment Station, is given by the exponential formula

$$X^n = (KH_d^{n-1}) Y \qquad (9.12)$$

where X and Y are, respectively, the horizontal and the vertical coordinates of the spillway, with their origins at the highest point of the crest, as shown in Figure 9.17, and H_d is the design head upstream, excluding the kinetic energy of the approach velocity. The *design head* corresponds to the design discharge, the maximum expected flow over the spillway. The coefficients K and n depend on the upstream slope of the spillway. They are given in Table 9.3.

---

[2]The design discharge is defined by hydrologic considerations and is commonly specified by governmental regulations.

[3]Henri Emile Bazin (France, 1829–1917) was Darcy's assistant. He continued Darcy's experimental work on open channel resistance in the water supply of Paris.

TABLE 9.2 Bazin-Craeger Coordinates for Contoured Spillway and Nappe Configuration

Horizontal Coordinate x	Lower Nappe Surface y_1	Spillway Surface y_2	Upper Nappe Surface y_3	Horizontal Coordinate x	Lower Nappe Surface y_1	Spillway Surface y_2	Upper Nappe Surface y_3
0	-0.126	-0.126	-0.831	2.1	1.456	1.369	0.834
0.1	-0.036	-0.036	-0.803	2.2	1.609	1.508	0.975
0.2	-0.007	-0.007	-0.772	2.3	1.769	1.654	1.140
0.3	0.000	0.000	-0.740	2.4	1.936	1.804	1.310
0.4	0.007	0.006	-0.702	2.5	2.111	1.960	1.500
0.5	0.027	0.025	-0.655	2.6	2.293	2.122	1.686
0.6	0.063	0.060	-0.620	2.7	2.482	2.289	1.880
0.7	0.103	0.098	-0.560	2.8	2.679	2.463	2.120
0.8	0.153	0.147	-0.511	2.9	2.883	2.640	3.390
0.9	0.206	0.198	-0.450	3.0	3.094	2.824	2.500
$X = H_d \rightarrow 1.0$	0.267	0.256	-0.380	3.1	3.313	3.013	2.70
1.1	0.355	0.322	-0.290	3.2	3.539	3.207	2.92
1.2	0.410	0.393	-0.219	3.3	3.772	3.405	3.16
1.3	0.497	0.477	-0.100	3.4	4.013	3.609	3.40
1.4	0.591	0.565	-0.030	3.5	4.261	3.818	3.66
1.5	0.693	0.662	+0.090	3.6	4.516	4.031	3.88
1.6	0.800	0.764	+0.200	3.7	4.779	4.249	4.15
1.7	0.918	0.873	+0.305	3.8	5.049	4.471	4.40
1.8	1.041	0.987	+0.405	3.9	5.326	4.699	4.65
1.9	1.172	1.108	+0.540	4.0	5.610	4.930	5.00
2.0	1.310	1.255	+0.693	4.5	17.50	6.460	6.54

To get the x and y values, multiply tabulated data by H_d. y_1, y_2, and y_3 are measured vertically from the crest.

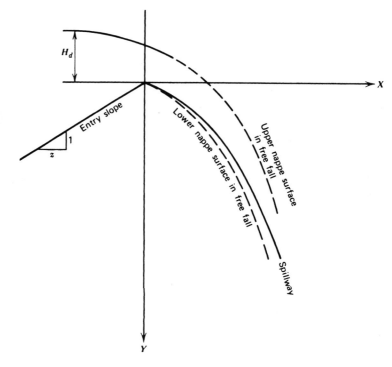

FIGURE 9.17 Interpretation of the notations for flow over spillways (Table 9.2).

TABLE 9.3 Coefficients for Equation 9.12

Upstream Slope, z	K	n
0	2.000	1.850
0.33	1.936	1.836
0.66	1.939	1.810
1.0	1.873	1.776

For small dams, spillways are often made in a straight, flat shape, particularly in the case of earthen dams where the spillway's concrete mat is laid on the top of the earth fill.

At the bottom of the dam the spillway is often built with an outward curvature to direct the flow downstream. The curvature of this portion, called the *bucket*, depends on the crest height and on the design head over the weir. Table 9.4 gives the radius of curvature, in meters, that is recommended for various crest heights and heads.

9.6 ENERGY DISSIPATORS

The energy gained by the water falling over the spillway can cause significant damage downstream if it is not abated. The velocity of the flow at the toe of the spillway can be computed by

$$v_1 = \sqrt{2\,g(P + H - y_1)} \tag{9.13}$$

where P is the crest height, H is the head over the weir including the kinetic energy head, and y_1 is the depth of the nappe at the toe. This velocity is usually very high. As such, it carries a large amount of kinetic energy that can cause heavy scouring. The situation is identical in the case of flow under a gate when the downstream water elevation is low or in the case of a culvert exit. To prevent a gradual erosion of the downstream channel and the possible undermining and destruction of the structure itself, the excess energy must be dissipated. Assuming that the downstream channel is resistant to scouring and that it is of a near-horizontal slope of constant cross section and, furthermore, that no particular effort is made to dissipate the energy of the water leaving the spillway, the following will occur: The flow will shoot out with an initial velocity according to Equation 9.13 into the channel, which here is called the *apron* or *tailrace*. The energy is gradually lost because of channel friction. Because this flow is supercritical, its energy loss rate will be considerable, resulting in a gradual decrease of velocity and a corresponding increase in depth. The flow can be computed with reasonable accuracy by using the Chézy equation, although the flow profile is somewhat curved, being a type of backwater curve. Applying Equation 8.13 and knowing the roughness coefficient, we can determine the rate of energy loss. As the depth of flow increases, the flow will reach a level at which the given discharge Q may flow in the given channel, defined by its slope S and roughness n, either at supercritical or subcritical velocity, according to the concepts described in Chapter 8. The corresponding depths involved here are the conjugate depths. According to the law of minimum energy, nature will select the flow condition with the lower rate of energy loss; that is, the flow will become subcritical. For this reason a hydraulic jump must occur. The energy loss in the hydraulic jump is computed from Equation 8.10.

TABLE 9.4 Recommended Radii of Curvature for Buckets Below Spillways

Crest Height, P m	H (m)						
	0.6	1.5	3.0	4.5	6.0	7.5	9.0
	Radius of Curvature, R m						
6	1.8	3.0	4.5	6.0	7.8	9.0	10.5
9	2.3	3.4	5.1	7.0	8.4	10.0	11.5
12	2.4	4.2	6.0	7.5	9.0	10.7	12.4
15	2.7	4.5	6.6	8.5	9.8	11.4	13.1
30	2.8	6.3	9.6	11.9	13.6	15.0	16.6
45	2.9	7.4	12.0	14.2	16.9	18.7	20.0
60	3.0	7.5	15.0	17.5	20.0	21.6	23.4

Source: From the book of standards of the Research Institute of Water Resources, Budapest, Hungary, written by O. Starosolszky and L. Muszkalay, entitled *Mütárgyhidraulikai Zsebkönyv* (Handbook of Hydraulic Structures, in Hungarian), Vols. 1 and 2, published by the Technical Bookpublisher Co., Budapest, Hungary, 1961.

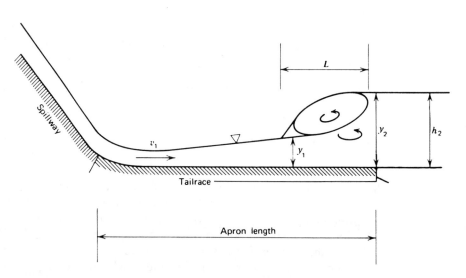

FIGURE 9.18 Notations for a hydraulic jump below a spillway.

If for a given discharge the downstream channel slope is mild, the downstream water level may be assumed to be of normal depth. As shown in Figure 9.18, this level will define the downstream total energy content of the flow after the jump. It will in turn also control the location of the jump. At the downstream end of the jump where the depth of the tail water y_n equals y_2 is the subcritical conjugate depth. At the beginning of the jump the supercritical conjugate depth is then y_1. Between these depths the length of the jump depends on the Froude number at the upstream end. With d_1 and Q known, this Froude number may be computed. The length L then can be obtained from Figure 9.19. This graph is based on experimental data obtained by the U.S. Bureau of Reclamation. The remaining uncertainty concerns the distance between the toe of the spillway (or the location of a gate) and the beginning of the jump. From the beginning of the tailrace, the water level will rise until it reaches the magnitude of y_1. For small discharges this length may be zero, and the jump occurs right at the bottom of the spillway. As the discharge increases, the jump will move away from the dam. The length of this distance is very sensitive to the roughness of the channel. Because the construction cost of downstream aprons depends on their required length, designers should strive to minimize it. This necessitates that the jump be localized by absorbing the excess energy in the supercritical flow, causing the jump to occur as close to the structure as possible. This end is served by the design and construction of *energy dissipators*.

A great many different energy dissipators have been built at various types of hydraulic structures. Reports on experimental studies and design recommendations are abundant in the hydraulic literature. Standard designs of energy dissipators used by agencies of the U.S. government exist. A typical such standard design,

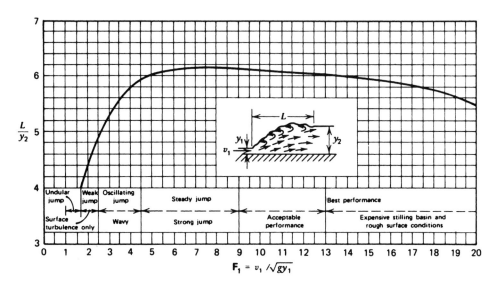

FIGURE 9.19 Ranges of various types of hydraulic jumps. (Courtesy of Bureau of Reclamation)

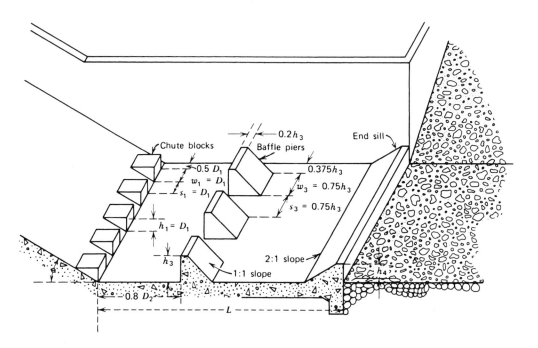

FIGURE 9.20   The U.S. Bureau of Reclamation's Type III energy dissipator. (Courtesy of Bureau of Reclamation)

incorporating several basic energy-absorbing features, is the U.S. Bureau of Reclamation's Type III basin, which is reviewed here as an example of an efficient energy dissipator. The design was developed for relatively small structures with the configuration shown in Figure 9.20. The sizes of the energy dissipator components of this stilling basin are shown in Figure 9.21 as a function of the Froude number and conjugate depth at the beginning (supercritical portion) of the hydraulic jump. The length of the basin, L, is recommended to be about $2.8d_2$, where d_2 is the subcritical conjugate depth of the jump. This design confines the hydraulic jump within the length of the basin. It also tends to shorten the length of the hydraulic jump by about 50 percent.

Even when the water passes an energy dissipator, it is still full of swirls and vortices. Because of this it is recommended that the channel below be protected against erosive currents by crushed stone called *riprap*. The necessary stone size as a function of the expected erosion velocity is presented in Figure 9.22. The curve is based on actual field data collected by the U.S. Bureau of Reclamation. The riprap should be composed of a well-graded mixture, but most of the stones should be of the size indicated by the curve. Riprap should be placed over a filter blanket, a protected filter cloth, or a bed of graded gravel in a layer 1.5 times or more as thick as the largest stone diameter. Erosion control in a riprap-covered channel

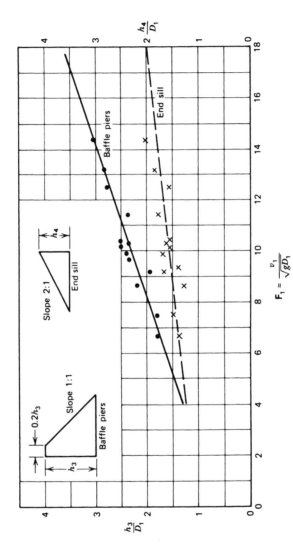

FIGURE 9.21 Sizing chart of energy dissipator components shown in Figure 9.20. (Courtesy of Bureau of Reclamation)

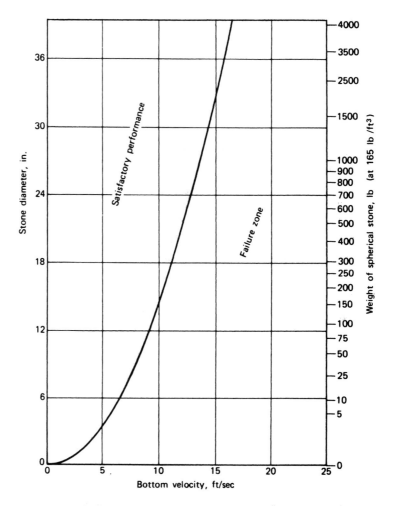

FIGURE 9.22   Recommended size of riprap for erosion control (After Bureau of Reclamation).

section following an energy dissipator may be further improved by allowing the channel to flare out to a size several times its normal width. Figure 9.23 shows the exit of a circular concrete culvert with an erosion control blanket of riprap.

9.7 CULVERTS

Highways and railroads traversing the land cut across individual watersheds. To allow the flow from each watershed across the embankment, culverts are built at the lowest points of the valleys. Although culverts are simple in appearance, their hydraulic design is no easy matter. The operation of culverts under the various possible discharge conditions presents a somewhat complex problem that cannot

FIGURE 9.23 Concrete culvert exit with riprap erosion protection. (Courtesy of Ohio Department of Transportation)

be classified either as flow under pressure or as free surface flow. The actual conditions involve both of these basic concepts.

The fundamental objective of hydraulic design of culverts is to determine the *most economic diameter at which the design discharge is passed without exceeding the allowable headwater elevation.* The major components of a culvert are its inlet, the culvert pipe barrel itself, and its outlet with the exit energy dissipator, if any. Each of these components has a definite capacity to deliver a discharge. The component having the least discharge delivery capacity will control the hydraulic performance of the whole structure.

One speaks of *inlet control* if, under given circumstances, the discharge of a culvert is dependent only on the headwater above the invert at the entrance, the size of the pipe, and the geometry of the entrance. When the inlet conditions control the flow, the slope, length, and roughness of the culvert pipe have no influence on the discharge. In this case, the pipe is never full, even though the headwater may exceed the top of the pipe entrance, causing the flow to enter the pipe under pressure. The flow, therefore, is open channel flow. Figure 9.24 shows a typical nomograph by which the discharge Q can be determined for a culvert of D diameter under a headwater depth HW.[4] The nomograph is for a square-edged entrance in a headwall. Similar nomographs are found in governmental and trade literature for many other entrance conditions. Short culverts with relatively negligible tailwater elevations almost always operate under inlet control.

Outlet control occurs when the discharge is dependent on all hydraulic variables of the structure. Figure 9.25 shows the notations relative to these variables. These include the slope S_0, length L, diameter D, roughness n, tailwater depth TW, and

---

[4]Yet another example of the enduring nature of antique engineering notations.

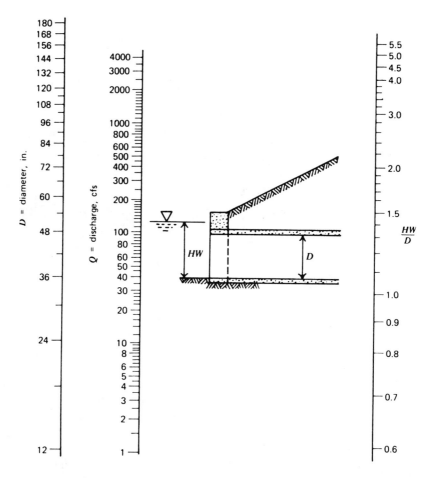

FIGURE 9.24 Typical nomograph for inlet-controlled culvert design. Square-edged entrance. (From *Handbook of Concrete Culvert Pipe Hydraulics,* Portland Cement Association, 1964)

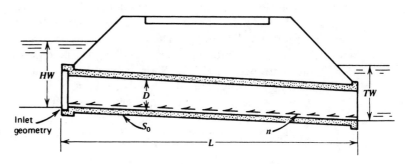

FIGURE 9.25 Notations for culvert analysis.

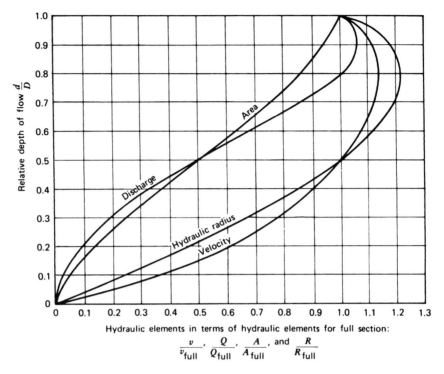

FIGURE 9.26 Hydraulic parameters of circular pipes flowing partially full. (From *Handbook of Concrete Culvert Pipe Hydraulics*, Portland Cement Association, 1964)

headwater depth *HW*. Unless the tailwater level is above the top of the culvert exit, the pipe will be only partly full. This means that the flow in the pipe, again, will be open channel flow.

Flow in partially full circular pipes may be analyzed in the manner described in Chapter 8. The depth of the flow *d* relative to the pipe diameter *D* determines the area of the flow as well as its hydraulic radius. Figure 9.26 aids in the computation of the area, hydraulic radius, velocity and discharge in relation to those at full flow. From these data the critical depth d_c may be obtained for any discharge.

Because of the vena contracta at the inlet created by the entrance conditions under the headwater *HW*, the flow in most cases starts in a supercritical condition in the culvert. The friction on the pipe wall, and the turbulence in the flow, gradually dissipate the energy, causing the depth of the flow to increase. Depending on the tailwater elevation, the supercritical flow may convert into subcritical flow through a hydraulic jump. Because the critical depth is characterized by the fact that it allows the maximum possible discharge to pass under the prevailing available hydraulic energy, it is the *essence* of desirable culvert operation that the flow be under

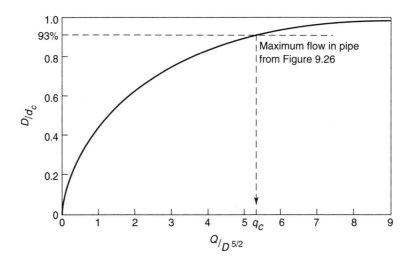

FIGURE 9.27 Critical depth graph for circular pipes flowing partially full. (From *Handbook of Concrete Culvert Pipe Hydraulics,* Portland Cement Association, 1964)

critical conditions. The total available energy, or specific energy, in a pipe is

$$H_0 = \frac{Q^2}{2\,g\,A^2} + d \tag{9.14}$$

where d is the depth of flow. This equation may be rewritten to express conditions for a partially full pipe as

$$a = \frac{\text{actual flow area}}{\text{area of pipe}} \tag{9.15}$$

To simplify the notations, a dimensional equivalent to the Froude number, called the *discharge factor,* was introduced into culvert hydraulics:

$$q_c = \frac{Q}{D^{5/2}} \tag{9.16}$$

It enables the determination of the critical depth in partially full flow in a circular pipe by using Figure 9.27. Introducing the discharge factor and dividing both sides of Equation 9.14 by the diameter D, we may write the specific energy equation in conventional American units as

$$\frac{H_0}{D} = \frac{0.025\,q_c^2}{a^2} + \frac{d}{D} \tag{9.17}$$

This equation is plotted in a dimensionless manner in Figure 9.28. The points at which the flow occurs with minimum energy are marked on this graph.

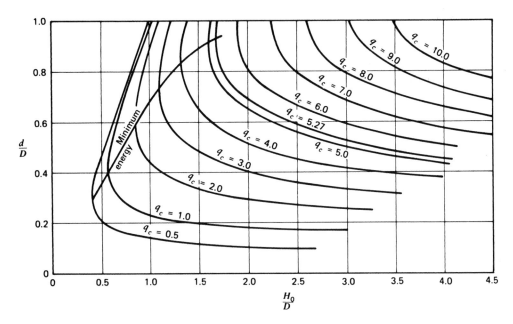

FIGURE 9.28 Specific energy graphs for circular pipes flowing partially full.

The most common materials used for culverts are concrete and corrugated steel. The roughness in both cases is usually assumed to be constant for any flow depth. Common values used in practice are $n = 0.012$ for concrete pipe and 0.024 for corrugated steel pipe. Although the selection of culvert materials in practice is made on a competitive economic basis, the fact that the roughness of corrugated steel pipes is twice that of concrete pipes suggests that their selection can be based on hydraulic considerations also. On hilly terrains where the culvert slope is expected to be relatively steep and the flow through the culvert gains considerable energy, corrugated steel pipes offer energy-dissipating advantages. On flat terrains energy loss through a culvert is undesirable; hence, concrete pipes are more suitable.

In Figure 9.26 the maximum discharge in a pipe flowing under free surface conditions occurs at a depth that is 93 percent of the pipe diameter. For the given energy conditions the *optimum discharge* is found at the critical depth, which is 93 percent of the diameter. On the vertical coordinate of Figure 9.27 this value gives $q_c = 5.27$ at the minimum energy line. Consequently, by the discharge factor formula, Equation 9.16, the optimum discharge in a culvert is

$$Q_{\text{optimum}} = 5.27\, D^{5/2} \tag{9.18}$$

in conventional American units. Substituting this value into the Chézy-Manning formula, Equation 8.19, we compute the *optimum (critical) slope* of a culvert to be

$$S_{\text{optimum}} = 111\, \frac{n^2}{D^{1/3}} \tag{9.19}$$

The reader may note the strong resemblance between this equation and Equation

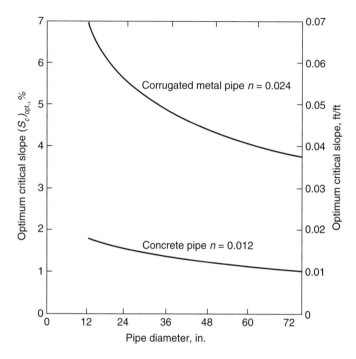

FIGURE 9.29 Graph of optimum critical slope of concrete and corrugated pipe culverts. (From *Handbook of Concrete Culvert Pipe Hydraulics,* Portland Cement Association, 1964)

8.26, developed for wide rectangular channels. For the commonly used roughness values for concrete and corrugated steel pipes this equation may be plotted in the form of Figure 9.29. As long as the depth of the flow in the culvert is less than critical the structure will operate under inlet control. If the depth reaches its critical value, outlet control will take over.

To aid in these computations, the U.S. Bureau of Public Roads developed a series of graphs that are widely available in design offices. One of these is reproduced in Figure 9.30. In this graph the dashed lines indicate the limiting outlet control conditions; the solid lines indicate the inlet control conditions for the culvert diameters indicated. For example, the graph indicates that a 54 in. diameter corrugated metal pipe will always work under inlet control as long as its length L is equal to or less than $50/100S_0$ in feet. For longer lengths it will operate under outlet control, provided that the same culvert slope S_0 is retained. The steplike dotted lines in the graphs indicate the upper limit within which the curves give very precise solutions for the problems. They represent HW values that are twice the heights of the pipes. In these zones the $L/100S_0$ index values can be approximated in a linear manner. Above these dotted lines the headwaters are high enough to allow the designer to use full flow nomographs.

To select the appropriate culvert size using graphs of the type shown in Figure 9.30 requires the following data: flood discharge Q in cubic feet per second, outlet

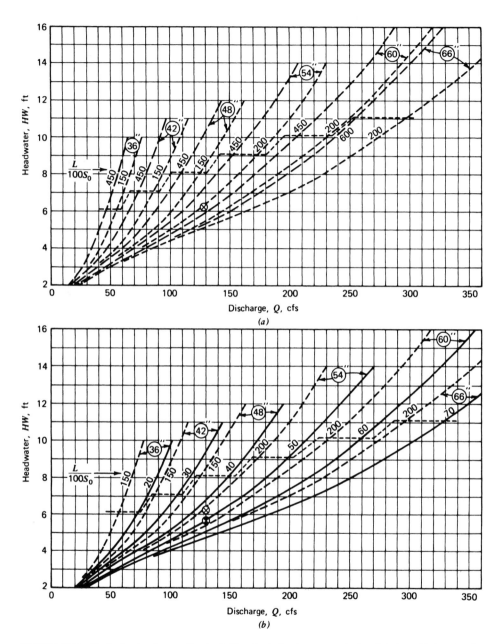

FIGURE 9.30 Typical culvert capacity chart. Standard circular corrugated metal pipe with headwall entrance. (From *Handbook of Concrete Culvert Pipe Hydraulics*, Portland Cement Association, 1964)

channel depth in relation to the critical depth in the culvert (to determine whether the tailwater exerts control over the flow), allowable headwater in feet, length of the culvert L in feet, slope of the culvert S_0, and the resulting index number $L/100S_0$. Once these basic data are established and the type of headwall and barrel are selected, the appropriate chart (like the one shown in Figure 9.31) may be entered at the proper Q value. Moving up to the first solid line in Figure 9.30b, we find the actual headwater for inlet-controlled culverts of any design that have an index number less than that indicated by the solid curve. If our culvert is longer or if it is laid on a flatter slope, resulting in an index number larger than the one indicated by the solid line, we can use linear approximation between the appropriate dotted and solid lines relating to the pipe diameter indicated.

When the tailwater depth exceeds the top of the exit pipe, the culvert will be *flowing full*. Similarly, with efficient entrance geometry a culvert may flow under pressure, at least in the initial portion of the pipe. Flow conditions for culverts flowing full may be determined by the nomograph shown in Figure 9.31. It is typical of several such nomographs made by the U.S. Bureau of Public Roads.

At the exit point of a culvert the velocity of the water is considerable, because culverts are designed to operate near critical velocity. As a result, it is often necessary to provide for energy dissipators at the outlet.

9.8 SEWER DESIGN

There are three main types of sewers: *Sanitary sewers* carry household or industrial sewage. *Storm sewers* collect rainwater in urban areas. *Combined sewers* do both, but they are not recommended anymore because of environmental considerations.

Most sewers, particularly smaller ones, are circular in cross section. Other shapes are also used; these generally provide for efficient conveyance of low flows by a small curvature on the bottom and by large cross sections on the higher parts for carrying larger discharges. The minimum recommended sewer diameter is 8 in., although in some cases 6 in. diameters are also permitted.

The horizontal alignment of sewers is controlled by their purpose and by considerations related to maintenance. Sanitary sewers are usually laid along the middle of the street to allow equal distances for connections to houses on both sides. Storm sewers are usually off center, in most cases located along the edge of the pavement. Sewers are composed of straight sections connecting manholes that are rarely farther apart than 100 or 150 m. This permits relatively easy cleaning in case of blockage.

Vertical alignment of sewers is guided by a number of rules, which on occasion may be broken. Sanitary sewers are to be placed deep enough to permit gravity drainage from all basements of residences served. In exceptionally deep positions some basements may have to be provided with pumps that raise the sewage to the level of the main sewer, as the extra expense of excavating deeper for the main sewer may not be warranted. In planning the depth of the sewer, one should consider the need for turns and the slope of the sewer line leading out of basements. Attempts should be made to follow the slope of the ground surface with the slope of the

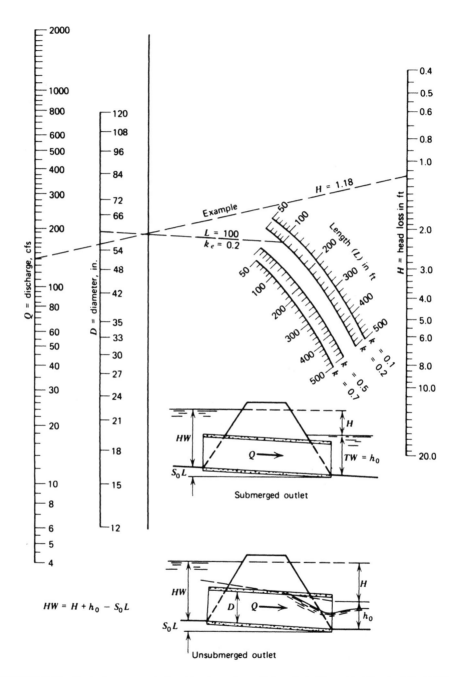

FIGURE 9.31 Typical nomograph for culverts under outlet control. (From *Handbook of Concrete Culvert Pipe Hydraulics,* Portland Cement Association, 1964) Head loss for concrete pipe culverts flowing full; $n = 0.012$. (Adapted from Bureau of Public Roads chart 1051.1)

sewer; this will minimize the depth of the excavation required. When the slope of the sewer is selected as a part of the hydraulic design, the velocity should be kept at full flow within 0.5 to 3 m/s. The smaller velocities will lead to sedimentation and clogging; higher velocities will cause erosion of the pipe walls. The designer should avoid the temptation to use a larger pipe than necessary to carry the design discharge. Larger pipes would allow flatter slopes, hence less excavation, but the resultant lower velocities would lead to clogging of the pipe, particularly during low flows. Changes of pipe diameter should always occur at manholes. The tops of the two pipes are usually placed at the same elevation, or the exiting pipe is located below the incoming one. Contents of a larger pipe should never discharge into a smaller one, as it always leads to blocking.

The required discharge capacity of sanitary sewers is dependent on the drainage area served and its population density; 400 liters (or 100 gal) per person per day is a reasonable value. More specific information is available in textbooks on sanitary engineering and in local codes. In addition to the sewage carried, the sanitary sewers are always subject to groundwater infiltration, regardless of the degree of care with which they are constructed. Storm sewer discharge capacity is dependent on the individual drainage area of each manhole and storm inlet. The discharge for each of these drainage areas is determined by the methods discussed in Chapter 7. For smaller projects the *rational method* furnishes acceptable results. In addition to the resulting discharge, the time of concentration must also be determined. Both factors will determine the times at which the peak runoff will reach the storm sewer at each manhole and the main trunk sewer at the junction of each tributary. To determine the magnitude of the peak flow in the various portions of the sewer line along the way, we must find the time of travel in each pipe length by dividing the length by the velocity of the flow. These values, when added to the time of concentration on the surface, will give an indication of how the peaks from the various tributaries will add up.

The *hydraulic design* of sewerage systems is relatively simple. The basic principle is to design the pipe as an open channel flowing just full. The normal flow formula is used, in which the hydraulic gradient is the slope of the sewer pipe. Knowing the available commercial pipe sizes and the roughness of the pipe, the designer can determine the velocity and the corresponding slope for each pipe section. As long as the velocity is within allowable limits, and all other fundamental design rules are met, the pipe may be drawn with the slope computed. The design and the profile of the sewer are usually drawn together. Figure 9.32 shows such a profile with all necessary information labeled.

Commercial computer programs are available for the design of storm as well as sanitary sewers. One of them, Storm Water Management Model, or SWMM, was already mentioned in Chapter 7, because it includes runoff analysis. It contains extensive routines for analyzing in-conduit storage, downstream control effects, pressure flow, flow reversal, surcharge, branches, loops, pumps, weirs, and regulators. Other commercial computer programs, such as StormCAD™ for Windows, and SewerCAD™ for Windows, offered by Haestad Methods, Inc., features a computer-

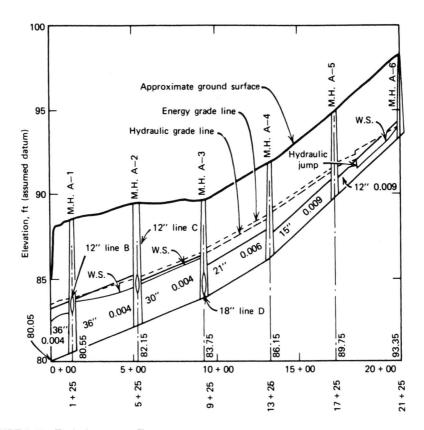

FIGURE 9.32 Typical sewer profile.

aided design type of graphical interface that allows drag-and-drop construction of systems. Introduction of such programs is beyond the scope of this text.

EXAMPLE PROBLEMS

Example 9.1   Determine the required diameter of a 3 ft long horizontal drain pipe located at the bottom of a concrete storage tank, 15 ft below the water level, under the condition that the discharge through the pipe be 9000 gpm. The drain pipe is smooth with a sharp entrance.

Solution. By Equation 9.4 the pipe diameter may be computed as

$$D = \left(\frac{4Q}{c\pi\sqrt{2gh}} \right)^{1/2}$$

where

$$Q = \frac{9000}{449} = 20 \text{ cfs}$$

$$h = 15 \text{ ft}$$

From Figure 9.6 we assume the value of c to be 0.84. Therefore,

$$D = \left(\frac{4(20)}{0.84\pi\sqrt{64.4(15)}}\right)^{1/2} = \left(\frac{80}{82.0}\right)^{1/2} = 1.0 \text{ ft}$$

$$l/D = 3/1.0 = 3$$

From Figure 9.6 $c = 0.84$, so the assumption was OK. ☐

Example 9.2

A 6 ft wide vertical gate on the top of a spillway withholds 4 ft deep water. Determine the discharge under the gate if the gate is raised by 1 ft.

Solution. Using Equation 9.7 with

$$b = 6.0 \text{ ft}$$

$$a = 1.0 \text{ ft}$$

$$H_0 = 4.0 \text{ ft}$$

we have

$$Q = 6(1.0)c\sqrt{64.4}\left(\frac{4}{\sqrt{4 + \psi}}\right)$$

The coefficients c and ψ may be obtained from Figure 9.8 with

$$\frac{H_0}{a} = \frac{4}{1} = 4$$

which gives

$$\psi = 0.624$$
$$c = 0.604$$

Hence,

$$Q = 6(0.604)8.025\frac{4}{\sqrt{4.624}} = 54 \text{ cfs}$$ ☐

Example 9.3

Determine the required opening of the gate described in Example 9.2 for a discharge of 90 cfs.

Solution. Entering our known values into Equation 9.7, we have

$$90 = 6ac\sqrt{64.4}\frac{4}{\sqrt{4 + \psi a}}$$

Simplifying and rearranging, we have

$$\frac{90}{24\sqrt{64.4}} = 0.47 = \frac{ac}{\sqrt{4 + \psi a}}$$

From Figure 9.8 we note that a trial-and-error solution is possible by assuming the value of a. Assuming $a = 2.0$, $H_0/a = 2$, and $\psi = 0.630$, then $c \simeq 0.5965$. Hence,

$$\frac{2.0\,(0.5965)}{\sqrt{4 + 2.0(0.630)}} = 0.520 > 0.47$$

Therefore, a must be further reduced. For our second trial let's take $a = 1.8$ so that

$$\frac{H_0}{a} = \frac{4}{1.8} = 2.22$$

and the values of the coefficients may be assumed to be $\psi = 0.628$ and $c = 0.596$. Therefore,

$$\frac{1.8(0.596)}{\sqrt{4 + 1.8(0.628)}} = 0.47$$

At $a = 1.8$ ft the gate delivers 90 cfs discharge.   ☐

Example 9.4

A 10 ft wide vertical gate discharges into a pool in which the water level is 6 ft. The upstream water level is 8 ft, and the gate opening is 1.6 ft. Determine the discharge through the structure.

Solution. From Figure 9.9 we find the type of flow condition existing.

$$\frac{H_2}{a} = \frac{6}{1.6} = 3.75 \qquad \frac{H_0}{a} = \frac{8}{1.6} = 5.0$$

Observing the location of the point described, we note that the outflow is influenced by the downstream level.

For partially retarded discharge we have to use Equation 9.8. Q is to be determined by Equation 9.7 and corrected by a coefficient k from Figure 9.10. The correction factor is $k = 0.6$.

$$Q = 10(1.6)c\sqrt{64.4}\left(\frac{8}{\sqrt{8 + \psi 1.6}}\right)$$

and, from Figure 9.8,

$$\psi = 0.624 \qquad \text{and} \qquad c = 0.6065$$

Therefore, the free discharge is

$$Q = 1027 \times \frac{0.6065}{\sqrt{8 + 1.6(0.624)}} = 208 \text{ cfs}$$

The retarded discharge is then

$$Q_r = kQ = 0.6(208) = 125 \text{ cfs}$$ ☐

Example 9.5

A 4 m wide gate in an irrigation canal is raised to 1.17 m. The water level is 3 m upstream and 2.7 m downstream. Determine the discharge passing through the gate.

Solution. From Figure 9.7 $H_1 = \psi a$. As a conservative estimate let $H_1 = 0.6(1.17) = 0.7$.

$$H_0 - H_1 = 3.0 - 0.7 = 2.3 < H_2 \ (H_2 = 2.7 \text{ m})$$

which indicates absolute downstream control. Therefore, Figure 9.11 may be used with the following variables:

$$\frac{H_2}{H_0} = \frac{2.7}{3.0} = 0.9$$

$$\frac{a}{H_0} = \frac{1.17}{3.0} = 0.39$$

Entering the graph as shown results in

$$\frac{Q}{B} = 2.5 \text{ m}^3/\text{s/m width.}$$

For the 4 m wide gate the discharge is

$$Q = 10 \text{ m}^3/\text{s} \qquad \qquad \square$$

Example 9.6

A rectangular broad-crested weir is 30 ft long and is known to have a discharge coefficient of 0.7. Determine the discharge if the upstream water level is 2 ft over the crest.

Solution. Equation 9.10 applies. Substituting the variables, we have

$$\begin{aligned} Q &= cb\sqrt{2g}\,H_1^{3/2} \\ &= 0.7(30)\sqrt{64.4}\,(2^{3/2}) \\ &= 477 \text{ cfs} \qquad \qquad \square \end{aligned}$$

Example 9.7

A weir 10 m long with a cross-sectional design similar to case 4 in Table 9.1 is flowing with an upstream head of 45 cm. Determine the discharge.

Solution. From Table 9.1 the metric discharge coefficient at 45 cm upstream head is $M = 1.77$. Using Figure 9.14 with $b = 10$ m, $H = 0.45$ m, and $M = 1.77$, we find the discharge to be

$$Q \simeq 6 \text{ m}^3/\text{s}$$

From Equation 9.11

$$\begin{aligned} Q &= MbH^{3/2} \\ &= 1.77(10)(0.45)^{3/2} \\ &= 5.3 \text{ m}^3/\text{s} \qquad \qquad \square \end{aligned}$$

Example 9.8

The crown of a broad-crested suppressed weir has the shape of case 12 in Table 9.1. The width of the weir is 10.5 m. Establish a rating curve (Q versus H) for the flow over the weir.

Solution. The first two rows of the following computational table show the data for case 12 in Table 9.1. The third row is the $H^{1.5}$ value. The last row is $bMH^{3/2} = Q$, by Equation 9.11.

$H = 0.30$	0.45	0.60	0.75	0.90	1.20	meter
$M = 1.88$	1.96	2.01	2.04	2.05	2.05	m$^{1/2}$/s
$H^{3/2} = 0.1643$	0.3019	0.4648	0.6495	0.8538	1.3145	
$Q = 3.24$	6.21	9.81	13.91	18.38	28.29	m^3/s

Example 9.9

A 20 ft long weir was measured independently to carry a 50 cfs discharge when the crest is overtopped by 8 in. of water. Determine the discharge coefficient of the weir.

Solution. Rearranging Equation 9.10 in the form of

$$c = \frac{Q}{b\sqrt{2g}H^{3/2}}$$

and substituting the known variables, we obtain

$$c = \frac{50}{20\sqrt{64.4}\,(8/12)^{1.5}}$$
$$= 0.57 \qquad \text{for } H = 0.67 \text{ ft}$$

which is the discharge coefficient sought.

Example 9.10

A dam's weir is designed to have a crest height of 16 ft. The maximum upstream head over the crest should not exceed 7 ft. However, the length of the weir is to be selected such that the downstream water level does not influence the upstream conditions and, therefore, does not retard the flow during critical floods. What is the allowable maximum downstream water level under these conditions?

Solution. Figure 9.16 may be used with

$$\frac{H_1}{P} = \frac{7}{16}$$
$$= 0.44$$

The left bounding line of the region of true hydraulic jump represents the limiting condition for no downstream influence. In our case, it is

$$\frac{H_2}{P} = 0.55$$

Therefore, with $P = 16$ the limiting value is

$$H_2 = 8.8 \text{ ft}$$

Example 9.11

A weir is to be placed in a channel to measure the discharge. A rating curve will be developed for it after construction by independent discharge measurements. The normal flow in the channel under extreme conditions will not exceed 4 m. What should be the minimum crest height of the weir to avoid downstream influence?

Solution. By using Figure 9.16 we consider the limit of the free fall condition, the uppermost curve in the graph. The curve may be approximated by a line. Developing the equation for this line, we determine that its slope is about 1.04 and its intercept at $H_2/P = 0$ is -0.13; therefore,

$$\frac{H_1}{P} = -0.13 + 1.04 \frac{H_2}{P}$$

from which $P = 8.0H_2 - 7.7H_1$.

Next, we establish the normal depth of the flow in the downstream channel under the maximum expected discharge, which gives our maximum value for H_2. In our case, H_2 equals 4 m. Therefore, the minimum required crest height is

$$P = 32.0 - 7.7H_1$$

where H_1 is determined by Equation 9.10 by using the maximum discharge expected. To minimize the required P crest height or to keep H_1 within a prescribed limit, the crest width b may be suitably selected in Equation 9.10.   □

Example 9.12   Design a spillway such that at a maximum flood of $Q = 14,000$ cfs the water elevation in the reservoir does not exceed 15 ft over the crest height, which is to be 75 ft. The discharge coefficient is assumed to be 0.62. The entrance slope of the spillway on the upstream side is to be $z = 0.33$. The curvature of the spillway should conform to the U.S. Army Corps of Engineers formula given in Equation 9.12, connecting to a downstream slope of $z = 0.2$.

Solution. First, we calculate the required spillway width for Equation 9.10,

$$Q = cb\sqrt{2g}H^{3/2}$$

with

$$Q = 14,000 \text{ cfs}$$
$$c = 0.62$$
$$H = 15 \text{ ft}$$

Hence,

$$b = \frac{14,000}{(0.62)\sqrt{64.4}(15)^{1.5}} = 48.4 \text{ ft}$$

For the curvature of the spillway we use Equation 9.12 and the parameters from Table 9.3,

$$K = 1.936 \quad \text{and} \quad n = 1.836$$

which corresponds to an entry slope of $z = 0.33$. Then,

$$X^{1.836} = [1.936(15)^{1.836-1}]Y$$

or

$$\frac{X^{1.836}}{18.63} = Y$$

Computing for a sequence of X values, we get

$X = 0$	2	4	6	8	10	12	14	16	18	20						
$Y = 0$	0.19	0.68	1.44	2.44	3.68	5.14	6.82	8.72	10.83	13.14						
$X = 22$	24	26	28	30	32	34	36	38	40	42	44	46	48	50	52
$Y = 15.65$	18.36	21.27	24.37	27.66	31.13	34.80	38.65	42.68	46.90	51.29	55.87	60.62	65.55	70.65	75.92

The curvature of the toe of this spillway may be determined from Table 9.4. Converting H and P into meters, we get

$$15\,\text{ft} = 4.6\,\text{m} \qquad 75\,\text{ft} = 22.9\,\text{m}$$

From Table 9.4 the radius of curvature is 10.4 m, which is

$$r = 34\,\text{ft}$$

This is connected to a straight portion with a slope $z = 0.2$.   □

Example 9.13   A spillway with a 5 ft crest height operates under a 15 ft upstream design head. Using the coordinates recommended by Bazin and Craeger (Table 9.2), determine the location of the outside and inside surfaces of the nappe and the recommended location of the spillway surface at one point, 25 ft below the crest.

Solution. The coordinates shown in Table 9.2 may be converted to actual measurements by multiplying them by the design head H_d, which in our case is 15 ft. To get the x coordinates needed, we find y_1, y_2, and y_3 values such that

$$H_x y = 25\,\text{ft}$$
$$y = \frac{25}{15} = 1.667$$

These values from Table 9.2 correspond to x values as follows:

$$\text{for } y_1 = 1.667 \qquad x_1 = 2.24$$
$$\text{for } y_2 = 1.667 \qquad x_2 = 2.31$$
$$\text{for } y_3 = 1.667 \qquad x_3 = 2.59$$

Multiplying these x coordinates by $H_d = 15$ ft, we obtain

$$X_1 = 2.24(15) = 33.6\,\text{ft}$$

which is the inside surface of the nappe for free fall and

$$X_3 = 2.59(15) = 38.9\,\text{ft}$$

which is the outside surface of the nappe for free fall.
Hence, the horizontal width of the overfalling sheet of water is

$$X_3 - X_1 = 38.9 - 33.6 = 5.3\,\text{ft}$$

at a location 25 ft below crest.

The location of the spillway surface at this point, recommended by Bazin and Craeger, is

$$X_2 = 2.31(15) = 34.7 \text{ ft}$$

measured horizontally away from the crest and 25 ft vertically below the crest. Since X_1, the inside nappe surface, is less than the position of the spillway, it is shown that the spillway is supporting the nappe to some extent.

For a complete spillway design similar values may be computed for other depths below the crest. The position of the outer nappe surface is required in the judicious design of the required side walls bordering the spillway. ☐

Example 9.14

A 12 ft wide spillway delivers a discharge of 250 cfs such that the depth of the water at the toe of the spillway is 1.2 ft, and the depth of the downstream water is 4 ft (Figure E9.14). Determine the required size of the downstream apron.

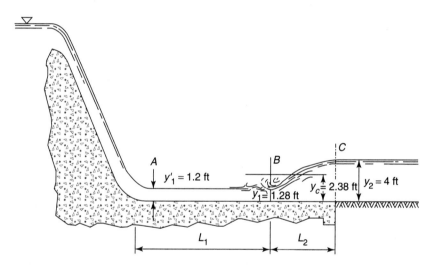

FIGURE E9.14

Solution. We assume that the spillway is rectangular and identical throughout the apron. From $Q = 250$ cfs, therefore, we can calculate the critical depth y_c. The discharge per unit width $= 250/12 = 20.83$ cfs/ft. Then using Example 8.7 as a guide, we get

$$y_c = \left[\frac{q^2}{g} \right]^{1/3} = \left[\frac{(20.83)^2}{32.2} \right]^{1/3} = 2.38 \text{ ft}$$

To find the conjugate depth of $y_2 = 4$ ft, we see that

$$\frac{q^2}{g} = \frac{1}{2} y_1 y_2 (y_1 + y_2)$$

$$= \frac{(20.83)^2}{32.2} = \frac{1}{2}(y_1)4(y_1 + 4)$$

$$y_1 = 1.28 \text{ ft}$$

Thus, at point B, right in front of the jump the velocity is

$$v_1 = \frac{250}{(12)1.28} = 16.3 \text{ fps}$$

The hydraulic radius is

$$R_1 = \frac{12 \times 1.28}{12 + 2 \times 1.28} = 1.05 \text{ ft}$$

At point A, the beginning of the apron, $y_1' = 1.2$ ft and

$$v_1' = \frac{250}{12 \times 1.2} = 17.4 \text{ fps} \qquad R_1' = \frac{12 \times 1.2}{12 + 2 \times 1.2} = 1 \text{ ft}$$

The average velocity between A and B is

$$v_{\text{ave}} = (17.4 + 16.3)/2 = 16.9 \text{ fps}$$

and

$$R_{\text{ave}} = \frac{1 + 1.05}{2} = 1.03 \text{ ft}$$

Now, to find the slope of the energy grade line S slope, we assume that the friction coefficient of the apron is $n = 0.014$, and we write Equation 8.19 as

$$v = \frac{Q}{A} = \frac{1.49 R^{2/3}}{n} S^{1/2}$$

$$16.9 = \frac{1.49}{0.014}(1.03)^{2/3} S^{1/2}$$

$$S = \left[\frac{16.9}{(1.03)^{2/3}} \times \frac{0.014}{1.49} \right]^2 = 0.024$$

Hence, the length of the tailrace before the jump is

$$(S_0 - S)L_{AB} = \left(\frac{v_1^2}{2g} + y_1 \right) - \left(\frac{v_1'^2}{2g} + y_1' \right)$$

$$= \left[\frac{(16.3)^2}{2 \times 32.2} + 1.28 \right] - \left[\frac{(17.4)^2}{2 \times 32.2} + 1.20 \right]$$

$$L_{AB} = \frac{5.41 - 5.90}{0 - 0.024} = \frac{-0.49}{-0.024} = 20.4 \text{ ft}$$

The Froude number at B is

$$\mathbf{F}_1 = \frac{v_1}{\sqrt{gy_1}} = \frac{16.3}{\sqrt{32.2 \times 1.28}} = 2.54$$

From Figure 9.19 the length of the jump $L_{BC} = 4.9 \times 4 = 19.6$ ft. Then, the total apron length is $20.4 + 19.6 = 40.0$ ft. □

Example 9.15   Determine the required stone size of the riprap to be placed after the apron designed in Example 9.14.

Solution. Assuming that the velocity immediately below the hydraulic jump is uniform, and computing this value from

$$Q = 250 \, \text{cfs}$$
$$\text{depth} = 4 \, \text{ft}$$
$$\text{width} = 12 \, \text{ft}$$
$$v_{\text{ave}} = \frac{250}{4(12)} = 5.21 \, \frac{\text{ft}}{\text{sec}}$$

Using Figure 9.22 with $V = 5.21$ fps, we find that the required stone diameter is 4 in.   □

Example 9.16   A 60 in. diameter concrete culvert is 100 ft long, and its entrance loss coefficient is $k_e = 0.2$ (Figure E9.16). Determine the head loss in the culvert if the discharge is 140 cfs and the pipe is flowing such that both ends are submerged.

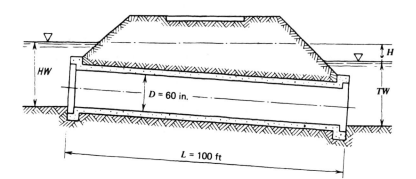

FIGURE E9.16

Solution. The culvert is flowing full. We assume the concrete roughness $n = 0.012$:

$$k_e = 0.2$$
$$L = 100 \, \text{ft}$$
$$Q = 140 \, \text{cfs}$$
$$D = \frac{60}{12} = 5 \, \text{ft}$$

From the nomograph shown in Figure 9.31, we use L, k_e, and D to get the turning point. Then we use Q and the turning point to get $H = 1.18$ ft.   □

Example 9.17   A corrugated metal culvert with a head wall entrance has a diameter of 48 in. and is 150 ft in length; it is laid on a slope of $S_0 = 0.02$ (Figure E9.17). For a headwater elevation of

$HW = 6$ ft, determine the discharge in the culvert, and state whether the flow is under inlet or outlet control.

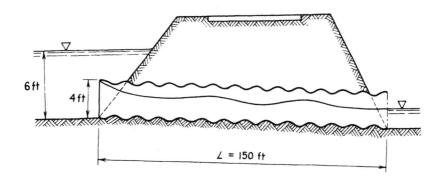

FIGURE E9.17

Solution. For a corrugated steel culvert,

$$n = 0.024$$
$$S_0 = 0.02$$

We assume it is partially full; we check by Equation 9.19:

$$S_{optimum} = \frac{111n^2}{D^{1/3}}$$
$$= \frac{111(0.024)^2}{(4)^{1/3}} = 0.04$$

Therefore, the culvert slope $S_0 < (S)_{optimum}$.

$$\frac{L}{100S_0} = \frac{150}{100(0.02)} = 75 > 40 \qquad \text{for a 48 in. pipe}$$

in Figure 9.30b. Thus, the culvert is operated under outlet control.

From Figure 9.30b with $L/100S_0 = 75$ and with $HW = 6$ ft for a 48 in. pipe, we obtain $Q = 112$ cfs.   ☐

PROBLEMS

9.1 A 12 ft long, 12 in. diameter rough galvanized pipe with a rounded entrance is opened to allow flow through a concrete dam. The depth of water above the pipe entrance is 10 ft. What is the discharge through the pipe?

9.2 A 5 m wide rectangular sluice gate is opened to a height of 1.5 m. The upstream water level is 4.0 m, and the downstream level is 2.0 m. Calculate the discharge under the gate.

9.3 What will be the discharge under the gate in Problem 9.2 if the downstream water level is 3.65 m?

9.4   A 15 ft wide rectangular sluice gate is opened to a height of 1 ft. The upstream water level is 10 ft, and the downstream level is 6 ft. Calculate the discharge under the gate.

9.5   A 4 m wide rectangular suppressed weir is overtopped by 1.5 m of water. For a discharge of 10 m³/s determine the metric discharge coefficient.

9.6   A 1 mi long flood wall with a rectangular cross section is overtopped by a flood. The depth of water over the wall is 1.5 ft. The wall is 5 ft high and 3 ft thick. Estimate the volume of water flowing over the wall for a 6-hr period.

9.7   The depth of flow over a weir is 1.5 ft above the weir crest. The weir is 1.5 ft high, and the water depth on the downstream side of the weir is 1.8 ft. Use Figure 9.16 to determine the type of flow across the weir.

9.8   A broad-crested weir is the shape of case 11 in Table 9.1. The width of the weir is 4.9 m. Establish the discharge rating curve for the weir.

9.9   On a horizontal downstream apron the supercritical flow before the jump is characterized by a Froude number of 9.5. The downstream water depth is 6.4 ft. How long will the hydraulic jump be? How is the jump classified according to Figure 9.19?

9.10 For a 1.5 ft water level over the crest of a spillway, compute the shape of the nappe if the downstream water level is 5 ft below the spillway crest.

9.11 A 5 ft diameter culvert is flowing 60 percent full. The discharge is 300 cfs. Determine the total energy of the flow.

9.12 What is the optimum discharge for a 5 ft diameter culvert?

9.13 Compute the optimum slope for the culvert in Problem 9.12 if the pipe material is corrugated metal. If the pipe was made of concrete, how would the optimum slope change?

9.14 A 250 ft long concrete sewer pipe section is to carry a maximum discharge of 11.0 cfs when 80 percent full. Select a slope and a pipe diameter such that the velocity is within the allowable limits for sewers (use 3.5 fps).

MULTIPLE CHOICE QUESTIONS

9.15 Torricelli's equation is an expression of the
   A. Kinetic energy head.
   B. Second law of Newton.
   C. Momentum-impulse theorem.
   D. Chézy formula.
   E. Darcy-Weisbach formula.

9.16 The discharge coefficient of the orifice formula is
   A. Less than 1/2.
   B. Independent of the viscosity of the fluid.
   C. Dependent on the vena contracta.
   D. Proportional to the velocity.
   E. Based on experimental data.

9.17 In flow under gates the downstream influence occurs when the
   A. Downstream level is less than the gate opening.
   B. Downstream level equals the gate opening.
   C. Downstream level exceeds the depth of flow under the gate, H_1.

D. All of the above.

E. None of the above.

9.18 Absolute downstream control in flow under gates means that the

A. Jet is fully submerged.

B. Upstream water level may be disregarded.

C. Partial downstream control coefficient is zero.

D. Discharge coefficient becomes constant for all depths.

E. Downstream depth determines the discharge.

9.19 Discharge under gates depends (in part) on the energy available raised to the nth power. What is n?

A. 2.0.

B. 3/2.

C. 5/3.

D. 1/2.

E. 2/3.

9.20 Discharge over weirs depends on the available energy level raised to the nth power. What is the value n?

A. 1/2.

B. 5/3.

C. 0.66.

D. 2.5.

E. 1.5.

9.21 A suppressed weir is one that is

A. Under downstream control.

B. Broad-crested.

C. Without side contraction.

D. Useless for discharge measurement.

E. Not perpendicular to the flow.

9.22 In flow over weirs the downstream water level must be below the weir crest to have a true hydraulic jump occur.

A. True.

B. False.

9.23 The Bazin-Craeger coordinates are based on

A. Experiments by the U.S. Bureau of Reclamation.

B. U.S. Army Corps of Engineers studies.

C. Dynamics of free fall.

D. Supercritical flow.

E. Practical experience of designers.

9.24 The upstream slope of the dam does not influence the recommended shape of spillways.

A. True.

B. False.

9.25 The need to minimize the length of the apron of a spillway results from

A. Economic conditions.

B. The danger of erosion.

C. Supercritical conditions.

D. The instability of the jump formation.

E. Downstream influences.

9.26 The length of a culvert is immaterial if it is under inlet control.
   A. True.
   B. False.

9.27 The friction factor of a culvert pipe is immaterial if it is under inlet control.
   A. True.
   B. False.

9.28 The recommended distance between sewer manholes is
   A. 50–60 ft.
   B. 60–80 ft.
   C. 30–50 m.
   D. 50–100 m.
   E. 100–150 m.

9.29 Allowable velocities in sewers must not exceed
   A. 20 m/s.
   B. 30 m/s.
   C. 0.5 m/s.
   D. 3 m/s.
   E. 12 m/s.

10

Coastal Hydraulics

As more and more of the American population move near the coastal areas hydraulic problems related to the seashore take on increasing importance. The generaton of waves and their travel across the oceans; tides; sediment transport; and erosion of the seashore are reviewed. The functions of various coastal structures and the basic design concepts of sea walls, jetties, breakwaters, and groins are discussed.

10.1 WAVE MECHANICS

Coastal hydraulics is basically the study of the effects of waves. The theoretical analysis of wave motion was one of the favorite subjects of nineteenth-century mathematicians. Among the many notable contributors,[1] George Biddle Airy (England, 1801–1892) was the first to present a workable formula.[2] His solution, based on a linear simplification of the otherwise complex partial differential equation, describes wave motion as sinusoidal, progressive, and periodic motion in which wave crests move with a celerity c. Such oscillatory waves can generally be described by their length, height, and period. The celerity, also called *phase velocity*, is

$$c = \frac{L}{T} \tag{10.1}$$

where L is the *wavelength*, and T is the *wave period*. These variables are shown in Figure 10.1. The basic assumptions of Airy's solution was that the amplitude—the

---

[1]Newton, Bernoulli, Laplace, Lagrange, Cauchy, Stokes, et al.

[2]In "Tides and Waves," *Encyclopaedia Metropolitana*, London, 1845.

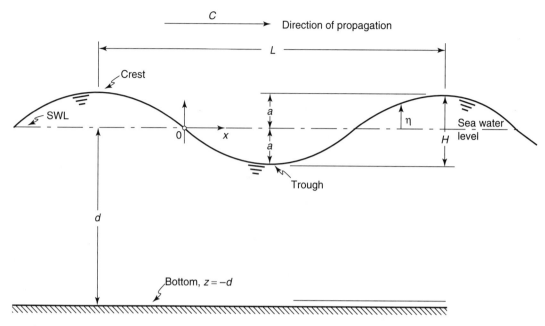

FIGURE 10.1 Definition of variables of a sinusoidal wave.

height of the waves, H—is small, the viscosity is neglected, the movement is irrotational, and the water is deep. In practice, depths exceeding $d = L/4$ are considered to satisfy this requirement. Airy's formula for the water surface is

$$\eta(x, t) = \frac{H}{2}\cos(kx - \sigma t) \qquad (10.2)$$

where η is measured from the horizontal still water level in terms of x horizontal dimension and time t. The term k in Equation 10.2 is an abbreviation for $2\pi/L$ and is called the *wave number*. The other term, σ, stands for $2\pi/T$. If we define the inverse of the wave period, $1/T = f$, as *wave frequency*, then σ is the *angular frequency*. Although the crest of the periodic waves appears to move with the celerity, the water particles under the surface are in orbital motion, as shown in the bottom part of Figure 10.2. In theory, the orbit diameter is largest for the water particles on the surface, while at the bottom of the water there is no movement. In deep water the orbits are circular, whereas in shallow waters they become increasingly ellipsoidal as the depth decreases.

In practice, the water is considered shallow if the depth is less than $L/20$. For *shallow waters* the celerity formula is

$$c_{\text{shallow water}} = \sqrt{gd} \qquad (10.3)$$

as we know from open channel flow. This equation, incidentally, was suggested by John Scott Russell (Scotland, 1808–1882) on the basis of extensive field studies

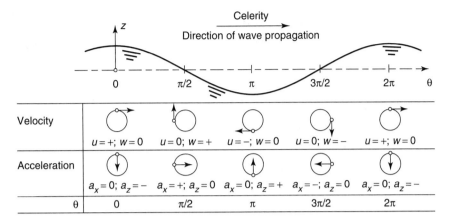

FIGURE 10.2 Particle velocities and accelerations in a wave.

concerning the design of ships. It was, however, formally derived earlier by Joseph Louis Lagrange (France, 1736–1813).[3]

Although theoretically deep waters are ones whose depth approaches infinity, in reality when d exceeds $L/4$ we speak of *deep waters*. For the transitional zone between deep and shallow waters the celerity can be derived from Equation 10.2:

$$c_{\text{transitional}} = \frac{gT}{2\pi} \tanh\left(\frac{2\pi d}{L}\right) \qquad (10.4)$$

where the tanh expression is the hyperbolic tangent of the bracketed term. Its value approaches unity with increasing depth. Hence, in the case when the depth can be taken as infinity, the celerity of the waves is

$$c_{\text{deep water}} = \frac{g}{2\pi} T \qquad (10.5)$$

It can be mathematically proved that this value is about twice as big as the celerity in shallow water. As the wave approaches the shore the depth of the water decreases, which causes a decrease in celerity. At the crest of the wave the depth of the water is the greatest, hence the celerity at that point is the highest. At one point, as the wave nears the shore, the top of the wave outraces the rest and the wave *breaks*. The breaker zone is generally considered to be the outer boundary of the concerns of coastal hydraulics.

---

[3]Lagrange wrote, "The velocity of propagation of waves will be the same as that which a weight would acquire in falling from a height equal to half the depth of water in the canal." He made the mistake, however, of presuming that the same relationship would apply to the upper stratum of deep water as well.

FIGURE 10.3   Plunging breaking wave on sand beach.

Using Equation 10.1, we can rearrange Equation 10.4 to express the wavelength as

$$L = \frac{gT^2}{2\pi} \tanh\left(\frac{2\pi d}{L}\right)$$ (10.6)

The problem with this formula is that L appears on both sides. Tabular solutions are available in the literature.[4]

Figure 10.3 shows a breaking wave rolling up a sandy beach. It contains quite a lot of energy. The energy carried by a wave is the sum of the kinetic energy inherent in the movement of the water particles and the potential energy of the particles that are located above the trough, all the way to the wave crest. Airy's theory quantifies this energy as

$$E = E_{\text{kinetic}} + E_{\text{potential}} = \gamma L \frac{H^2}{8}$$ (10.7)

where H is the wave height.

It is more convenient to express this value for a unit surface area, dividing by L to obtain the *energy density*,

$$E_{\text{d}} = \gamma \frac{H^2}{8}$$ (10.8)

---

[4]*Shore Protection Manual*, Vol. 2, U.S. Army Corps of Engineers, 1984.

The entire energy transmitted across a vertical plane, all the way to the bottom perpendicular to the wave advance and in a unit width is given by

$$P = ncE_d \qquad (10.9)$$

which is called *wave power*. The term n in this equation is

$$n = \frac{1}{2}\left[1 + \frac{4\pi\,d/L}{\sinh(4\pi\,d/L)}\right] \qquad (10.10)$$

For deep water n takes the value of $\frac{1}{2}$, and for shallow depths it is 1. Thus, we conclude that the wave power for deep waters is

$$P_{\text{deep water}} = \frac{c}{16}\gamma H^2 \qquad (10.11)$$

and for shallow waters it is

$$P_{\text{shallow water}} = \frac{c}{8}\gamma H^2 \qquad (10.12)$$

where the celerity is defined by Equation 10.3.

In nature, periodic waves appear with an infinite variety of frequencies and lengths. In linear wave theory, waves of different lengths and periods can be superimposed. If two wave trains of slightly different lengths and periods travel together, the superimposed wave forms create an envelope with the appearance of a wave. Because of this interference, where the two waves cancel each other, the envelope has zero value. Where the two waves are in phase, the composite wave's height doubles. Thus, two superimposed wave trains create an apparently new wave that has a velocity of

$$c_g = nc \qquad (10.13)$$

where c_g is called *group velocity*, and n is as defined by Equation 10.10. Because n equals $\frac{1}{2}$ in deep waters, the group velocity is half the celerity. In shallow water, with n being unity, the two terms are equal. Group velocity is important because it is the velocity by which the wave power is propagated. Wave power is also called *energy flux*. In the analysis of coastal problems like beach formation, erosion, and generation of longshore currents, the application of the energy conservation equation is an important tool.

Airy's wave theory works quite well in deep water. However, it becomes quite inadequate in the shallows of coastal areas. Higher-order solutions exist in the literature. One of these is the Stokes wave formula. Another is the cnoidal formula.[5]

---

[5]This name comes from the term *cn*, which is a Jacobian elliptic function.

Both of these describe shallow periodic waves that are typical of the widely spaced shallow waves seen running up on beaches. These are beyond our scope.

10.2 WAVE GENERATION

Wave actions on coastal regions are caused by wind-induced waves and tidal waves. Generally, waves induced by the wind are referred to as seas or swells. *Seas* are waves induced by air pressure variations at the surface due to turbulence, eddies, and vortices in the winds during storms. *Swells* are periodic waves transporting the storm-induced energy to great distances over the oceans. Commonly, they are the waves that are encountered in beach processes. For practical purposes, the periods of swells range between 5 and 15 sec. Their heights depend on the magnitude of the storms that caused them.

During storms part of the momentum and energy in the atmosphere is transferred to the water by some mechanism that is not yet fully understood. This energy transfer causes random motions of the water surface. Such random motion may be analyzed by separating it into components of waves differentiated by their frequencies and intensities. This mathematical process is called *spectral analysis*. Waves with smaller frequencies when superimposed over waves of larger frequencies are washed out in the process. They transfer their momentum to the larger waves. To describe such random wave motion, the concept of *significant wave height* was introduced. It is defined as the average of the highest third of the random waves measured during a period, usually 20 min.

Studies have shown that there appears to be a relationship between the total range of periods and the maximum sustained wind velocity. From this it has been concluded that the significant wave height derived from a wave spectrum is proportional to the wind speed raised to the $\frac{5}{2}$ power.

The magnitude of the waves—hence, the energy density developed by a storm—depends on its *duration*, wind velocity, and geographic extent. The last factor is measured by the *fetch*, the distance measured on the water surface in the direction of the wind across the area of the storm. Each wind velocity generates a well-defined wave period with a well-defined maximum. Given an unlimited fetch and duration, the storm causes the seas to be fully developed. Table 10.1 shows some characteristics of fully developed seas. These data are based on studies applying the techniques of spectrum analysis. The wind velocity here is given in *knots*, which means nautical miles[6] per hour. One knot equals 1.15 miles[7] per hour.

Once the waves propagated by a storm move out of the storm area and are no longer influenced by the wind, they travel in the form of swells. The direction of the wind causing the storm determines the direction of wave propagation. In the swell the waves become more regular by dispersing waves of smaller frequencies. The wave train travels with the group velocity in the direction of its original propa-

---

[6]One nautical mile equals 1013 fathoms. Three fathoms equal a league.

[7]Statutory miles, not nautical miles in this case.

TABLE 10.1   Characteristics of Fully Developed Seas As Determined by Spectrum Analysis

Wind Velocity knots	Peak Period s	Significant Wave Height m
10	4.0	0.43
20	8.1	2.44
30	12.1	6.58
40	16.1	13.8
50	20.2	23.8
56	22.6	31.4

Source: Excerpted from Pierson, Neumann, and James, Observing and forecasting ocean waves by means of wave spectra and statistics. U.S. Navy, Hydrograph. Office. Publ. No. 603, 1955.

gating wind. The train's width is about equal to the width of the storm. Its length equals the product of the duration of the storm and the group velocity. Swells traveling in such wave trains may travel thousands of miles across the open oceans without a significant reduction of energy. Winter storms off South Africa, for instance, can cause heavy wave action on the beaches of New Jersey during the summer months. The physical processes outlined here are useful in forecasting wave action based on weather reports from thousands of miles away.

In addition to seas and swells, coastal areas are also subjected to tidal waves. *Tides* are caused by the mutual attraction of the masses of Earth and the Moon and, to a lesser extent, the Sun. The force of attraction is defined by Newton's law of gravitation.[8] The rotation of the Moon around Earth causes tides to rise and fall twice a day. The period of the tidal wave is hence about 12 hr. These are called *semidiurnal* tides. There are exceptions to this: At some locations on Earth the tides are *diurnal*; their period is about 24 hr. Because of the varying relative position of the three bodies, the magnitude and the time of cresting of the tides also vary. Since the lunar day is some 50 minutes longer than the solar day, peaks occur 50 minutes later in successive days. At times when the Sun, Earth, and the Moon are approximately in line—the times of full moon and new moon—tides of extreme magnitude occur. These are called *spring tides.*[9]

There are great variations in the magnitude of tides around the world. Around the United States spring tides range from 2 ft at Key West, Florida, to 23 ft at Calais, Maine. On the Pacific Coast, diurnal tides range from 5 ft at Point Loma, California, to 15 ft at Puget Sound, Washington. Generally, tides vary substantially from place to place. Daily tidal information is a regular feature of newspapers in coastal areas. These predictions are issued by the National Oceanic and Atmospheric Administration (NOAA).

The velocity of travel of tides depends on the depth, according to Equation 10.3. Near the coast the tidal wave height increases. When tides enter or exit bays

---

[8]In fact, it was Newton who first published a clear explanation of the mechanism of tidal motion in 1686.

[9]From the Anglo-Saxon word *springan*, meaning "rising of the water."

FIGURE 10.4 Damaging wave action at Oxnard, California. (Courtesy of U.S. Army Corps of Engineers, Wave Dynamics Division)

and estuaries they create strong currents. Tidal currents in coastal areas contribute to sediment transport and erosion. During storms, or when heavy swells are encountered, high tides are significant contributors to major coastal erosion. Figure 10.4, from Oxnard, California, shows such an effect.

The behavior of tides in ocean basins or large lakes is often analyzed with the concept of *forced waves*. Standing or stationary waves may be observed in enclosed basins. If in a basin of length l and depth h a solitary wave oscillates by reflecting off the boundaries, the period of oscillation, T_n, is the time it takes the wave to travel a distance $2l$ with celerity c. Using Equation 10.3, we find that the period of forced oscillation is then

$$T_n = \frac{2l}{\sqrt{gd}}$$ (10.14)

where d is the depth of the basin and l is the width. Periodic oscillations may be caused by wind or atmospheric pressure variations, and not necessarily by tides. Such low-frequency oscillations in water bodies are called *seiches*. These fluctuations have been measured to be as much as +8.8 feet in Lake Erie at Buffalo, New York, and −8.9 feet at Toledo, Ohio, also on Lake Erie.

10.3 LITTORAL PROCESSES

Geologists refer to the coastal zone within which coastal sediment may be moved by waves and currents as the *littoral zone*. Seaward from this littoral zone the sea

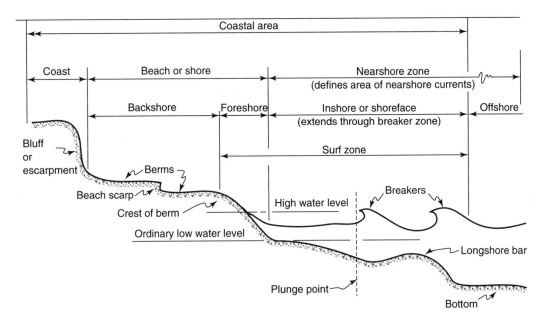

FIGURE 10.5 Definition of terms in a typical beach profile.

bottom remains undisturbed. Figure 10.5 shows the various components of the littoral zone. The *inshore* zone, shown here, begins where the breakers appear. This is the area of reduced depth where the top of the waves overrun the wave itself. This happens when the wave height is about 80 to 125 percent of the depth. Hence, larger waves break in deeper water than smaller waves. In the breaker zone there usually is an underwater sandbar, the *longshore bar*. The *near-shore* zone contains near-shore currents that contribute to the transport of sediments. The *foreshore* ranges from the upper limit of the wave swash at high tide to the low water mark of the backrush of the wave swash at low tide. The slope of the beach in this zone is inversely related to the average grain size of the sediment. Sand and silt beaches are flatter than those with pebbles and rocks. Beaches where the incoming wave energy is high are usually steeper than low-energy beaches. The *back shore* extends from the starting line of the vegetation or the coastal bluff, to the edge of the highest point exposed to water. Within the back shore one or more nearly horizontal *berms* may be found.

The shape of the cross section of the littoral zone is not permanent. In the Northern Hemisphere, summer swells caused by winter storms in the Southern Hemisphere build up the beaches. During winter storms, coastal beaches are often eroded.

Landward wind tends to blow away the dried sand from the berm and create *dunes.* The formation of dunes is encouraged by vegetation or fencing. Dunes are nature's protective structures. They build up between storms but are often washed away during major storms, causing coastal flooding. In some cases coastal dunes are not fully recovered because the sand is deposited in the offshore water too deep to be moved back onshore by regular wave action. Also, if coastal dunes are covered by parking lots and condominiums, their beneficial effects of maintaining the beaches are lost.

The momentum flux of the waves causes the water level shoreward from the breaker line to rise toward the shore. The existence of this *wave setup* has been proved theoretically as well as experimentally. It is defined as the superelevation of the mean water level caused by wave action alone. The wave setup is balanced by a *wave setdown* seaward from the breaker zone. The wave setup in the *surf zone* depends on local conditions and on the magnitude of the waves. Because the shore is not uniform along the beach, the local differences in the wave setup cause *longshore currents* that carry the excess water and, of course, sediments. At certain locations where conditions are favorable, the excess water in longshore currents returns to the sea through *riptides.*

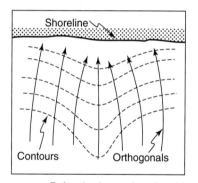 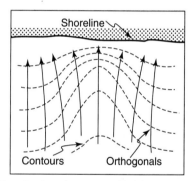

Refraction by a submarine ridge (left) and submarine canyon (right).

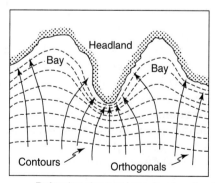 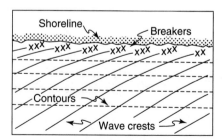

Refraction along an irregular shoreline (left) and along a straight beach with parallel bottom contours (right).

FIGURE 10.6 Effects of underwater topography on wave refraction.

FIGURE 10.7 Wave diffraction behind a breakwater.

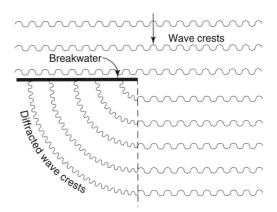

When the waves approach the shore at an oblique angle, the part of each wave that is still traveling in deeper water will move faster than the part in the shallows. The reason for this is that the celerity decreases with decreasing depth. As the celerity decreases, the wavelength decreases proportionally. The result is that the incoming waves bend toward the shoreline. This bending effect is called *refraction*, and it is analogous to refraction in light or sound waves. Figure 10.6 shows typical coastal features that cause refraction of waves. Refraction greatly influences the relative magnitude of the waves arriving at different points along the coast as they converge or diverge toward various locations. As a result, there are significant variations in the amount of wave energy and wave forces on the beach. Refraction alters erosion and deposition of beach materials along the shore and greatly influences the potential development of rip currents. In practice, refraction computations can be made by computer as long as bathymetric information is available and the underwater terrain is not too complex.

Even if a bay or a protected harbor is not subject to direct wave action, indirectly wave energy can enter it laterally by diffraction. The effect of *wave diffraction* in the protected zone behind a breakwater is shown in Figure 10.7. If the breakwater does not absorb the energy of the waves completely, it creates a wave *reflection* on the seaward side. The determination of wave diffraction and reflection effects is important in the design of some coastal structures. Diffraction of waves also plays a part in coastal sedimentation and erosion. As an example, Figure 10.8 shows an aerial photograph of the Lake Erie coast at Lakeview Park, Ohio, where the diffraction effect of segmented breakwaters helps build the beach by encouraging sediment deposition in the protected zone. Direct and diffracted waves can be seen quite well in this picture.

Waves arriving obliquely at the shore create longitudinal currents along the coast. This current brings about a *littoral sediment transport* that is very important

FIGURE 10.8   Direct waves and diffracted waves behind segmented offshore breakwater at Lakeview Park, Ohio. (Courtesy of U.S. Army Corps of Engineers, Wave Dynamics Division)

in the stability of beaches. It is determined by the longshore component of the wave power:

$$P_l = (P)_b \sin \alpha_b \cos \alpha_b \tag{10.15}$$

where α is the angle the breaking wave crests make with the shoreline. The subscript b indicates that the terms are to be evaluated at the breaker zone. P is the energy flux determined from Equation 10.9. Based on field measurements[10] of transported sediments the littoral transport rate may be expressed by the formula

$$I_l = 0.77 \, P_l \tag{10.16}$$

where I_l is called the *immersed weight transport rate* in dynes per second and which is defined as

$$I_l = (\rho_s - \rho) \, g a' S_l \tag{10.17}$$

where ρ_s is the dry density of the sediment, ρ is the density of the sea water, a' is a correction factor accounting for the porosity of the sediment (about 0.6 for beach deposits), and S_l is the *longshore volume transport rate*. The dimensions of S_l is volume per unit time, for example, cubic feet per day, or cubic meters per year.

---

[10]From Paul D. Komar, *Beach Processes and Sedimentation*, Prentice Hall, 1976.

FIGURE 10.9 River inlet on Lake Michigan, protected by jetties. Note sediment deposits alongside the jetties. (Courtesy of Illinois Department of Transportation)

The magnitude of this littoral drift of sediment can be as much as 306,000 m³/year at Ocean City, New Jersey, or as little as 22,600 m³/year at Atlantic Beach, North Carolina. The dominant wind direction and the availability of sand are some of the variables determining littoral drift.

When the mechanism of littoral drift is interrupted either by a jetty protecting a harbor entrance or by a groin retaining a beach, the transported sand will be deposited on the updrift side, and the downdrift side will encounter beach erosion. At Cape Canaveral, for example, jetties built during the 1950s caused a beach loss of 400 ft south of the jetties, endangering coastal homes. In such cases the sand may be bypassed by pumps, dredging, or other means to the downstream side for

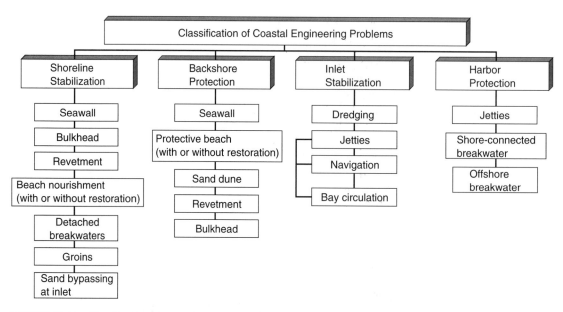

FIGURE 10.10 Classification of coastal engineering problems.

FIGURE 10.11 Breakwater-protected harbor. (Courtesy of U.S. Army Corps of Engineers, Wave Dynamics Division)

beach renourishment. At Pompano Beach, Florida, for instance, 100,000 yd^3 of sand is moved each year to maintain the beaches deprived of drifting sand by a jetty. The effect of jetties on lateral sediment movement is illustrated in Figure 10.9, showing a river inlet protected by jetties. The upper portion of the photograph shows that the jetty retains much of the sand drifting along the coast.

10.4 COASTAL STRUCTURES

There are two primary purposes for coastal structures: One is to improve navigation by preventing shoaling in navigation channels and harbors. The other is to prevent the erosion of the coast. These may be further divided into coastal engineering problems as shown in Figure 10.10. In each of the problems listed design considerations generally encompass hydraulics as well as sedimentation analysis, structural analysis and design, navigation, environmental concerns, legal and economic matters, and maintenance problems.

FIGURE 10.12 Concrete seawall at Galveston Beach, Texas. Note riprap blanket in front. (Courtesy of U.S. Army Corps of Engineers, Wave Dynamics Division)

FIGURE 10.13 Concrete revetment at Del Ray Beach, Florida. (Courtesy of U.S. Army Corps of Engineers, Wave Dynamics Division)

Jetties—or *wharfs* and *piers* as they are called in Britain—as mentioned in the previous section, are structures built from the shore that extend into a body of water into the littoral zone. Their main purpose is to prevent the longshore littoral drift from depositing in a navigable channel and causing frequent need to dredge. Placed at the mouths of rivers, jetties direct and confine the stream or tidal flow and allow the entering water to carry its sediment into deeper water. Such jetties are called *training walls.*

Structurally, *breakwaters* are similar to jetties, but their purpose is to provide protection from incoming waves. They may be connected to the coast, such as the one shown in Figure 10.11, protecting a harbor, or they may be entirely offshore. The segmented breakwater, shown in Figure 10.8, provides partial protection, allowing some near-shore circulation of water, hence littoral sand movement.

Sea walls protect the coast from wave action. After disastrous hurricane damage, the wall shown in Figure 10.12 was built at Galveston Beach, Texas. It is a recurving concrete wall protected in front by a rubble mound. Design of such structures includes the consideration of static and dynamic forces due to waves, the potential effect of pumping action on the sand behind and underneath, and the effect of

FIGURE 10.14 Groins protecting sand beach below protected inlet at Keansburg, New Jersey. (Courtesy of U.S. Army Corps of Engineers, Wave Dynamics Division)

wave runup and overtopping. In the design of major structures hydraulic laboratory studies of a great variety of designs are not uncommon. The dynamic force of waves acting on sea walls depends on the wave type. An unbroken deep-water wave fluctuating at the wall from crest to trough with a certain period results in a fluctuating hydrostatic force. But when the wave is broken, the force transferred to the wall may be much greater. Spilling or plunging breakers hitting a wall result in sudden impulse forces that significantly exceed the hydrostatic force that corresponds to the wave height.

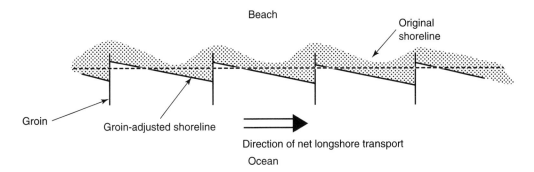

FIGURE 10.15 Effect of groins on the shoreline.

Revetments are protective blankets of a variety of designs, laid directly on the beach, such as the concrete mat at Del Ray Beach, Florida, shown in Figure 10.13. The concrete blocks serve as energy dissipators for the oncoming waves.

Groins are low walls built perpendicular to the shore line at intervals to prevent, at least partially, the littoral movement of sand. Their purpose is to prevent beach erosion. Two are shown in Figure 10.14, at a distance away from the protected inlet at Keansburg, New Jersey. Groins may be made of sheet piles, concrete panels held by steel or precast concrete pilings, or even wood. They do not extend too far into the breaker zone, in order to allow some lateral sand movement. The way groins influence the alignment of the shoreline is depicted in Figure 10.15.

EXAMPLE PROBLEMS

Example 10.1

A wave with a period of 10 sec is propagated at a depth of 600 ft. Find the wave celerity and wavelength.

Solution. Combining Equations 10.1 and 10.5 we get

$$L = \frac{g}{2\pi} T^2 = \frac{32.2}{2\pi}(10)^2 = 512 \text{ ft}$$

$(L/4 = 128) < (d = 600)$, so Equation 10.5 was the correct choice.

$$c_{\text{deep water}} = \frac{g}{2\pi} T = \frac{32.2}{2\pi}(10) = 51.2 \text{ ft/s} \qquad \square$$

Example 10.2

Repeat Example 10.1 for a depth of 6 ft.

Solution. Combining Equations 10.1 and 10.3 we get

$$L = T\sqrt{gd} = 10\sqrt{32.2(6)} = 139 \text{ ft}$$

$(L/20 = 7) > (d = 6)$, so Equation 10.3 was the correct choice.

$$c_{\text{shallow water}} = \sqrt{gd} = \sqrt{32.2(6)} = 13.9 \text{ ft/s} \qquad \square$$

Example 10.3

For the wave in Example 10.1 with a height of 5 ft, determine the rate at which energy per unit crest width is transported toward the shoreline in horsepower per foot of wave crest.

Solution. From Equation 10.11 with a specific weight of seawater of 64 lb/ft³

$$P_{\text{deep water}} = \frac{c}{16}\gamma H^2 = \frac{51.2}{16}(64)(5)^2 = 5120 \frac{\text{ft-lb}}{\text{sec}} \text{ per ft of wave crest}$$

To arrive at horsepower, we divide the result by 550 (Equation 2.18), for an answer of 9.3 HP per foot of wave crest. $\qquad \square$

Example 10.4

For a given wave power, what angle must the breaking waves make with the shoreline to maximize the littoral sediment transport?

Solution. We know that the littoral sediment transport is a function of $(P)_b \sin \alpha_b \cos \alpha_b$ (Equation 10.15). Since the wave power is constant, we must maximize $\sin \alpha_b \cos \alpha_b$. Figure E10.4 is a graph of $\sin \alpha \cos \alpha$ versus α. It shows that the littoral sediment transport will be maximized at 45°.

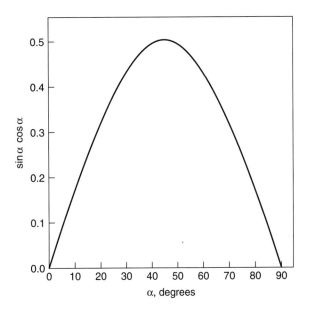

FIGURE E10.4

Example 10.5

Determine the longshore component of the wave power for waves like that described in Example 10.2 with a height of 3 ft. The waves make a 30° angle with the shoreline.

Solution. From Equation 10.12 we have

$$P_{\text{shallow water}} = \frac{c}{8} \gamma H^2 = \frac{13.9}{8} (64)(3)^2 = 1000 \text{ ft-lb/sec per ft of wave crest}$$

From Equation 10.15 the longshore component of the wave power equals the answer to the preceding equation multiplied by $\sin \alpha_b \cos \alpha_b$ or

$$P_l = 1000 \sin (30) \cos (30) = 433 \text{ ft-lb/sec per ft of wave crest}$$

PROBLEMS

10.1 A wave with a period of 10 sec moves through water at a depth of 2 m. What is its wave length and the celerity?

10.2 What is the wave number, wave frequency, and angular frequency for the wave in Problem 10.1?

10.3 A wave with a period of 10 sec moves through water at a depth of 4 m. What is its wave length and the celerity?

10.4 What is the wave power for the wave in Problem 10.1 if the wave is in sea water ($\rho = 1024$ kg/m^3) and has a height of 1.5 m?

10.5 What is the period of forced oscillation for a basin with a depth of 6.5 ft and a width of 10 miles?

10.6 For Example 10.5, how much greater would the longshore component of the wave power be if the waves made a 45° angle with the shore?

10.7 For Problem 10.4, what would be the immersed weight transport rate in dynes/sec if the waves made a 3° angle with the shoreline?

Appendices

APPENDIX A

Appendix A contains the following tables:

Conversion Factors and Important Quantities
Volume Conversion Factors
Discharge Conversion Factors
Rainfall and Runoff Conversion Factors
Conversion Factors for Kinematic Viscosity
Bakhmeteff's Gradually Varied Flow Functions

TABLE A.1 Conversion Factors and Important Quantities

	To Convert Conventional American Unit	Multiply by	To Obtain Metric (SI) Unit
Acceleration	ft/sec^2	0.3048	m/s^2
Area	in.^{2}	645.2	mm^2
	ft^2	0.0929	m^2
	acre	0.4047	hectare (ha)
Density	slug/ft^3	515.4	kg/m^3
Energy	ft/lb	1.356	joule (J) = N/m
	ft/lb	3.77×10^{-7}	kWhr
	BTU = 778 ft/lb	1055	joule (J) = N/m
Flow rate	cfs	0.0283	m^3/s = 10^3 l/s
	mgd = 1.55 cfs	0.0438	m^3/s = 10^3 l/s
	1000 gpm = 2.23 cfs	0.0631	m^3/s = 10^3 l/s

TABLE A.1 *(continued)*

To Convert Conventional American Unit	Multiply by	To Obtain Metric (SI) Unit
Force	lb	4.448	newton (N)
Kinematic viscosity	ft²/sec	0.0929	m²/s
Length	in.	25.4	mm
	ft	0.3048	m
	mile	1.609	km
Mass	slug	14.59	kg
	lb (mass)	453.6	g (mass)
Power	ft-lb/sec	1.356	W = J/s = N·m/s
	hp = 550 ft-lb/sec	745.7	W
Pressure	psi	6895	N/m² (Pa)
	lb/ft²	47.88	N/m²
Specific heat	ft-lb (slug)(°R)	0.1672	N/m (kg)(K)
Velocity	fps	0.3048	m/s
	mph	1.609	km/hr
Viscosity	lb sec/ft²	47.88	N·s/m² = 10 P
Volume	ft³	0.0283	m³
	U.S. gallon = 0.1337 ft³	3.785	liter = 10^{-3} m³

	Conventional American Unit	SI Unit
Acceleration of gravity	32.2 ft/sec²	9.81 m/s²
Density of water (39.4°F, 4°C)	1.94 slug/ft³ = 1.94 lb/ft² sec^{-4}	1000 kg/m³ = g/cm³ or 1.0 Mg/m³
Specific weight of water (50°F, 15°C)	62.4 lb/ft³	~9810 N/m³ or 9.81 kN/m³
Standard sea-level atmosphere	⎰ 14.7 psia ⎱ 29.92 in. Hg ⎰ 33.9 ft water	⎰ 101.32 kN/m², abs ⎱ 760 mm Hg ⎰ 10.33 m water

TABLE A.2 Volume Conversion Factors

	Cubic Inch	U.S. Gallon	Imperial Gallon	Cubic Foot	Cubic Yard	Cubic Meter	Acre-Foot	Second-Foot-Day
Cubic inch	1	0.00433	0.00361	5.79×10^{-4}	2.14×10^{-5}	1.64×10^{-5}	1.33×10^{-8}	6.70×10^{-9}
U.S. gallon	231	1	0.833	0.134	0.00495	0.00379	3.07×10^{-6}	1.55×10^{-6}
Imperial gallon	277	1.20	1	0.161	0.00595	0.00455	3.68×10^{-6}	1.86×10^{-6}
Cubic foot	1728	7.48	6.23	1	0.0370	0.0283	2.30×10^{-5}	1.16×10^{-5}
Cubic yard	46,656	202	168	27	1	0.765	6.20×10^{-4}	3.12×10^{-4}
Cubic meter	61,000	264	220	35.3	1.31	1	8.11×10^{-4}	4.09×10^{-4}
Acre-foot	7.53×10^{7}	3.26×10^{5}	2.71×10^{5}	43,560	1610	1230	1	0.504
Second-foot-day	1.49×10^{8}	6.46×10^{5}	5.38×10^{5}	86,400	3200	2450	1.98	1

TABLE A.3 Discharge Conversion Factors

	Million Gallons/Day	Gallons/Minute	Cubic Feet/Second	Cubic Feet/Day	Acre-Feet/Day	Meter³/Second	Meter³/Hour	Meter³/Day	Millimeters/Year/Kilometer²	Inches/Year/Mile²
Million gallons/day	1	694	1.55	134×10^3	307	43.8×10^{-3}	157.68	3784	1381	20.98
Gallons/minute	1.44×10^{-3}	1	2.227×10^{-3}	193	4.42×10^{-3}	6.31×10^{-5}	0.227	5.448	198.8	0.03
Cubic feet/second	0.646	449	1	86,400	1.98	0.0283	101.88	2445	892.4	13.56
Cubic feet/day	7.48×10^{-6}	5.19×10^{-3}	1.16×10^{-5}	1	2.30×10^{-5}	3.28×10^{-7}	1.18×10^{-3}	0.028	0.0102	1.55×10^{-4}
Acre-feet/day	0.326	226	0.504	43,560	1	0.0143	51.48	1235	450	6.84
Meter³/second	22.8	15,800	35.3	3.05×10^6	70	1	3600	86,400	31,536	479
Meter³/hour	6.34×10^{-3}	4.40	9.81×10^{-3}	847.5	0.0194	2.77×10^{-4}	1	24	8.76	0.133
Meter³/day	2.64×10^{-4}	0.184	4.09×10^{-4}	35.7	8.1×10^{-4}	1.157×10^{-5}	41.6×10^{-3}	1	0.365	5.55×10^{-3}
Millimeters/year/kilometer²	7.24×10^{-4}	5.03×10^{-3}	1.12×10^{-3}	98	2.22×10^{-3}	3.17×10^{-5}	0.114	2.74	1	1.52×10^{-4}
Inches/year/mile²	0.047	33.33	0.074	6451	0.146	2.1×10^{-3}	7.5	180	6579	1

TABLE A.4 Rainfall and Runoff Conversion Factors

	Millimeters/Year	Inches/Year	Liters/Second/Kilometer²	Meter³/Year/Kilometer²	Feet³/Second/Mile²
Millimeters/year	1	0.0394	0.03169	1000	0.3468
Inches/year	25.4	1	8.05×10^{-3}	25,400	8.85
Liters/second/kilometer²	31.55	124.2	1	31,550	10.94
Meter³/year/kilometer²	0.001	0.392×10^{-4}	0.316×10^{-4}	1	3.47×10^{-4}
Feet²/second/mile²	2.88	0.113	0.0914	2880	1

TABLE A.5 Conversion Factors for Kinematic Viscosity

ν	meter²/second	meter²/hour	centimeters²/second	feet²/second	feet³/hour
m²/s	1	3600	1×10^{4}	10.7639	3.875×10^{4}
m²/hr	277.8×10^{-6}	1	2.778	2.99×10^{-3}	10.7639
cm²/s (Stoke)	1×10^{-4}	0.36	1	10.7639×10^{-4}	3.875
ft²/sec	0.092903	334.45	929.03	1	360
ft²/hr	25.806×10^{-6}	0.092903	0.25806	2.77×10^{-3}	1

TABLE A.6   Bakhmeteff's Varied Flow Functions

u, v \ N, J	2.2	2.4	2.6	2.8	3.0	3.2	3.4	3.6	3.8	4.0
0.00	0.000	0.000	0.000	0.000	0.000	0.000	0.000	0.000	0.000	0.000
0.02	0.020	0.020	0.020	0.020	0.020	0.020	0.020	0.020	0.020	0.020
0.04	0.040	0.040	0.040	0.040	0.040	0.040	0.040	0.040	0.040	0.040
0.06	0.060	0.060	0.060	0.060	0.060	0.060	0.060	0.060	0.060	0.060
0.08	0.080	0.080	0.080	0.080	0.080	0.080	0.080	0.080	0.080	0.080
0.10	0.100	0.100	0.100	0.100	0.100	0.100	0.100	0.100	0.100	0.100
0.12	0.120	0.120	0.120	0.120	0.120	0.120	0.120	0.120	0.120	0.120
0.14	0.140	0.140	0.140	0.140	0.140	0.140	0.140	0.140	0.140	0.140
0.16	0.161	0.161	0.160	0.160	0.160	0.160	0.160	0.160	0.160	0.160
0.18	0.181	0.181	0.181	0.180	0.180	0.180	0.180	0.180	0.180	0.180
0.20	0.202	0.201	0.201	0.201	0.200	0.200	0.200	0.200	0.200	0.200
0.22	0.223	0.222	0.221	0.221	0.221	0.220	0.220	0.220	0.220	0.220
0.24	0.244	0.243	0.242	0.241	0.241	0.241	0.240	0.240	0.240	0.240
0.26	0.265	0.263	0.262	0.262	0.261	0.261	0.261	0.260	0.260	0.260
0.28	0.286	0.284	0.283	0.282	0.282	0.281	0.281	0.281	0.280	0.280
0.30	0.307	0.305	0.304	0.303	0.302	0.302	0.301	0.301	0.301	0.300
0.32	0.329	0.326	0.325	0.324	0.323	0.322	0.322	0.321	0.321	0.321
0.34	0.351	0.348	0.346	0.344	0.343	0.343	0.342	0.342	0.341	0.341
0.36	0.372	0.369	0.367	0.366	0.364	0.363	0.363	0.362	0.362	0.361
0.38	0.395	0.392	0.389	0.387	0.385	0.384	0.383	0.383	0.382	0.382
0.40	0.418	0.414	0.411	0.408	0.407	0.405	0.404	0.403	0.403	0.402
0.42	0.442	0.437	0.433	0.430	0.428	0.426	0.425	0.424	0.423	0.423
0.44	0.465	0.460	0.456	0.452	0.450	0.448	0.446	0.445	0.444	0.443
0.46	0.489	0.483	0.479	0.475	0.472	0.470	0.468	0.466	0.465	0.464
0.48	0.514	0.507	0.502	0.497	0.494	0.492	0.489	0.488	0.486	0.485
0.50	0.539	0.531	0.525	0.521	0.517	0.514	0.511	0.509	0.508	0.506
0.52	0.565	0.557	0.550	0.544	0.540	0.536	0.534	0.531	0.529	0.528
0.54	0.592	0.582	0.574	0.568	0.563	0.559	0.556	0.554	0.551	0.550
0.56	0.619	0.608	0.599	0.593	0.587	0.583	0.579	0.576	0.574	0.572
0.58	0.648	0.635	0.626	0.618	0.612	0.607	0.603	0.599	0.596	0.594
0.60	0.676	0.663	0.653	0.644	0.637	0.631	0.627	0.623	0.620	0.617
0.61	0.691	0.678	0.667	0.657	0.650	0.644	0.639	0.635	0.631	0.628
0.62	0.706	0.692	0.680	0.671	0.663	0.657	0.651	0.647	0.643	0.640
0.63	0.722	0.707	0.694	0.684	0.676	0.669	0.664	0.659	0.655	0.652
0.64	0.738	0.722	0.709	0.698	0.690	0.683	0.677	0.672	0.667	0.664
0.65	0.754	0.737	0.724	0.712	0.703	0.696	0.689	0.684	0.680	0.676
0.66	0.771	0.753	0.738	0.727	0.717	0.709	0.703	0.697	0.692	0.688
0.67	0.787	0.769	0.754	0.742	0.731	0.723	0.716	0.710	0.705	0.701
0.68	0.804	0.785	0.769	0.757	0.746	0.737	0.729	0.723	0.718	0.713
0.69	0.822	0.804	0.785	0.772	0.761	0.751	0.743	0.737	0.731	0.726
0.70	0.840	0.819	0.802	0.787	0.776	0.766	0.757	0.750	0.744	0.739
0.71	0.858	0.836	0.819	0.804	0.791	0.781	0.772	0.764	0.758	0.752
0.72	0.878	0.855	0.836	0.820	0.807	0.796	0.786	0.779	0.772	0.766
0.73	0.898	0.874	0.854	0.837	0.823	0.811	0.802	0.793	0.786	0.780
0.74	0.918	0.892	0.868	0.854	0.840	0.827	0.817	0.808	0.800	0.794

TABLE A.6   (*continued*)

u, v \ N, J	2.2	2.4	2.6	2.8	3.0	3.2	3.4	3.6	3.8	4.0
0.75	0.940	0.913	0.890	0.872	0.857	0.844	0.833	0.823	0.815	0.808
0.76	0.961	0.933	0.909	0.890	0.874	0.861	0.849	0.839	0.830	0.823
0.77	0.985	0.954	0.930	0.909	0.892	0.878	0.866	0.855	0.846	0.838
0.78	1.007	0.976	0.950	0.929	0.911	0.896	0.883	0.872	0.862	0.854
0.79	1.031	0.998	0.971	0.949	0.930	0.914	0.901	0.889	0.879	0.870
0.80	1.056	1.022	0.994	0.970	0.950	0.934	0.919	0.907	0.896	0.887
0.81	1.083	1.046	1.017	0.992	0.971	0.954	0.938	0.925	0.914	0.904
0.82	1.110	1.072	1.044	1.015	0.993	0.974	0.958	0.945	0.932	0.922
0.83	1.139	1.099	1.067	1.039	1.016	0.996	0.979	0.965	0.952	0.940
0.84	1.171	1.129	1.094	1.064	1.040	1.019	1.001	0.985	0.972	0.960
0.85	1.201	1.157	1.121	1.091	1.065	1.043	1.024	1.007	0.993	0.980
0.86	1.238	1.192	1.153	1.119	1.092	1.068	1.048	1.031	1.015	1.002
0.87	1.272	1.223	1.182	1.149	1.120	1.095	1.074	1.055	1.039	1.025
0.88	1.314	1.262	1.228	1.181	1.151	1.124	1.101	1.081	1.064	1.049
0.89	1.357	1.302	1.255	1.216	1.183	1.155	1.131	1.110	1.091	1.075
0.90	1.401	1.343	1.294	1.253	1.218	1.189	1.163	1.140	1.120	1.103
0.91	1.452	1.389	1.338	1.294	1.257	1.225	1.197	1.173	1.152	1.133
0.92	1.505	1.438	1.351	1.340	1.300	1.266	1.236	1.210	1.187	1.166
0.93	1.564	1.493	1.435	1.391	1.348	1.311	1.279	1.251	1.226	1.204
0.94	1.645	1.568	1.504	1.449	1.403	1.363	1.328	1.297	1.270	1.246
0.950	1.737	1.652	1.582	1.518	1.467	1.423	1.385	1.352	1.322	1.296
0.960	1.833	1.741	1.665	1.601	1.545	1.497	1.454	1.417	1.385	1.355
0.970	1.969	1.866	1.780	1.707	1.644	1.590	1.543	1.501	1.464	1.431
0.975	2.055	1.945	1.853	1.773	1.707	1.649	1.598	1.554	1.514	1.479
0.980	2.164	2.045	1.946	1.855	1.783	1.720	1.666	1.617	1.575	1.536
0.985	2.294	2.165	2.056	1.959	1.880	1.812	1.752	1.699	1.652	1.610
0.990	2.477	2.333	2.212	2.106	2.017	1.910	1.873	1.814	1.761	1.714
0.995	2.792	2.621	2.478	2.355	2.250	2.159	2.079	2.008	1.945	1.889
0.999	3.523	3.292	3.097	2.931	2.788	2.663	2.554	2.457	2.370	2.293
1.000	∞	∞	∞	∞	∞	∞	∞	∞	∞	∞
1.001	3.317	2.931	2.640	2.399	2.184	2.008	1.856	1.725	1.610	1.508
1.005	2.587	2.266	2.022	1.818	1.649	1.506	1.384	1.279	1.188	1.107
1.010	2.273	1.977	1.757	1.572	1.419	1.291	1.182	1.089	1.007	0.936
1.015	2.090	1.807	1.602	1.428	1.286	1.166	1.065	0.978	0.902	0.836
1.020	1.961	1.711	1.493	1.327	1.191	1.078	0.982	0.900	0.828	0.766
1.03	1.779	1.531	1.340	1.186	1.060	0.955	0.866	0.790	0.725	0.668
1.04	1.651	1.410	1.232	1.086	0.967	0.868	0.785	0.714	0.653	0.600
1.05	1.552	1.334	1.150	1.010	0.896	0.802	0.723	0.656	0.598	0.548
1.06	1.472	1.250	1.082	0.948	0.838	0.748	0.672	0.608	0.553	0.506
1.07	1.404	1.195	1.026	0.896	0.790	0.703	0.630	0.569	0.516	0.471
1.08	1.346	1.139	0.978	0.851	0.749	0.665	0.595	0.535	0.485	0.441
1.09	1.295	1.089	0.935	0.812	0.713	0.631	0.563	0.506	0.457	0.415
1.10	1.250	1.050	0.897	0.777	0.681	0.601	0.536	0.480	0.433	0.392
1.11	1.209	1.014	0.864	0.746	0.652	0.575	0.511	0.457	0.411	0.372
1.12	1.172	0.981	0.833	0.718	0.626	0.551	0.488	0.436	0.392	0.354

TABLE A.6 *(continued)*

u, v $\diagdown$ N, J	2.2	2.4	2.6	2.8	3.0	3.2	3.4	3.6	3.8	4.0
1.13	1.138	0.950	0.805	0.692	0.602	0.520	0.468	0.417	0.371	0.337
1.14	1.107	0.921	0.780	0.669	0.581	0.509	0.450	0.400	0.358	0.322
1.15	1.078	0.892	0.756	0.647	0.561	0.490	0.432	0.384	0.343	0.308
1.16	1.052	0.870	0.734	0.627	0.542	0.473	0.417	0.369	0.329	0.295
1.17	1.027	0.850	0.713	0.608	0.525	0.458	0.402	0.356	0.317	0.283
1.18	1.003	0.825	0.694	0.591	0.509	0.443	0.388	0.343	0.305	0.272
1.19	0.981	0.810	0.676	0.574	0.494	0.429	0.375	0.331	0.294	0.262
1.20	0.960	0.787	0.659	0.559	0.480	0.416	0.363	0.320	0.283	0.252
1.22	0.922	0.755	0.628	0.531	0.454	0.392	0.344	0.299	0.264	0.235
1.24	0.887	0.725	0.600	0.505	0.431	0.371	0.322	0.281	0.248	0.219
1.26	0.855	0.692	0.574	0.482	0.410	0.351	0.304	0.265	0.233	0.205
1.28	0.827	0.666	0.551	0.461	0.391	0.334	0.288	0.250	0.219	0.193
1.30	0.800	0.644	0.530	0.442	0.373	0.318	0.274	0.237	0.207	0.181
1.32	0.775	0.625	0.510	0.424	0.357	0.304	0.260	0.225	0.196	0.171
1.34	0.752	0.605	0.492	0.408	0.342	0.290	0.248	0.214	0.185	0.162
1.36	0.731	0.588	0.475	0.393	0.329	0.278	0.237	0.204	0.176	0.153
1.38	0.711	0.567	0.459	0.378	0.316	0.266	0.226	0.194	0.167	0.145
1.40	0.692	0.548	0.444	0.365	0.304	0.256	0.217	0.185	0.159	0.138
1.42	0.674	0.533	0.431	0.353	0.293	0.246	0.208	0.177	0.152	0.131
1.44	0.658	0.517	0.417	0.341	0.282	0.236	0.199	0.169	0.145	0.125
1.46	0.642	0.505	0.405	0.330	0.273	0.227	0.191	0.162	0.139	0.119
1.48	0.627	0.493	0.394	0.320	0.263	0.219	0.184	0.156	0.133	0.113
1.50	0.613	0.480	0.383	0.310	0.255	0.211	0.177	0.149	0.127	0.108
1.55	0.580	0.451	0.358	0.288	0.235	0.194	0.161	0.135	0.114	0.097
1.60	0.551	0.425	0.335	0.269	0.218	0.179	0.148	0.123	0.103	0.087
1.65	0.525	0.402	0.316	0.251	0.203	0.165	0.136	0.113	0.094	0.079
1.70	0.501	0.381	0.298	0.236	0.189	0.153	0.125	0.103	0.086	0.072
1.75	0.480	0.362	0.282	0.222	0.177	0.143	0.116	0.095	0.079	0.065
1.80	0.460	0.349	0.267	0.209	0.166	0.133	0.108	0.088	0.072	0.060
1.85	0.442	0.332	0.254	0.198	0.156	0.125	0.100	0.082	0.067	0.055
1.90	0.425	0.315	0.242	0.188	0.147	0.117	0.094	0.076	0.062	0.050
1.95	0.409	0.304	0.231	0.178	0.139	0.110	0.088	0.070	0.057	0.046
2.00	0.395	0.292	0.221	0.169	0.132	0.104	0.082	0.066	0.053	0.043
2.10	0.369	0.273	0.202	0.154	0.119	0.092	0.073	0.058	0.046	0.037
2.20	0.346	0.253	0.186	0.141	0.107	0.083	0.065	0.051	0.040	0.032
2.3	0.326	0.235	0.173	0.129	0.098	0.075	0.058	0.045	0.035	0.028
2.4	0.308	0.220	0.160	0.119	0.089	0.068	0.052	0.040	0.031	0.024
2.5	0.292	0.207	0.150	0.110	0.082	0.062	0.047	0.036	0.028	0.022
2.6	0.277	0.197	0.140	0.102	0.076	0.057	0.043	0.033	0.025	0.019
2.7	0.264	0.188	0.131	0.095	0.070	0.052	0.039	0.029	0.022	0.017
2.8	0.252	0.176	0.124	0.089	0.065	0.048	0.036	0.027	0.020	0.015
2.9	0.241	0.166	0.117	0.083	0.060	0.044	0.033	0.024	0.018	0.014
3.0	0.230	0.159	0.110	0.078	0.056	0.041	0.030	0.022	0.017	0.012
3.5	0.190	0.126	0.085	0.059	0.041	0.029	0.021	0.015	0.011	0.008
4.0	0.161	0.104	0.069	0.046	0.031	0.022	0.015	0.010	0.007	0.005

TABLE A.6   (*continued*)

u, v \ N, J	2.2	2.4	2.6	2.8	3.0	3.2	3.4	3.6	3.8	4.0
4.5	0.139	0.087	0.057	0.037	0.025	0.017	0.011	0.008	0.005	0.004
5.0	0.122	0.076	0.048	0.031	0.020	0.013	0.009	0.006	0.004	0.003
6.0	0.098	0.060	0.036	0.022	0.014	0.009	0.006	0.004	0.002	0.002
7.0	0.081	0.048	0.028	0.017	0.010	0.006	0.004	0.002	0.002	0.001
8.0	0.069	0.040	0.022	0.013	0.008	0.005	0.003	0.002	0.001	0.001
9.0	0.060	0.034	0.019	0.011	0.006	0.004	0.002	0.001	0.001	0.000
10.0	0.053	0.028	0.016	0.009	0.005	0.003	0.002	0.001	0.001	0.000
20.0	0.023	0.018	0.011	0.006	0.002	0.001	0.001	0.000	0.000	0.000

TABLE A.6 (*continued*)

u, v \ N, J	4.2	4.6	5.0	5.4	5.8	6.2	6.6	7.0	7.4	7.8
0.00	0.000	0.000	0.000	0.000	0.000	0.000	0.000	0.000	0.000	0.000
0.02	0.020	0.020	0.020	0.020	0.020	0.020	0.020	0.020	0.020	0.020
0.04	0.040	0.040	0.040	0.040	0.040	0.040	0.040	0.040	0.040	0.040
0.06	0.060	0.060	0.060	0.060	0.060	0.060	0.060	0.060	0.060	0.060
0.08	0.080	0.080	0.080	0.080	0.080	0.080	0.080	0.080	0.080	0.080
0.10	0.100	0.100	0.100	0.100	0.100	0.100	0.100	0.100	0.100	0.100
0.12	0.120	0.120	0.120	0.120	0.120	0.120	0.120	0.120	0.120	0.120
0.14	0.140	0.140	0.140	0.140	0.140	0.140	0.140	0.140	0.140	0.140
0.16	0.160	0.160	0.160	0.160	0.160	0.160	0.160	0.160	0.160	0.160
0.18	0.180	0.180	0.180	0.180	0.180	0.180	0.180	0.180	0.180	0.180
0.20	0.200	0.200	0.200	0.200	0.200	0.200	0.200	0.200	0.200	0.200
0.22	0.220	0.220	0.220	0.220	0.220	0.220	0.220	0.220	0.220	0.220
0.24	0.240	0.240	0.240	0.240	0.240	0.240	0.240	0.240	0.240	0.240
0.26	0.260	0.260	0.260	0.260	0.260	0.260	0.260	0.260	0.260	0.260
0.28	0.280	0.280	0.280	0.280	0.280	0.280	0.280	0.280	0.280	0.280
0.30	0.300	0.300	0.300	0.300	0.300	0.300	0.300	0.300	0.300	0.300
0.32	0.321	0.320	0.320	0.320	0.320	0.320	0.320	0.320	0.320	0.320
0.34	0.341	0.340	0.340	0.340	0.340	0.340	0.340	0.340	0.340	0.340
0.36	0.361	0.361	0.360	0.360	0.360	0.360	0.360	0.360	0.360	0.360
0.38	0.381	0.381	0.381	0.380	0.380	0.380	0.380	0.380	0.380	0.380
0.40	0.402	0.401	0.401	0.400	0.400	0.400	0.400	0.400	0.400	0.400
0.42	0.422	0.421	0.421	0.421	0.420	0.420	0.420	0.420	0.420	0.420
0.44	0.443	0.442	0.441	0.441	0.441	0.440	0.440	0.440	0.440	0.440
0.46	0.463	0.462	0.462	0.461	0.461	0.461	0.460	0.460	0.460	0.460
0.48	0.484	0.483	0.482	0.481	0.481	0.481	0.480	0.480	0.480	0.480
0.50	0.505	0.504	0.503	0.502	0.501	0.501	0.501	0.500	0.500	0.500
0.52	0.527	0.525	0.523	0.522	0.522	0.521	0.521	0.521	0.520	0.520
0.54	0.548	0.546	0.544	0.543	0.542	0.542	0.541	0.541	0.541	0.541
0.56	0.570	0.567	0.565	0.564	0.563	0.562	0.562	0.561	0.561	0.561
0.58	0.592	0.589	0.587	0.585	0.583	0.583	0.582	0.582	0.581	0.581
0.60	0.614	0.611	0.608	0.606	0.605	0.604	0.603	0.602	0.602	0.601
0.61	0.626	0.622	0.619	0.617	0.615	0.614	0.613	0.612	0.612	0.611
0.62	0.637	0.633	0.630	0.628	0.626	0.625	0.624	0.623	0.622	0.622
0.63	0.649	0.644	0.641	0.638	0.636	0.635	0.634	0.633	0.632	0.632
0.64	0.661	0.656	0.652	0.649	0.647	0.646	0.645	0.644	0.643	0.642
0.65	0.673	0.667	0.663	0.660	0.658	0.656	0.655	0.654	0.653	0.653
0.66	0.685	0.679	0.675	0.672	0.669	0.667	0.666	0.665	0.664	0.663
0.67	0.697	0.691	0.686	0.683	0.680	0.678	0.676	0.675	0.674	0.673
0.68	0.709	0.703	0.698	0.694	0.691	0.689	0.687	0.686	0.685	0.684
0.69	0.722	0.715	0.710	0.706	0.703	0.700	0.698	0.696	0.695	0.694
0.70	0.735	0.727	0.722	0.717	0.714	0.712	0.710	0.708	0.706	0.705
0.71	0.748	0.740	0.734	0.729	0.726	0.723	0.721	0.719	0.717	0.716
0.72	0.761	0.752	0.746	0.741	0.737	0.734	0.732	0.730	0.728	0.727
0.73	0.774	0.765	0.759	0.753	0.749	0.746	0.743	0.741	0.739	0.737
0.74	0.788	0.779	0.771	0.766	0.761	0.757	0.754	0.752	0.750	0.748

TABLE A.6   (*continued*)

u, v \ N, J	4.2	4.6	5.0	5.4	5.8	6.2	6.6	7.0	7.4	7.8
0.75	0.802	0.792	0.784	0.778	0.773	0.769	0.766	0.763	0.761	0.759
0.76	0.817	0.806	0.798	0.791	0.786	0.782	0.778	0.775	0.773	0.771
0.77	0.831	0.820	0.811	0.804	0.798	0.794	0.790	0.787	0.784	0.782
0.78	0.847	0.834	0.825	0.817	0.811	0.806	0.802	0.799	0.796	0.794
0.79	0.862	0.849	0.839	0.831	0.824	0.819	0.815	0.811	0.808	0.805
0.80	0.878	0.865	0.854	0.845	0.838	0.832	0.828	0.823	0.820	0.818
0.81	0.895	0.881	0.869	0.860	0.852	0.846	0.841	0.836	0.833	0.830
0.82	0.913	0.897	0.885	0.875	0.866	0.860	0.854	0.850	0.846	0.842
0.83	0.931	0.914	0.901	0.890	0.881	0.874	0.868	0.863	0.859	0.855
0.84	0.949	0.932	0.918	0.906	0.897	0.889	0.882	0.877	0.872	0.868
0.85	0.969	0.950	0.935	0.923	0.912	0.905	0.898	0.891	0.887	0.882
0.86	0.990	0.970	0.954	0.940	0.930	0.921	0.913	0.906	0.901	0.896
0.87	1.012	0.990	0.973	0.959	0.947	0.937	0.929	0.922	0.916	0.911
0.88	1.035	1.012	0.994	0.978	0.966	0.955	0.946	0.938	0.932	0.927
0.89	1.060	1.035	1.015	0.999	0.986	0.974	0.964	0.956	0.949	0.943
0.90	1.087	1.060	1.039	1.021	1.007	0.994	0.984	0.974	0.967	0.960
0.91	1.116	1.088	1.064	1.045	1.029	1.016	1.003	0.995	0.986	0.979
0.92	1.148	1.117	1.092	1.072	1.054	1.039	1.027	1.016	1.006	0.999
0.93	1.184	1.151	1.123	1.101	1.081	1.065	1.050	1.040	1.029	1.021
0.94	1.225	1.188	1.158	1.134	1.113	1.095	1.080	1.066	1.054	1.044
0.950	1.272	1.232	1.199	1.172	1.148	1.128	1.111	1.097	1.084	1.073
0.960	1.329	1.285	1.248	1.217	1.188	1.167	1.149	1.133	1.119	1.106
0.970	1.402	1.351	1.310	1.275	1.246	1.219	1.197	1.179	1.162	1.148
0.975	1.447	1.393	1.348	1.311	1.280	1.250	1.227	1.207	1.190	1.173
0.980	1.502	1.443	1.395	1.354	1.339	1.288	1.262	1.241	1.221	1.204
0.985	1.573	1.508	1.454	1.409	1.372	1.337	1.309	1.284	1.263	1.243
0.990	1.671	1.598	1.537	1.487	1.444	1.404	1.373	1.344	1.319	1.297
0.995	1.838	1.751	1.678	1.617	1.565	1.519	1.479	1.451	1.419	1.388
0.999	2.223	2.102	2.002	1.917	1.845	1.780	1.725	1.678	1.635	1.596
1.000	∞	∞	∞	∞	∞	∞	∞	∞	∞	∞
1.001	1.417	1.264	1.138	1.033	0.951	0.870	0.803	0.746	0.697	0.651
1.005	1.036	0.915	0.817	0.737	0.669	0.612	0.553	0.526	0.481	0.447
1.010	0.873	0.766	0.681	0.610	0.551	0.502	0.459	0.422	0.389	0.360
1.015	0.778	0.680	0.602	0.537	0.483	0.440	0.399	0.366	0.336	0.310
1.02	0.711	0.620	0.546	0.486	0.436	0.394	0.358	0.327	0.300	0.276
1.03	0.618	0.535	0.469	0.415	0.370	0.333	0.300	0.272	0.250	0.228
1.04	0.554	0.477	0.415	0.365	0.324	0.290	0.262	0.236	0.215	0.195
1.05	0.504	0.432	0.374	0.328	0.289	0.259	0.231	0.208	0.189	0.174
1.06	0.464	0.396	0.342	0.298	0.262	0.233	0.209	0.187	0.170	0.154
1.07	0.431	0.366	0.315	0.273	0.239	0.212	0.191	0.168	0.154	0.136
1.08	0.403	0.341	0.292	0.252	0.220	0.194	0.172	0.153	0.137	0.123
1.09	0.379	0.319	0.272	0.234	0.204	0.179	0.158	0.140	0.125	0.112
1.10	0.357	0.299	0.254	0.218	0.189	0.165	0.146	0.129	0.114	0.102
1.11	0.338	0.282	0.239	0.201	0.176	0.154	0.135	0.119	0.105	0.094
1.12	0.321	0.267	0.225	0.192	0.165	0.143	0.125	0.110	0.097	0.080

TABLE A.6 (*continued*)

u, v N, J	4.2	4.6	5.0	5.4	5.8	6.2	6.6	7.0	7.4	7.8
1.13	0.305	0.253	0.212	0.181	0.155	0.135	0.117	0.102	0.099	0.080
1.14	0.291	0.240	0.201	0.170	0.146	0.126	0.109	0.095	0.084	0.074
1.15	0.278	0.229	0.191	0.161	0.137	0.118	0.102	0.089	0.078	0.068
1.16	0.266	0.218	0.181	0.153	0.130	0.111	0.096	0.084	0.072	0.064
1.17	0.255	0.208	0.173	0.145	0.123	0.105	0.090	0.078	0.068	0.060
1.18	0.244	0.199	0.165	0.138	0.116	0.099	0.085	0.073	0.063	0.055
1.19	0.235	0.191	0.157	0.131	0.110	0.094	0.080	0.068	0.059	0.051
1.20	0.226	0.183	0.150	0.125	0.105	0.088	0.076	0.064	0.056	0.048
1.22	0.209	0.168	0.138	0.114	0.095	0.080	0.068	0.057	0.049	0.042
1.24	0.195	0.156	0.127	0.104	0.086	0.072	0.060	0.051	0.044	0.038
1.26	0.182	0.145	0.117	0.095	0.079	0.065	0.055	0.046	0.039	0.033
1.28	0.170	0.135	0.108	0.088	0.072	0.060	0.050	0.041	0.035	0.030
1.30	0.160	0.126	0.100	0.081	0.066	0.054	0.045	0.037	0.031	0.026
1.32	0.150	0.118	0.093	0.075	0.061	0.050	0.041	0.034	0.028	0.024
1.34	0.142	0.110	0.087	0.069	0.056	0.045	0.037	0.030	0.025	0.021
1.36	0.134	0.103	0.081	0.064	0.052	0.042	0.034	0.028	0.023	0.019
1.38	0.127	0.097	0.076	0.060	0.048	0.038	0.032	0.026	0.021	0.017
1.40	0.120	0.092	0.074	0.056	0.044	0.036	0.028	0.023	0.019	0.016
1.42	0.114	0.087	0.067	0.052	0.041	0.033	0.026	0.021	0.017	0.014
1.44	0.108	0.082	0.063	0.049	0.038	0.030	0.024	0.019	0.016	0.013
1.46	0.103	0.077	0.059	0.046	0.036	0.028	0.022	0.018	0.014	0.012
1.48	0.098	0.073	0.056	0.043	0.033	0.026	0.021	0.017	0.013	0.010
1.50	0.093	0.069	0.053	0.040	0.031	0.024	0.020	0.015	0.012	0.009
1.55	0.083	0.061	0.046	0.035	0.026	0.020	0.016	0.012	0.010	0.008
1.60	0.074	0.054	0.040	0.030	0.023	0.017	0.013	0.010	0.008	0.006
1.65	0.067	0.048	0.035	0.026	0.019	0.014	0.011	0.008	0.006	0.005
1.70	0.060	0.043	0.031	0.023	0.016	0.012	0.009	0.007	0.005	0.004
1.75	0.054	0.038	0.027	0.020	0.014	0.010	0.008	0.006	0.004	0.003
1.80	0.019	0.034	0.024	0.017	0.012	0.009	0.007	0.005	0.004	0.003
1.85	0.045	0.031	0.022	0.015	0.011	0.008	0.006	0.004	0.003	0.002
1.90	0.041	0.028	0.020	0.014	0.010	0.007	0.005	0.004	0.003	0.002
1.95	0.038	0.026	0.018	0.012	0.008	0.006	0.004	0.003	0.002	0.002
2.00	0.035	0.023	0.016	0.011	0.007	0.005	0.004	0.003	0.002	0.001
2.10	0.030	0.019	0.013	0.009	0.006	0.004	0.003	0.002	0.001	0.001
2.20	0.025	0.016	0.011	0.007	0.005	0.004	0.002	0.001	0.001	0.001
2.3	0.022	0.014	0.009	0.006	0.004	0.003	0.002	0.001	0.001	0.001
2.4	0.019	0.012	0.008	0.005	0.003	0.002	0.001	0.001	0.001	0.001
2.5	0.017	0.010	0.006	0.004	0.003	0.002	0.001	0.001	0.000	0.000
2.6	0.015	0.009	0.005	0.003	0.002	0.001	0.001	0.001	0.000	0.000
2.7	0.013	0.008	0.005	0.003	0.002	0.001	0.001	0.000	0.000	0.000
2.8	0.012	0.007	0.004	0.002	0.001	0.001	0.001	0.000	0.000	0.000
2.9	0.010	0.006	0.004	0.002	0.001	0.001	0.000	0.000	0.000	0.000
3.0	0.009	0.005	0.003	0.002	0.001	0.001	0.000	0.000	0.000	0.000
3.5	0.006	0.003	0.002	0.001	0.001	0.000	0.000	0.000	0.000	0.000
4.0	0.004	0.002	0.001	0.000	0.000	0.000	0.000	0.000	0.000	0.000

TABLE A.6   (*continued*)

u, v \ N, J	4.2	4.6	5.0	5.4	5.8	6.2	6.6	7.0	7.4	7.8
4.5	0.003	0.001	0.001	0.000	0.000	0.000	0.000	0.000	0.000	0.000
5.0	0.002	0.001	0.000	0.000	0.000	0.000	0.000	0.000	0.000	0.000
6.0	0.001	0.000	0.000	0.000	0.000	0.000	0.000	0.000	0.000	0.000
7.0	0.001	0.000	0.000	0.000	0.000	0.000	0.000	0.000	0.000	0.000
8.0	0.000	0.000	0.000	0.000	0.000	0.000	0.000	0.000	0.000	0.000
9.0	0.000	0.000	0.000	0.000	0.000	0.000	0.000	0.000	0.000	0.000
10.0	0.000	0.000	0.000	0.000	0.000	0.000	0.000	0.000	0.000	0.000
20.0	0.000	0.000	0.000	0.000	0.000	0.000	0.000	0.000	0.000	0.000

TABLE A.6　(*continued*)

u, v \ N, J	8.2	8.6	9.0	9.4	9.8
0.85	0.878	0.875	0.873	0.870	0.868
0.86	0.892	0.889	0.886	0.883	0.881
0.87	0.907	0.903	0.900	0.897	0.894
0.88	0.921	0.918	0.914	0.911	0.908
0.89	0.937	0.933	0.929	0.925	0.922
0.90	0.954	0.949	0.944	0.940	0.937
0.91	0.972	0.967	0.961	0.957	0.953
0.92	0.991	0.986	0.980	0.975	0.970
0.93	1.012	1.006	0.999	0.994	0.989
0.94	1.036	1.029	1.022	1.016	1.010
0.950	1.062	1.055	1.047	1.040	1.033
0.960	1.097	1.085	1.074	1.063	1.053
0.970	1.136	1.124	1.112	1.100	1.087
0.975	1.157	1.147	1.134	1.122	1.108
0.980	1.187	1.175	1.160	1.150	1.132
0.985	1.224	1.210	1.196	1.183	1.165
0.990	1.275	1.260	1.243	1.228	1.208
0.995	1.363	1.342	1.320	1.302	1.280
0.999	1.560	1.530	1.500	1.476	1.447
1.000	∞	∞	∞	∞	∞
1.001	0.614	0.577	0.546	0.519	0.494
1.005	0.420	0.391	0.368	0.350	0.331
1.010	0.337	0.313	0.294	0.278	0.262
1.015	0.289	0.269	0.255	0.237	0.223
1.020	0.257	0.237	0.221	0.209	0.196
1.03	0.212	0.195	0.181	0.170	0.159
1.04	0.173	0.165	0.152	0.143	0.134
1.05	0.158	0.143	0.132	0.124	0.115
1.06	0.140	0.127	0.116	0.106	0.098
1.07	0.123	0.112	0.102	0.094	0.086
1.08	0.111	0.101	0.092	0.084	0.077
1.09	0.101	0.091	0.082	0.075	0.069
1.10	0.092	0.083	0.074	0.067	0.062
1.11	0.084	0.075	0.067	0.060	0.055
1.12	0.077	0.069	0.062	0.055	0.050
1.13	0.071	0.063	0.056	0.050	0.045
1.14	0.065	0.058	0.052	0.046	0.041
1.15	0.061	0.054	0.048	0.043	0.038
1.16	0.056	0.050	0.045	0.040	0.035
1.17	0.052	0.046	0.041	0.036	0.032
1.18	0.048	0.042	0.037	0.033	0.029
1.19	0.045	0.039	0.034	0.030	0.027
1.20	0.043	0.037	0.032	0.028	0.025
1.22	0.037	0.032	0.028	0.024	0.021
1.24	0.032	0.028	0.024	0.021	0.018

TABLE A.6   (*continued*)

u, v \ N, J	8.2	8.6	9.0	9.4	9.8
1.26	0.028	0.024	0.021	0.018	0.016
1.28	0.025	0.021	0.018	0.016	0.014
1.30	0.022	0.019	0.016	0.014	0.012
1.32	0.020	0.017	0.014	0.012	0.010
1.34	0.018	0.015	0.012	0.010	0.009
1.36	0.016	0.013	0.011	0.009	0.008
1.38	0.014	0.012	0.010	0.008	0.007
1.40	0.013	0.011	0.009	0.007	0.006
1.42	0.011	0.009	0.008	0.006	0.005
1.44	0.010	0.008	0.007	0.006	0.005
1.46	0.009	0.008	0.006	0.005	0.004
1.48	0.009	0.007	0.005	0.004	0.004
1.50	0.008	0.006	0.005	0.004	0.003
1.55	0.006	0.005	0.004	0.003	0.003
1.60	0.005	0.004	0.003	0.002	0.002
1.65	0.004	0.003	0.002	0.002	0.001
1.70	0.003	0.002	0.002	0.001	0.001
1.75	0.002	0.002	0.002	0.001	0.001
1.80	0.002	0.001	0.001	0.001	0.001
1.85	0.002	0.001	0.001	0.001	0.001
1.90	0.001	0.001	0.001	0.001	0.000
1.95	0.001	0.001	0.001	0.000	0.000
2.00	0.001	0.001	0.000	0.000	0.000
2.10	0.001	0.000	0.000	0.000	0.000
2.20	0.000	0.000	0.000	0.000	0.000
0.00	0.000	0.000	0.000	0.000	0.000
0.02	0.020	0.020	0.020	0.020	0.020
0.04	0.040	0.040	0.040	0.040	0.040
0.06	0.060	0.060	0.060	0.060	0.060
0.08	0.080	0.080	0.080	0.080	0.080
0.10	0.100	0.100	0.100	0.100	0.100
0.12	0.120	0.120	0.120	0.120	0.120
0.14	0.140	0.140	0.140	0.140	0.140
0.16	0.160	0.160	0.160	0.160	0.160
0.18	0.180	0.180	0.180	0.180	0.180
0.20	0.200	0.200	0.200	0.200	0.200
0.22	0.220	0.220	0.220	0.220	0.220
0.24	0.240	0.240	0.240	0.240	0.240
0.26	0.260	0.260	0.260	0.260	0.260
0.28	0.280	0.280	0.280	0.280	0.280
0.30	0.300	0.300	0.300	0.300	0.300
0.32	0.320	0.320	0.320	0.320	0.320
0.34	0.340	0.340	0.340	0.340	0.340
0.36	0.360	0.360	0.360	0.360	0.360
0.38	0.380	0.380	0.380	0.380	0.380

TABLE A.6   (*continued*)

u, v	N, J 8.2	8.6	9.0	9.4	9.8
0.40	0.400	0.400	0.400	0.400	0.400
0.42	0.420	0.420	0.420	0.420	0.420
0.44	0.440	0.440	0.440	0.440	0.440
0.46	0.460	0.460	0.460	0.460	0.460
0.48	0.480	0.480	0.480	0.480	0.480
0.50	0.500	0.500	0.500	0.500	0.500
0.52	0.520	0.520	0.520	0.520	0.520
0.54	0.540	0.540	0.540	0.540	0.540
0.56	0.561	0.560	0.560	0.560	0.560
0.58	0.581	0.581	0.580	0.580	0.580
0.60	0.601	0.601	0.601	0.600	0.600
0.61	0.611	0.611	0.611	0.611	0.610
0.62	0.621	0.621	0.621	0.621	0.621
0.63	0.632	0.631	0.631	0.631	0.631
0.64	0.642	0.641	0.641	0.641	0.641
0.65	0.652	0.652	0.651	0.651	0.651
0.66	0.662	0.662	0.662	0.661	0.661
0.67	0.673	0.672	0.672	0.672	0.671
0.68	0.683	0.683	0.682	0.682	0.681
0.69	0.694	0.693	0.692	0.692	0.692
0.70	0.704	0.704	0.703	0.702	0.702
0.71	0.715	0.714	0.713	0.713	0.712
0.72	0.726	0.725	0.724	0.723	0.723
0.73	0.736	0.735	0.734	0.734	0.733
0.74	0.747	0.746	0.745	0.744	0.744
0.75	0.758	0.757	0.756	0.755	0.754
0.76	0.769	0.768	0.767	0.766	0.765
0.77	0.780	0.779	0.778	0.777	0.776
0.78	0.792	0.790	0.789	0.788	0.787
0.79	0.804	0.802	0.800	0.799	0.798
0.80	0.815	0.813	0.811	0.810	0.809
0.81	0.827	0.825	0.823	0.822	0.820
0.82	0.839	0.837	0.835	0.833	0.831
0.83	0.852	0.849	0.847	0.845	0.844
0.84	0.865	0.862	0.860	0.858	0.856

(Reproduced by permission from Ven Te Chow: *Integrating the Equation of Gradually Varied Flow, Proceedings American Society of Civil Engineers,* Vol. 81, Paper 838, pp. 1–32, November 1955)

APPENDIX B   SOLUTIONS TO PROBLEMS

Chapter 1

1.1 4.16×10^6 BTU

1.2 -12.89 psi at $20°$

1.3 1.29 lb less at the higher temperature

1.4 42 percent reduction of the original height

1.5 $+0.043$ m³

1.6 310 psi

1.8 3.79×10^{-6} ft²/sec

1.9 Increases 370 percent

1.10 0.15 percent

1.11 Increases 424 percent

1.12 1.089×10^7 J

1.13 0.64 m/s

1.14 69.98 lb

1.15 Decreases 5 percent

1.16 A

1.17 D

1.18 A

1.19 E

1.20 A

1.21 D

1.22 E

1.23 C

1.24 A

1.25 B

1.26 E

1.27 B

1.28 A

1.29 D

1.30 A

1.31 A

1.32 E

1.33 C

1.34 A

1.35 D

Chapter 2

2.1 $\mathbf{M} = 96$ m³, $\mathbf{I_c} = 597$ m⁴

2.2 562,000 lb

2.3 154 N

2.4 Force = 9000 lb

2.5 Force = 2760 lb

2.6 $y_a = 11.3$ ft

2.7 $y_a = 1.47$ m

2.8 $e = 0.082$ ft

2.9 Resultant horizontal force = 2164 lb/ft

2.10 Thickness = 2.9 ft

2.11 13 drums

2.12 $Q = 39$ cfs

2.13 31.2 min

2.14 15.4 lb

2.15 102 N

2.16 $\Delta M = 2.33$ N

2.17 9.8 fps

2.18 136.1 HP

2.19 Total energy at both locations = 8.15 m

2.20 7.3 HP

2.21 Maximum elevation of hill = 120.6 m

2.22 6.4 m

2.23 319.6 ft of head

2.24 Turbulent

2.25 Sand grain has a terminal velocity 5.6 times greater in the water

2.26 540 N

2.27 C

2.28 A

2.29 A

2.30 E

2.31 B

2.32 D

2.33 D

2.34 A

2.35 A

2.36 D

2.37 A

2.38 C

2.39 B

2.40 D

2.41 C

2.42 D

2.43 E

2.44 B

2.45 D

2.46 C

2.47 B

2.48 A

2.49 E

2.50 D

2.51 A

Chapter 3

3.1 4.73 fps

3.2 $K = 0.0324$ ft^4/(sec-lb$^{0.5}$)

3.3 15.6 ft assuming the measured fluid is water

3.4 1.42 ft

3.5 20.4 psf

3.6 Yes

3.7 0.60 m/s

3.8 14.4 cfs

3.9 13.2 cfs

3.10 $(\rho g L^2)/(\mu v)$

3.11 $(\mu v)/(\Delta p L)$

3.12 3.1×10^5

3.13 0.43

3.14 One dimensionless variable is $\mu/(\rho v d)$

3.15 $P_p = 109 \, P_m$

3.16 $T_R = L_R \, \rho_R^{0.5} \, K_R^{-0.5}$

3.17 $L_R = \sigma_R \, v_R^{-2} \, \rho_R^{-1}$

3.18 Diameter of model culvert = 4 in.

3.19 μ_R (required) = 0.1128; glycerin is unsatisfactory

3.20 C

3.21 A

3.22 D

3.23 C

3.24 A

3.25 D

3.26 C

3.27 A

3.28 C

3.29 D

3.30 B

3.31 B

3.32 D

3.33 D

3.34 A

3.35 E

Chapter 4

4.1 6 in.

4.2 2.6 m

4.3 0.021

4.4 0.18 m

4.5 6 in.

4.6 0.03 m^3/s

4.7 0.063 ft/ft

4.8 0.048 ft/ft

4.9 Total head loss = 1.66 m

4.10 $Q_1 = 2.287$ cfs

4.11 1340 psf

4.12 6.80×10^6 g

4.13 Head loss = 0.03 m

4.14 E

4.15 D

4.16 B

4.17 C

4.18 D

4.19 C

4.20 B

4.21 B	**4.27** D
4.22 A	**4.28** E
4.23 A	**4.29** A
4.24 E	**4.30** D
4.25 A	**4.31** A
4.26 E	

Chapter 5

5.1 Specific speed = 1410	**5.16** E
5.2 4.48 ft-lb	**5.17** E
5.3 For case A the brake HP is 52	**5.18** B
5.4 Change in head is 65 ft	**5.19** D
5.5 50.8 kW	**5.20** A
5.6 New discharge = 1250 gpm	**5.21** C
5.7 New torque = 14.5 ft-lb	**5.22** E
5.8 −17.9 ft	**5.23** E
5.10 Brake HP = 49	**5.24** B
5.11 In parallel operation the two pumps will deliver 3760 gpm at 100 ft of head	**5.25** C
	5.26 E
5.12 1200 gpm	**5.27** D
5.13 3900 gpm	**5.28** C
5.14 23.3 percent of the year	**5.29** B
5.15 A	**5.30** A

Chapter 6

6.1 18.7 lb/ft^3	**6.15** D
6.2 5.7 gpm	**6.16** E
6.3 1.3 × 10^{-11} m^2	**6.17** D
6.4 4.8 cm^3/s	**6.18** A
6.5 Depth = 25.5 ft	**6.19** E
6.6 2.9 × 10^{-5} cfs/ft	**6.20** E
6.7 998 psf	**6.21** E
6.8 Discharge = 5 × 10^{-6} m^3/s	**6.22** D
6.9 2.5 × 10^{-4} m^3/s	**6.23** D
6.10 1.15 m	**6.24** E
6.11 81.8 gpm	**6.25** D
6.12 0.034 cm/s	**6.26** C
6.13 6.3 × 10^{-8} m^3/s for a 100 m section	**6.27** C
6.14 9.2 m/day	**6.28** A
	6.29 D

Chapter 7

7.1 1.55 in.	**7.14** A
7.2 28.6 in.	**7.15** C
7.3 40 min	**7.16** C
7.4 9800 cfs	**7.17** E
7.5 75.5 cfs	**7.18** A
7.6 Peak discharge = 125 cfs	**7.19** D
7.7 230 cfs	**7.20** D
7.9 Peak discharge = 233 cfs	**7.21** A
7.10 46 cfs	**7.22** D
7.11 D	**7.23** B
7.12 D	**7.24** E
7.13 B	**7.25** C

Chapter 8

8.1 3.14 ft	**8.20** D
8.2 1.36 ft	**8.21** B
8.3 1.1 m^3/s	**8.22** E
8.4 198 m$^{2.5}$	**8.23** A
8.5 0.86 m	**8.24** C
8.6 39.8 cfs	**8.25** E
8.7 1.6 m	**8.26** E
8.8 Inlet Froude number = 4.1	**8.27** D
8.9 58.2 cfs	**8.28** D
8.10 2600 cfs	**8.29** B
8.11 0.014 ft/ft	**8.30** B
8.12 Tractive force = 0.61 N/m^2	**8.31** A
8.14 Channel is stable	**8.32** C
8.15 D	**8.33** C
8.16 C	**8.34** E
8.17 D	**8.35** E
8.18 E	**8.36** B
8.19 A	**8.37** A
	8.38 A

Chapter 9

9.1 16.7 cfs	**9.7** True hydraulic jump
9.2 35.7 m^3/s	**9.8** $H = 0.9$ m; $Q = 8.6$ m^3/s
9.3 16.0 m^3/s	**9.9** Length of jump is 39 ft
9.4 186 cfs	**9.11** 12.1 ft
9.5 1.36 m$^{0.5}$/s	**9.12** 295 cfs
9.6 607 × 10^6 ft^3	**9.13** Optimum slope for corrugated metal = 0.037 ft/ft

9.14 Diameter = 24 in.   **9.22** B
9.15 A   **9.23** C
9.16 E   **9.24** B
9.17 C   **9.25** A
9.18 A   **9.26** A
9.19 D   **9.27** A
9.20 E   **9.28** E
9.21 C   **9.29** D

Chapter 10
10.1 L = 44.3 m   **10.5** 2.0 hr
10.2 k = 0.142/m   **10.6** 15.5% greater
10.3 L = 64.4 m   **10.7** 5.0×10^7 dynes/sec
10.4 12,430 N · m/s per m of wave crest

APPENDIX C   QUESTIONS FOR A COMPREHENSIVE ORAL FINAL EXAMINATION

These questions cover the important concepts discussed in the text. Frequent reference to them while working through the book will help the reader master the information presented.

1. How would you solve a three-reservoir branching pipe problem?
2. Discuss the formation of hydraulic jumps.
3. What is a scaling ratio?
4. Explain the Froude model law and its use.
5. How is the H. Cross method used for pipe network analysis?
6. What are sluiceways and how are they designed?
7. What is the cavitation parameter?
8. What is the role of a hydraulic jump in an energy dissipator?
9. What are the variables influencing f in the Darcy-Weisbach equation?
10. Explain the development of Moody's diagram.
11. Discuss the hydraulic concepts related to flood routing.
12. What is a specific energy graph?
13. What do we mean by optimum critical slope in culverts?
14. What do we mean by equivalent sand roughness?
15. Explain the relationships among velocity, depth, energy, discharge, and Froude number in an open channel.
16. How are stable channels designed?
17. What percentage of the maximum possible precipitation is usually selected for the design of flood control projects?
18. Explain the drawdown lag in wells.
19. Define the following terms: work, energy, hydraulic energy, conjugate depths, power, and critical depth.

20. How do you determine hydrostatic forces on slanted plane surfaces?
21. Sketch and explain the operation of a Bourdon gage.
22. How would you determine the magnitude and direction of a hydrostatic force acting on a surface that is curved about a horizontal axis?
23. Explain the difference between temporal and spatial accelerations, and show examples from hydraulic applications.
24. Show the derivation of the conveyance method.
25. What do we mean by capillary discharge?
26. Discuss the law of momentum conservation, and interpret the term impulse force in hydraulics.
27. Criticize the Hazen-Williams formula.
28. Derive an equation for the pressure difference measured with an inverted differential manometer.
29. Describe the difference between confined and unconfined seepage.
30. Discuss the term viscosity in all of its interpretations.
31. What physical effects does the discharge coefficient include in orifice measurements?
32. Why do two pumps operating in parallel deliver less discharge through a piping network than twice the discharge of one pump?
33. What is the difference between critical, supercritical, and subcritical slope?
34. Explain Manning's equation.
35. Why does the celerity wave travel slower in an elastic pipe than in a rigid pipe?
36. What do we mean by time of concentration, and what physical parameters enter into it?
37. Explain the mechanism of longshore currents.
38. Describe what is meant by exit and outlet control of culverts.
39. What is the practical advantage of the conveyance method?
40. What is the essence of Prandtl's boundary layer theory?
41. What is the main advantage of Parshall flumes?
42. Explain the mechanism of capillarity.
43. What is the difference between contracted and suppressed weirs?
44. Is there an exit gradient associated with unconfined seepage? Why?
45. What is the net positive suction head?
46. Explain the operational principle of an LVDT.
47. How do we check a hand-drawn flow net for correctness?
48. Discuss the variables that may play a part in the magnitude of the C coefficient of empirical pipe flow equations.
49. List the main variables and their relative importance in the magnitude of floods.
50. How are channel-delivery graphs developed?
51. Explain how wave refraction and wave diffraction alters the deposition and erosion of the shoreline.
52. Explain the cavitation mechanism in pumps.

53. Explain the mechanism of the backwater curve, showing physical examples from hydraulics.
54. How does a flood hydrograph look for rains shorter than or exceeding the time of concentration?
55. What do we mean by instantaneous closure?
56. Explain the terms in the rational formula and criticize it.
57. What physical effects enter into the weir coefficient?
58. How will the system characteristics change if three pumps operate in a series instead of two?
59. Explain the mechanism of melting ice using the terminology of specific heat and latent heat.
60. Explain the difference between direct and indirect flow determination using practical examples.
61. Explain the law of minimum energy.
62. How do we use Buckingham's π theorem?
63. Explain what causes tides.
64. How does a magnetic flow meter operate?
65. What do we mean by potential function in seepage?
66. Discuss the term contact angle.
67. Within what range of flow is the conveyance method valid?
68. Describe a flow duration graph and its method of development for creeks and rivers.
69. What do we mean by time scale?
70. Describe the hydraulic effects you may observe on the top of a broad-crested weir if the downstream elevation is varied over a broad range.
71. Explain the difference between a mixed flow and a centrifugal pump.
72. What do we mean by quick condition?
73. Explain in detail the mechanism of water hammer.
74. On what mechanical concepts are the Bazin-Craeger coordinates based?
75. Draw a typical pump characteristics graph and explain all its terms.
76. What are the relative advantages of the graphs made by Li and Asthana over Moody's?
77. Explain the Reynolds model law and its use.
78. What do we mean by normal flow, and what are its basic assumptions?
79. What is the Froude number, and what does it represent?
80. Write the conveyance formula and explain its terms.
81. What types of similarities should be observed between models and prototypes?
82. What are the critical design considerations in the determination of the shape of a spillway?
83. Define the concept of exit gradient in groundwater flow.
84. What do we mean by specific speed and how do we use it?
85. With flow under gates, what are partial and absolute downstream controls?
86. Define the three main variables in rainfall statistics.

87. What are the affinity laws?
88. Sketch a Cipoletti weir and show the steps in deriving its discharge formula.
89. What do we mean by drag coefficient?
90. How will the magnitude of surface tension vary if we add salt to the water?
91. Explain the hydrostatics of buoyancy.
92. How is the salt wave method used for discharge measurements?
93. How do we construct a flow net if a free surface is present?
94. Define in detail all variables in Darcy's equation.
95. Discuss the concept of weight in SI and conventional American terminology and explain the slug unit.
96. How does a pressure transducer operate?
97. What is the difference between seas and swells?
98. Describe the process of reservoir siltation.
98. Explain the difference between Kaplan and Francis turbines.
100. Describe the difference between balancing of heads and balancing of discharges in pipe network analysis.

APPENDIX D   CAPSTONE PROJECTS

These projects are intended to integrate knowledge from this text and other courses in a technical or engineering curriculum toward the solution of practical problems. Because building materials, construction methods, and regulations vary from region to region, the projects are given in general form. Details for a particular locale will need to be supplied by instructors or determined by students. The goal, however, is the same: to provide creative, safe, cost-effective, reliable, and socially acceptable project designs. Projects should be directed by qualified instructors.

1. Design a constructed wetland. Factors to consider, include

   Soil types
   Aquatic habitat
   Water quality
   Inflow and outflow hydrographs
   Flood routing
   Sediment transport
   Hydraulic structures
   Embankments
   Seepage and evaporative losses
   Cost

2. Design an open channel conveyance system for irrigation. Factors to consider include:

   Source, quality, and quantity of water
   Reliablity of water supply
   Crops grown
   Soil types

Water demand
Hydraulic structures
Flow measurement
Channel shapes, slopes, and construction
Erosion and sedimentation
Seepage and evaporative losses
Cost

3. Design a water distribution system for a small community. Factors to consider, include

Source, quality, and quantity of water
Reliability of water supply
Water demand
Pipe network design
Water hammer
Pumping station design
Flow measurement
Fire protection
Cost

4. Design a dewatering system for a construction site. Factors to consider include

Hydraulic properties of geologic material
Piezometric surface
Design of wells
Design of pumps
Drawdown curves
Seepage
Design of retaining walls
Critical gradient
Safety
Cost

5. Design a small water power generation plant. Factors to consider include

Hydrostatic pressure forces
Reliability of water supply
Regional geology and geography
Turbine design
Flow measurement
Flood routing
Backwater computations
Power generation
Power demand
Water conveyance structures
Erosion and sedimentation
Cost

Bibliography

Addison, H. *A Treatise on Applied Hydraulics,* 5th ed. London: Chapman & Hall, 1964.

Albertson, M. L., J. R. Barton, and D. B. Simons. *Fluid Mechanics for Engineers.* Upper Saddle River, NJ: Prentice Hall, 1960.

American Concrete Pipe Association. *Concrete Pipe Design Manual,* 1st ed. Vienna, VA, 1970.

American Society of Civil Engineers. *Coastal Engineering*, Vols. 1 and 2. New York: American Society of Civil Engineers, 1984.

American Society of Civil Engineers. *Hydrology Handbook*, ASCE Manual of Engineering Practice No. 28. New York, 1963.

American Society of Civil Engineers. *Hydraulic Models,* ASCE Manual of Engineering Practice No. 25. New York, 1963.

Bean, H. S., ed. *Fluid Meters.* ASME Report, 6th ed. New York: American Society of Mechanical Engineers, 1971.

Bedient, P. B. *Hydrology & Floodplain Analysis.* Reading, MA: Addison-Wesley, 1988.

Benefield, L. D., J. F. Judkins, Jr., and A. D. Parr. *Treatment Plant Hydraulics for Environmental Engineers.* Upper Saddle River, NJ: Prentice Hall, 1984.

Bouwer, H. *Groundwater Hydrology*, 1st ed. New York: McGraw-Hill, 1978.

Brater, E. F., and H. W. King. *Handbook of Hydraulics*, 6th ed. New York: McGraw-Hill, 1976.

Cedergren, H. R. *Seepage, Drainage and Flow Nets,* 3d ed. New York: Wiley, 1989.

Cheremisinoff, N. P., and R. Gupta. *Handbook of Fluids in Motion.* Ann Arbor, MI: Ann Arbor Science Publishers, 1983.

Chow, V. T. *Open Channel Hydraulics.* New York: McGraw-Hill, 1959.

Chow, V. T., D. R. Maidment, and L. W. Mays. *Applied Hydrology.* New York: McGraw-Hill, 1988.

Dake, J. M. K. *Essentials of Engineering Hydraulics.* New York: Wiley-Interscience, 1972.

Daugherty, R. L., and J. B. Franzini. *Fluid Mechanics with Engineering Applications*, 8th ed. New York: McGraw-Hill, 1985.

Davis, C. V. *Handbook of Applied Hydraulics*, 2d ed. New York: McGraw-Hill, 1952.

Donkin, C. T. B. *Elementary Practical Hydraulics of Flow in Pipes.* New York: Oxford University Press, 1959.

Fair, G. M., and J. C. Geyer. *Water Supply and Waste-Water Disposal.* New York: Wiley, 1956.

Featherstone, R. E., and C. Nalluri. *Civil Engineering Hydraulics.* New York: American Society of Civil Engineers, 1995.

Finch, V. C. *Pump Handbook.* California: National Press, 1948.

Fox, R. W., and A. T. McDonald. *Introduction to Fluid Mechanics,* 4th ed. New York: Wiley, 1994.

French, R. H. *Open-Channel Hydraulics.* New York: McGraw-Hill, 1985.

Golze, Alfred A. *Handbook of Dam Engineering.* New York: Van Nostrand Reinhold, 1977.

Graf, W. H. *Hydraulics of Sediment Transport.* Littleton, CO: Water Resources Publications, 1984.

Harr, M. E. *Groundwater and Seepage.* New York: Dover, 1992.

Henderson, F. M. *Open Channel Flow.* New York, Macmillan: 1966.

Hicks, T. G., and T. W. Edwards. *Pump Application Engineering.* New York: McGraw-Hill, 1971.

Hwang, N. H., and C. E. Hita. *Fundamentals of Hydraulic Engineering Systems,* 2d ed. Upper Saddle River, NJ: Prentice Hall, 1987.

Jaeger, Charles. *Engineering Fluid Mechanics.* Translated from the German by P. O. Wolf. London: Blackie & Son, 1956.

Jensen, M. E. *Design and Operation of Farm Irrigation Systems.* St. Joseph, MI: American Society of Agricultural Engineers, 1980.

Jeppson, R. W. *Analysis of Flow in Pipe Networks.* Ann Arbor, MI: Science Publishers, 1977.

Johnstone, D., and W. P. Cross. *Elements of Applied Hydrology.* New York: Ronald Press, 1949.

Karassik, I. J., and W. C. Krutzsch. *Pump Handbook*, 2d ed. New York: McGraw-Hill, 1986.

Kaufmann, W. *Fluid Mechanics.* New York: McGraw-Hill, 1963.

King, H. W., and E. F. Brater. *Handbook of Hydraulics,* 5th ed. New York: McGraw-Hill, 1963.

Kinori, B. Z. *Manual of Surface Drainage Engineering,* Vol. 1. Amsterdam: Elsevier, 1970.

Komar, P. D. *Beach Processes and Sedimentation.* Upper Saddle River, NJ: Prentice Hall, 1976.

Lea, F. C. *Hydraulics for Engineers and Engineering Students,* 4th ed. London: Edward Arnold, 1923.

Leliavsky, S. *River and Canal Hydraulics.* London: Chapman & Hall, 1965.

Li, Wen-Hsiung. *Fluid Mechanics in Water Resources Engineering.* Upper Saddle River, NJ: Prentice Hall, 1983.

Linsley, R. K., J. B. Franzini, D. L. Freyberg, and G. Tchobanoglous. *Water Resources Engineering,* 4th ed. New York: McGraw-Hill, 1993.

Linsley, R. K., M. A. Kohler, and J. L. H. Paulus. *Hydrology for Engineers,* 3d. New York: McGraw-Hill, 1982.

Luthin, J. N. *Drainage Engineering,* rev. ed. Melbourne, FL: Krieger, 1978.

McCuen, R. H. *A Guide to Hydrologic Analysis Using SCS Methods.* Upper Saddle River, NJ: Prentice Hall, 1982.

McDaugh, F. W. *Elementary Hydraulics.* New York: Van Nostrand, 1924.

Metcalf and Eddy, Inc., and G. Tchobanoglous. *Wastewater Engineering: Treatment, Disposal and Reuse,* 3d ed. New York: McGraw-Hill, 1991.

Morris, H. M., and J. M. Waggert. *Applied Hydraulics in Engineering,* 2d ed. New York: Wiley, 1972.

Mosonyi, E. *Water Power Development,* Vols. 1 and 2. New York: State Mutual Book and Periodical Service, 1965.

O'Brien, M. P., and George H. Hickox. *Applied Fluid Mechanics.* New York: McGraw-Hill, 1937.

Owczarek, J. A. *Introduction to Fluid Mechanics.* Scranton, PA: International Textbook, 1968.

Parmakian, J. *Waterhammer Analysis.* Upper Saddle River, NJ: Prentice Hall, 1955.

Plate, E., ed. *Engineering Meteorology.* New York: Elsevier, 1982.

Portland Cement Association. *Handbook of Concrete Culvert Pipe Hydraulics.* Skokie, IL, 1964.

Powell, R. W. *An Elementary Text in Hydraulics and Fluid Mechanics.* New York: Macmillan, 1951.

Powell, R. W. *Mechanics of Liquids.* New York: Macmillan, 1940.

Raghunath, H. M. *Dimensional Analysis and Hydraulic Model Testing.* Cincinnati, OH: Asia Publishing House, 1967.

Rouse, H. *Engineering Hydraulics.* New York: Dover, 1978.

Russell, G. E. *Hydraulics,* 5th ed. New York: Holt, 1942.

Sellin, R. H. J. *Flow in Channels.* London: Gordon and Breach, 1970.

Silvester, R. *Coastal Engineering, I.* Amsterdam: Elsevier, 1974.

Silvester, R. *Coastal Engineering, II.* Amsterdam: Elsevier, 1974.

Streeter, V. L. *Handbook of Fluid Dynamics.* New York: McGraw-Hill, 1961.

Streeter, V. L., and W. E. Benjamin. *Fluid Mechanics,* 8th ed. New York: McGraw-Hill, 1985.

Tchobanoglous, G. *Wastewater Engineering: Collection and Pumping of Wastewater,* 1st ed. New York: McGraw-Hill, 1981.

Todd, D. K. *Ground Water Hydrology.* New York: Wiley, 1980.

U.S. Army, Corps of Engineers. *Shore Protection Manual,* Vols. 1, 2, and 3. Washington, DC: U.S. Army Coastal Engineering Research Center, 1984.

U.S. Department of Agriculture. *National Engineering Handbook.* Washington, DC: Soil Conservation Service, 1971.

Vennard, J. K., and R. L. Street. *Fluid Mechanics,* 6th ed. New York: Wiley, 1983.

Verruijt, A. *Theory of Groundwater Flow.* London: Gordon and Breach, 1970.

Viessman, W., Jr., and M. J. Hammer. *Water Supply and Pollution Control*, 5th ed. New York: HarperCollins, 1992.

Wood, A. M. M. *Coastal Hydraulics.* London: Macmillan, 1969.

Woodward, S. M., and C. J. Posey. *Hydraulics of Steady Flow in Open Channels.* New York: Wiley, 1941.

Ziparro, V. J., and H. Hasen. *Davis' Handbook of Applied Hydraulics*, 4th ed., rev., New York: McGraw-Hill, 1993.

Index